Liquid Materialities
A History of Milk, Science and the Law

PETER ATKINS
Durham University, UK

Routledge
Taylor & Francis Group

LONDON AND NEW YORK

First published 2010 by Ashgate Publishing

2 Park Square, Milton Park, Abingdon, Oxon OX14 4RN
711 Third Avenue, New York, NY 10017, USA

Routledge is an imprint of the Taylor & Francis Group, an informa business

First issued in paperback 2016

British Library Cataloguing in Publication Data
Atkins, Peter.
  Liquid materialities : a history of milk, science and the
  law. -- (Critical food studies)
  1. Milk--Social aspects. 2. Milk--Quality. 3. Dairy
  laws--History.
  I. Title II. Series
  641.3'7-dc22

Library of Congress Cataloging-in-Publication Data
Atkins, P. W. (Peter William), 1940-
  Liquid materialities : a history of milk, science, and the law / by Peter Atkins.
    p. cm.
  Includes bibliographical references and index.
  ISBN 978-0-7546-7921-9 (hardback)
  1. Milk. 2. Milk as food. I. Title.
  TX556.M5A85 2009
  641.3'71--dc22

                                                                    2009030146

ISBN 978-0-7546-7921-9 (hbk)
ISBN 978-1-138-26043-6 (pbk)

LIQUID MATERIALITIES

Critical Food Studies

Series Editor

Michael K. Goodman, Kings College London, UK

The study of food has seldom been more pressing or prescient. From the intensifying globalization of food, a world-wide food crisis and the continuing inequalities of its production and consumption, to food's exploding media presence, and its growing re-connections to places and people through 'alternative food movements', this series promotes critical explorations of contemporary food cultures and politics. Building on previous but disparate scholarship, its overall aims are to develop innovative and theoretical lenses and empirical material in order to contribute to – but also begin to more fully delineate – the confines and confluences of an agenda of critical food research and writing.

Of particular concern are original theoretical and empirical treatments of the materialisations of food politics, meanings and representations, the shifting political economies and ecologies of food production and consumption and the growing transgressions between alternative and corporatist food networks.

Contents

List of Figures and Tables   vii
Abbreviations   ix
Acknowledgements   xi
Preface   xiii

PART I: REMATERIALIZING FOOD HISTORY

Introduction to Part I   3
1   A Material World   11
2   Daniel Schrumpf: A Trial by Instruments   39

PART II: IN SEARCH OF MILK

Introduction to Part II   49
3   Seeking the Natural: Laboratories and the 'Knowability' of Milk   55
4   Expertise   91
5   Standards   115

PART III: DISCIPLINING MILK

Introduction to Part III   135
6   Moralizing Milk   137
7   Policing the Natural   159
8   Legal Ontologies and the Performative Realm of the Law   195

PART IV: IMPURITY AND DANGER

Introduction to Part IV   223
9   Dirty Milk and the Ontology of 'Clean'   225
10   The Material Politics of Milk   247
11   Conclusion   277

References   281

Index   331

List of Figures and Tables

Figures

3.1	Gerber butyrometers	80
III.1	Dairy products: percentage adulterated, 1877–1938	135
7.1	The fat content of Birmingham milk samples	177
7.2	Birmingham milk samples below standard, 1899–1930	177

Tables

1.1	The average percentage composition of different milks	21
1.2	Typical reactions to milk problems by actors in the period c. 1850–1940	27
3.1	The reduction of total solids with addition of water	63
3.2	The results (percentages) of 14 analysts asked to look at 15 samples of milk in 1884/5	71
3.3	Most popular chemical methods of milk fat analysis in Britain up to the 1960s	72
3.4	Summary of milk tests standard in the mid-twentieth century	86
4.1	Analytical disagreements (per cent of reference samples) between the government laboratory and local authority analysts	97
4.2	Somerset House analyses of milk composition, 1876	98
4.3	Expenses associated with analyses	106
4.4	Urban population per milk sample, 1893–1897	108
4.5	Factors affecting milk composition	112
5.1	The 'standard' of milk constituents current before the 1901 regulations	121
5.2	Witnesses called by the Mackenzie Committee, 1922	125
5.3	Weighted average composition of milk, 1901–1950	129
6.1	The proportion of milk sourced from cows kept within the urban boundaries	141
6.2	The average distances that milk was brought to London by selected railway companies in about 1900-1904	142
6.3	The virtues of prominent dairymen, 1850–1950, as noted in their obituaries	146
6.4	Orders of Worth in the British milk industry, *circa* 1930	149
7.1	Evidence of the adulteration of London milk before 1872	170
7.2	Samples collected by county and local authorities and the percentage that were judged to be adulterated	172

7.3	Average monthly composition of milk supplied to the Aylesbury Dairy Co., 1897–1916	175
7.4	The types of milk adulteration in the nineteenth and early twentieth centuries	180
7.5	Colouring matter in Lewisham milk samples	182
7.6	Results of a survey on milk cooling by the Milk Marketing Boards and the Ministry of Agriculture, 1950 (percentage of farmers)	183
7.7	Adulteration of milk, 1877–1913, by type of authority	190
7.8	A comparison of milk adulteration in 14 cities	191
8.1	Legal action nationwide against the adulteration of milk, 1887–1913	205
8.2	Percentage of Birmingham milk adulterated at different points in the chain	207
9.1	Sediment in Manchester's milk supply, 1896–1906 (mm of sediment separated by centrifuge)	229
9.2	Percentage of Yorkshire milk samples containing sediment, 1908	229
9.3	Sediment in London's milk supply, 1904/5: percentage of samples in each class (classes in parts per million by volume, measured by centrifuge)	231
10.1	The principal bacteria found in milk	248
10.2	Variations in bacterial counts according to plate media	250
10.3	The contamination of milk retailed in London, 1903–4 (million bacteria per cubic centimetre of milk sampled)	251
10.4	Bacteriological results of 70 samples in Manchester, 1916–17	251
10.5	The accuracy and speed of bacteriological tests	253
10.6	The objectives of the National Clean Milk Society	258
10.7	Licences for designated grades, England	260
10.8	The definitions of graded milk under the 1923 Order	263
10.9	Leicestershire County Council's monitoring scheme, percentage of milk samples in each category	269
10.10	Average bacterial count per cc of Midland Counties Dairy Ltd milk	271
10.11	The percentage of farmers supplying milk in each of three bacteriological standards to the Midland Counties Dairy, Birmingham	271
10.12	Grades of milk specified in the Milk (Special Designation) Orders and Drinking-Milk Regulations	275

Abbreviations

Cox CC	*Cox's Criminal Law Cases*
JC	*Justiciary Cases*
JP	*Justice of the Peace*
KB	*Law Reports: King's Bench*
LGR	*Local Government Reports*
LJKB	*Law Journal King's Bench Division*
LJMC	*Law Journal Magistrates' Cases*
LJQB	*Law Journal Queen's Bench*
LT	*Law Times Reports*
MAF	*Ministry of Agriculture and Fisheries*
NA	*National Archives*
NIRD	National Institute for Research in Dairying
PP	*British Parliamentary Papers*
QB	*Law Reports: Queen's Bench Division*
ScLR	*Scottish Law Reporter*
SC(J)	*Scotland, Courts of Session and Justiciary*
SJ & WR	*Solicitors' Journal and Weekly Reporter*
TLR	*Times Law Reports*
WR	*Weekly Reporter*

Acknowledgements

First of all I want to thank my wife, Liz, who for decades has put up with my crazy 80 hour working weeks. I dedicate this book to her, with love. Also my father (now deceased) and mother encouraged my early dabblings in this genre. Rowan and baby Ned; Miranda and Tim; Val and Kevin; and my extended family on all sides will all be surprised to hear that they kept me going, but they did, with smiles and kind words. Despite my detached and occasionally grumpy demeanour, I have never taken this support for granted. It is, after all, wonderful to be alive and to have the opportunity to be creative, so thank you all.

Thanks are also due to my colleagues in Durham University. Its Department of Geography not only has staff, support staff and students of the very highest calibre, but also has a happy and collegial atmosphere – a very rare combination that helps one to think clearly. Amongst those who have actively contributed to my enterprise through professional friendship are Nick Cox (who educated me about the history of statistics and drew Figures 7.1 and 7.2) and Christine Dunn, and also Bruce Scholten, a fellow dairy enthusiast. Hamzah Bin Muzaini briefly acted as a Research Assistant, for which I am grateful.

Beyond Durham, I must mention cheerful comrades Richard Hall, Glenn Mitchell and Paul Laxton. In the International Commission for Research on European Food History there are many who have heard preliminary versions of chapters at our biennial symposia. Among these, Derek Oddy and Peter Scholliers have been colleagues on several projects and they have given support when I needed it. In Paris my collaborators and friends Arouna Ouédraogo, Martin Bruegel and Alessandro Stanziani have all three been an inspiration in different ways. The same is true of the staff and students associated with the Institut Européen de l'Histoire et Culture de l'Alimentation in Tours and the Università delle Scienze Gastronomice at Colorno, where I teach occasional courses. My thanks also are due to Barbara Orland, who kindly discussed her ongoing work on milk with me, and to staff at the Milk Development Council and Dairy UK, such as Ed Komorowski and Mark Holden. In addition, felicitations to Mike Goodman, editor of the series in which this book is published, for suggesting a much better short title than I had in mind, and to Katy Low, my Ashgate editor, for her encouragement and her understanding of my strange way of working.

Research these days requires funding and the Wellcome Trust kindly provided this through their Research Leave Fellowship programme and their system of small grants for research expenses. I promised to write one book but they will in fact be getting two: buy one, get one free. Not bad value for their investment, although I admit that publication of this volume has taken longer than I expected. Financial support has also been received from the British Academy and Durham University, and indirectly through a second route from the Wellcome Trust to the

Northern Centre for the History of Medicine. My thanks for encouragement here are especially due to Holger Maehle of the Durham Philosophy Department and to Philip van der Eijk of Newcastle University.

My unreasonable and at times shrill demands of librarians and archivists have been borne with stoicism and goodwill. Particular thanks to the staffs of the Public Record Office, the British Library, the British Newspaper Library, the Wellcome Library and the Contemporary Medical Archives Centre, the Tyne & Wear Archives, and the University libraries in Durham, Manchester, Liverpool, Glasgow, University College London, and the London School of Economics. Once more I have been pleasantly surprised by the generosity of dozens of authors who have shared their work with me, including, in several cases, unpublished material. This access to the writing of others has been invaluable and has been boosted further in my recent experience of electronic versions of books available through the Internet. Discovering the Open Library's Internet Archive and Google Books was also a great help, so thanks to the many anonymous people who scanned the hundreds of books that I have downloaded – all of them out of copyright I hasten to add.

As noted earlier, this book has emerged from over 30 years of scholarship on food history. Its path has been far from straight and the argument you will find here was still forming in my mind as I wrote. This is a stressful way to work, and not necessarily to be recommended, but it has at least delivered a final product that is fresher than I had originally envisaged. A number of the theoretical points and a very brief version of some of the empirical material had an outing in a paper for *Environment & Planning D: Society & Space* in 2007. Parts of Chapters 9 and 10 were published in French, also in 2007, and I have drawn upon a number of my other papers.

Preface

In a general sense this book is a history of materials, a topic that until recently has received little attention. More specifically, I am interested in perishable foodstuffs, and particularly in milk. My narrative is not an economic history, a social history, nor even a cultural history, although these perspectives will play a part as it unfolds. Instead it is a history of the material that we call milk, a food that has been revered and ignored, respected and feared, in almost equal measure. I have chosen the last 200 years in Britain as a particularly dynamic and intriguing period of food history; one which sheds light upon the malleability and the resistance of materials to the ordering of modernity. I am hoping this story will make it impossible to see the materials in our diet in quite the same light again. They have biographies; they have careers. The mundane, taken-for-granted nature of words such as bread or milk shows that our linguistic and intellectual vocabulary hitherto has been inadequate to express the complex and changeable character of these foods. In this book I will suggest that an approach through the material itself will provide a way forward.

I do not remember being especially fascinated by milk as a child but something of my surroundings must have been absorbed by osmosis. My sister and I regularly used to visit Mr Rimmer's small dairy farm, which was in a field behind our house. His shippon (cowshed) seemed huge to me but no doubt its dimensions were exaggerated by childish perception. If I stood on the swing in our back garden I could see the cows in the fields, but not on Saturdays because then they were shooed indoors so that the field could be used as a rugby pitch. In the 1950s our part of the outskirts of north Liverpool was not yet fully developed and there were also still nurseries and small-holdings growing vegetables nearby. These, and Mr Rimmer, soon disappeared, to be replaced by semi-detached houses and bungalows.

The cows and the milk entered my life again in 1971 when my supervisor at Cambridge, Clifford Darby, encouraged me to start thinking about what he called 'a vertical theme'. I had first heard this term as an undergraduate in his third year specialist course, entitled 'Methods in Historical Geography'. Then as a new doctoral student I needed a topic and I can still hear Darby's sonorous tones as we debated whether I should take this route or his other proffered alternative, a 'cross-section'. If the choice was between a broad regional geography and a more specialized topic over a long period, for me it had to be the latter. So I set about in the first term of research looking for an area in the history of agriculture that had not already been intensively cultivated. Historical geography, agricultural history and economic history were the disciplinary contexts in which I moved at first since the most interesting scholarship seemed to be there. Alan Baker was also a formative influence, especially through his praise of Braudel and the Annales

School, and gradually it dawned on me that food would be my theme, especially those perishables whose production remained relatively close to their urban markets in the nineteenth century. This seemed geographical enough to justify an approach that made a nod to the type of locational analysis that was *de rigeur* in geography at that time.

So, in a sense, I have been writing this book since 1971. The objective of my work has remained the same: writing a history of the everyday that transcends disciplinary boundaries, in my case the common sphere of food studies. But my approach has changed markedly. I started with an interest in mapping change, for instance locations of production, and I had it in my mind to construct a dynamic model of transfer to market – a dendritic web of flows rather like a watershed. Over the years, the decades, my thinking has undergone several changes of direction, including a long period of interest in the development geographies of South Asia, but I see these as refinements of the overall quest. I may have rejected my early research questions and analytical perspectives but food history still feels like home. This is not too far from the spirit of John Dewey, for whom 'intellectual progress usually occurs through the sheer abandonment of questions ... an abandonment that results from their decreasing vitality and a change of urgent interest. We do not solve them: we get over them'.[1]

This particular book is the first volume in a quartet on the history of milk in Britain. The second will look at the association between milk and the spread of a number of diseases, especially bovine tuberculosis. This will reveal the origins of our concerns about animal diseases in the food supply, which, in the era of BSE, salmonella and E. coli O157, have come to undermine our confidence in what we eat. The third volume will be an historical political ecology of dairying in Britain, covering at regional and national scale the origins of specialist milk production and the evolution of the complex supply systems that kept city households provisioned. It will argue for a holistic study of this and other commodity chains. Finally, the fourth book addresses the issue of infant feeding in the late nineteenth and early twentieth centuries, arguing that the increased use of unclean cow's milk was responsible at that time for a heavy toll of morbidity and mortality.

My principal purpose with this quartet is to demonstrate that food, rather than occupying the register of the dull and the commonplace, is in fact 'continuously contested, imagined and re-imagined, transformed, negotiated'.[2] It provides a common ground for the hybridization of characteristics that might otherwise appear to be singularly natural or cultural. In other words, we have an opportunity to highlight the multiple lives of food.

I mentioned earlier the curious mixture of reverence and everyday forgetfulness with which we regard milk. In India it has traditionally been used in the worship of Hindu deities such as Ganesh, the elephant god, but it has also achieved an almost mythic status in western societies where it is seen to be clothed in a white

---

1   Dewey 1910: 19.
2   These are the words of Tully 1995: 11, writing about culture.

veil of innocence and as a vital gift of motherhood – our primal nourishment and a performed source of reassurance.[3] At the global scale there is a genetic divide into two groups. In the first, some individuals, particularly in Africa and East Asia, lose their ability to metabolize milk after weaning and the consumption of dairy products therefore tends to be low in these regions. In the second group, in Europe and the Americas, milk remains an important part of the diet throughout childhood and among adults and it has become integral to the econo-politics of food in those continents. In representational terms milk is also significant here for its reputed nutritional balance as 'the perfect food'. An early version of this belief was expounded in 1834 with quasi-religious enthusiasm by William Prout:

> All other matters appropriated by animals as food, exist for themselves; or for the use of the vegetable or animal of which they form a constituent part. But milk is designed and prepared by nature expressly as food; and it is the *only material* throughout the range of organization that is so prepared. In milk, therefore, we should expect to find a model of what an alimentary substance ought to be – a kind of prototype, as it were, of nutritious materials in general. Now, every sort of milk that is known, is a mixture of the three staminal principles we have described; that is to say, milk always contains a *saccharine* principle, a butyraceous or *oily* principle, and a caseous, or strictly speaking, an *albuminous principle,* Though, in the milk of different animals, these three staminal principles exist in endlessly modified forms, and in very different proportions; yet neither of the three is at present known to be entirely wanting in the milk of any animal.[4]

By way of introduction to the period of Prout and the decades that followed, it is worth reflecting that there was also a cynical, popular view in Britain that milk was what the retailer could get away with. By that time, as a foodstuff it had become an object of suspicion. One problem was that of perception, because milk accommodates a great deal of watering and creaming before a cognitive alert is sounded. It is easy to manipulate but any adulteration is difficult to detect. In addition, the compositional variations in cow's milk are substantial, from season to season, from animal to animal, from breed to breed, even from feed to feed. Identifying, analysing, measuring, qualifying, valuing, and sensing this material known as milk: nowhere and at no time has this been anything other than problematic, contested, and open-ended. Since nature refuses a straightforward revelation, human instinct has been, on the one hand, to impose norms of what ought to be considered natural, and, on the other, to exploit the uncertainty for profit. With consummate and ruthless skill, dairy capitalism operated on both tracks simultaneously from the late nineteenth century onwards.

---

3   Vernon 2000.
4   Prout 1834: 477.

From about 1850, 'drinking-milk' was commodified on a large scale in Britain for the first time. Before that, the rural dairy had been known for its butter and cheese, products that had a longer shelf-life than milk and which could be stored for months or even years.[5] Liquid milk was unsalable after as little as a week, or just a few days in hot weather. It was always thought of as highly perishable, along with food items such as fresh fish, fruit, and delicate green vegetables, and, as such, it was difficult to transport and store, and also relatively expensive as a source of sustenance. To most people it was a luxury, and only really essential for those infants who could not be nursed.

The decades of the 1850s, 1860s and 1870s saw changes due to a combination of factors. First, the rapid pace of early Victorian urbanization and a growth in disposable incomes created foci of demand that were difficult to satisfy from local resources. Under the primitive technological conditions of the early nineteenth century, liquid milk, in order to be fresh on delivery, had to be produced in cowsheds actually within the large urban areas and their immediate suburbs, close to its market. But the production costs of town milk eventually made it unprofitable. In the overcrowded cowsheds there was an expensive toll from endemic cattle diseases such as pleuro-pneumonia, and occasional epidemics of rinderpest and foot and mouth. To this was added the cost of meeting the sanitary requirements of the local authorities, starting with John Simon's 1853 rules and regulations in the City of London and later spreading throughout the rest of the capital and to other cities.[5]

From the human health point of view, one can readily appreciate the desire of urban administrators to be rid of the cowkeepers. The environmental hazard of the quarter of a million tonnes of manure produced by London's cows each year in the early 1860s was sufficient to make them the object of universal obloquy.[6] On the other hand, London's producers were closely regulated in the last three decades of the century and their milk is therefore less likely to have been contaminated with dirt and disease than the output of their unregulated rural colleagues. Town milk was also fresher and had a higher fat content.

> As a rule the milk in town dairies is richer in quality, because the town dairy farmers understand their business better; they keep their cows to the mark, and give them more nutritious food, and have an interest in not letting the cows go back, in order that they may be sold in good condition to the butchers when they cease to yield milk ....[6]

Second, the newly available technology of the railways was exploited from the 1840s by entrepreneurs looking to link rural producers with city markets. There were a number of problems to overcome, such as high freight rates and the lack of facilities at stations, but, from the 1860s onwards, railway milk flooded into

---

5   Holderness 2000.

6   Augustus Voelcker of the Royal Agricultural Society, in evidence to the Read Committee 1874: Q.5518.

London in particular. This was facilitated by middlemen such as George Barham of Express Dairies offering contracts to replace previous relationships based upon trust. The milk trade in London switched from a producer-driven to a buyer-driven system. The emergence of wholesale firms and large retail companies gradually led to the replacement of open and relatively loose market structures by the contractual subordination of suburban and country producers. These contracts were normally for periods of six months at a time and, by the end of the century, were a means of driving up quality standards and enforcing reliability of supply.

The third, a push factor, was competition from cheap imports from the 1870s for the manufactured dairy products market. As the globalization of trade in foodstuffs progressed, so, from the 1870s onwards, farm-based and artisan products, with their variable quality, felt the sharp edge of competition from North America, Australia and New Zealand.[7] The numbers of producers fell and the variety of British dairy products diminished even further, quite unlike the situation in, say, France, where consumer sentiment remained loyal to local foods and where capital made slower inroads into the dairy and wider food industries.[8] Farmers whose parents and grandparents before them had specialized in producing butter and cheese were now thrown back on selling liquid milk in order to make a living. For those close to a railway station with good access to a big city, this rapidly became the basis of their long-term prosperity.[9] Overall, the shift in the centre of gravity of dairy farming amounted to one of the most fundamental structural changes of British agriculture in the period of study, especially in the decades when the profitability of cereals was under threat from cheap imports and mixed farmers therefore leaned increasingly on the livestock side of their business.[10]

In terms of product range, the British market for liquid milk was even more restricted than that for butter and cheese. The consumer was presented with a remarkably limited choice with type or quality. Before the First World War there were essentially three products. First, and overwhelmingly the market leader, there was whole milk, for drinking and cooking. This was understood by all participants in the market to be 'milk as it came from the cow', nothing added and nothing taken away. There were other descriptions used by both retailer and consumer, such as 'new milk', as a means of insinuating some differentiation, but these tended to be descriptions of freshness rather than any real variations on the core understanding of the product. Second, the by-products of cream, butter and cheese-making – skim milk, butter milk and whey respectively – were marketed, often in rural areas and occasionally in cities.[11] These were much cheaper than whole milk and

---

7   Collins 2000: 199.
8   De Planhol and Claval 1994.
9   Atkins 1978.
10   Taylor 1976, 1987.
11   A nice distinction was sometimes made between 'skim milk' on the one hand, the result of hand skimming, and 'skimmed' or 'separated milk', on the other, which were the result of machine skimming. Richmond and Boseley 1893b.

were therefore popular with poorer consumers, who would otherwise have had to go without; but the quality was often poor. As a result of being set for the cream to rise, skim milk was frequently on the point of souring by the time it was sold.[12] It was used, for example, in Glasgow, especially among recently arrived migrants from rural areas who crowded into the slums of these cities in search of work.[13] Third, 'invalid' or 'nursery' milk was sold at a premium. Consumers seem to have assumed that this type of milk was particularly rich in butterfat and/or free from disease, and they were told that it came from 'special' animals kept nearby in urban cowsheds. Unfortunately the evidence suggests that fraud was particularly rife in this category, which was regarded by some dairymen as an opportunity to mark up the price of their normal milk for vulnerable and credulous customers.[14]

In 1900 drinking-milk was more readily available and more widely consumed than ever before. It was delivered to the doorstep by a system of increasing complexity which, when all of the components linked together and worked efficiently, was a wonder in its own right. Yet, apart from the adulteration that we have already mentioned, there was another problem. This was the so-called 'milk question' – one of the major issues of public health at the end of the nineteenth and beginning of the twentieth centuries. In short, milk was thought to be responsible for morbidity and mortality on a large scale in Europe and North America. Milton Rosenau, writing in 1912, asked 'why all this fuss about milk and milk products? We do not have a bread question, a grain question, a fruit question, or a vegetable question – these substances also represent standard articles of diet'.[15] The answer was the unique capacity of milk to transmit infections of various types, including epidemic diseases such as scarlet fever and typhoid, and slow-burn diseases like tuberculosis. This was coupled with the ever expanding use of raw cow's milk, for instance as an infant feed, thus increasing the risk. Along with Britain and America, France suffered very much the same challenges, which were summed up by Jacques Rennes as 'Le Lait Qui Tue. Le Lait Qui Sauve'.[16]

In Britain, local authorities often lacked the specialist knowledge and the resources to inspect and regulate milk production conditions adequately, and at the turn of the twentieth century the result was a spectrum of responses running from a limited number of progressive cities with good monitoring and local powers to intervene when necessary, to smaller authorities with different levels of concern, or none. Again, Rosenau summed up the contemporary view:

---

12   Foster Committee 1894: Q.1816, Bannister.

13   Rew 1892.

14   George Barham, founder of Express Dairies, evidence to the Read Committee 1874: Q.2437; Augustus Voelcker, Royal Agricultural Society, evidence to the Read Committee 1874: Q.5518; Foster Committee 1894: Q.2272 Long, Anon. 1894: 185.

15   Rosenau 1912: 1. For more on Rosenau, see Block 2005.

16   Rennes 1923.

There is probably no single problem in the whole realm of modern sanitation and hygiene that is so complex, so involved, so intricate, and so harassing. The difficulties sometimes appear impossible to overcome; the dangers too subtle to guard against.[17]

The story of the period roughly 1850 to 1950 is one of the introduction of various measures to deal with the milk question.[18] As a commodity, it became a site of politics as different groups vied to have their interests protected or their solutions implemented. The main emphasis in the first 70 years was sanitary, with attempts to squeeze cowkeepers out of cities, or at least to force them to reduce their environmental impact by requiring better buildings, the proper drainage of effluent and the removal of manure. This campaign was gradually extended to rural producers in the hope that cleaner production conditions would yield cleaner milk.

This sanitary drive in the milk trade was part of the much larger movement in society to cleanse the urban environment to make it fit the imagined ideal for a modern society. The stink of urban slaughterhouses and food markets added colour to the gothic realism of novels by Dickens in London and Baudelaire in Paris, and a newly energized hygienist movement lost no time in trying to purify the cities of such sights and smells. Thus the Victorians and Edwardians reshaped understandings of what constituted a 'good' city by reconceptualizing its components. This meant excluding the production and processing of animal foods, which in broader terms amounted to a purification and separation of the social realm of the urban from the activities and natural substances of the country.

In the 1920s, 1930s and 1940s the emphasis of milk discourse shifted from environmental hygiene to technological developments. The failure to eliminate dirt and disease from bulk supplies, despite significant investment of resources, encouraged public health officials and the commercial dairy companies to look to heat treatment as their silver bullet. Sterilized or pasteurized milk was consumed by more and more people from the 1920s onwards and it seemed, for a time at least, as though mastery over the environmental context of this key food had been achieved at last. This optimism proved to be illusory, however, because of very strong resistance from anti-pasteurization groups, particularly in the interwar years, but also right down to the present day.

Milk, then, was representative of efforts to redraw the boundaries between nature and society: first through the creation of purified cities and, second, by the redefinition of natural foods as having normative qualities. The 'true' essences of cities and food were being rediscovered and enforced. Bruno Latour argues that such moves were typical of Enlightenment thinking and were always doomed to failure in both metaphysical and practical terms.[19]

---

17 Rosenau 1912: 4.
18 For the American story, see DuPuis 2002.
19 Latour 1993.

Milk, despite its iconic life-giving status, resembled a mass-produced industrial product throughout the nineteenth and early twentieth centuries, with high volume, rapid turnover, but limited choice. It is only within living memory in Britain that drinking-milk, yoghurt and cream have become differentiated into dozens, maybe hundreds, of different products: reduced fat, sweetened, flavoured, or fortified with vitamins or minerals, and attractively and conveniently packaged in a range of sizes.[20] Some of these products perform previously unanticipated functions, such as probiotic yoghurt as a medicine, or ice cream as a meal course. What does this tell us about its material quality? The answer is that food quality, more broadly, even when defined in physical or organoleptic terms, has taken many forms and my purpose in this book is to demonstrate that most of these dimensions have historical origins and that their present-day application is best understood as part of a dynamic social process that was strongly influenced by the properties of the material. A hint of some of the themes that we will discuss is contained in a recent statement by Renard:

> Quality ... is not a condition inherent in a product. It must be constructed and then promoted in order to become a collective comparative advantage. A particular quality has to achieve market recognition, which in turn requires an organization to champion it. In sum, the construction of market niches revolving around specific quality definitions is a collective process that seeks consumer recognition through quality labels and (public and/or private) certification practices: the valorization of quality within a market is produced via certification processes.[21]

Milk was never the same materially, socially, culturally, economically, or politically after its entry into the networks that provisioned cities. These were not just systems of delivery but vast engines of transformation. They nourished bodies; they spread disease; they encouraged the make-over of agro-ecosystems and landscapes in the distant countryside; they enabled a re-imagining of cities as spaces without farming; they transformed food economies; and they encouraged a new form of food politics.

My story of milk is, in effect, an archaeology of qualities. So far we have mentioned production in relation to dirt and disease, but there were other qualities that were important in various ways to other system actors, including the consumer. I will come to these but first, in Chapter 1, I will reflect on the book's genealogy. It is written as food history but its links with the work of other food historians are tenuous. In order to show why my angle is new, I will sketch the boundaries of the territory on which we will trespass.

---

20 Gronow 2004: 39-40.
21 Renard 2005: 421.

The horseman serves the horse,
The neat-herd serves the neat,
The merchant serves the purse,
The eater serves his meat;
'Tis the day of the chattel,
Web to weave, and corn to grind,
Things are in the saddle,
And ride mankind.

Ralf Waldo Emerson 1867

PART I
Rematerializing Food History

Introduction to Part I

Food History

Recent historiographies of food history identify a new dynamism in its mission and a surge in the number and quality of publications.[1] This refortified enthusiasm is the result of, first, the rediscovery of links between food and ill-health following the food scares of the last two decades, such as the cholesterol/heart disease nexus; food poisoning from salmonella and E. coli O157; BSE; and, most recently, the increase of obesity.[2] Together these have encouraged a closer scrutiny of diet-disease history and a search for foods that will deliver better health, such as those in the Mediterranean dietary complex. Second, environmental concerns about carbon emissions, coupled with low levels of traceability and accountability in long-distance food transfers, have started a debate about the desirability of local food. Studies of gastronomic patrimony, terroir and of typical foods have flourished as a result, as have geographical works seeking to lay bare the evolving nature of food chains. Third, modern prosperity has meant that the portion of disposable incomes spent on food has been significantly reduced in the last 50 years or so in Europe and North America. Consumers are willing and able to prioritize eating out and the purchase of expensive items such as organic foods. Quality in all of its guises, including the material, has therefore become an issue, as never before. Fourth, there is the strange paradox that, while we are less and less willing to spend time cooking complicated dishes, there is more and more media exposure for celebrity chefs and stories about food. In other words, food has been transformed from a basic necessity to the raw material of entertainment and leisure.

Food historians at the beginning of the twenty-first century are certainly active. The International Commission for Research on European Food History was founded in 1989 and has so far published the proceedings of ten symposia. The Institut Européen d'Histoire et des Cultures de l'Alimentation was launched at Tours in France in 2002, and has already made its mark with a number of major international conferences and a new journal, *Food and History*. Various other organizations also accommodate an historical perspective, such as the International Commission for Ethnological Food Research, the Oxford Symposium on Food and Cookery, the Association for the Study of Food and Society, the Leeds Symposium on Food History and Traditions, and there are others. My own database of food historians worldwide now has over five hundred names and addresses, its largest number ever, although it has to be said that these are highly concentrated, with about half in France and Germany, the continuing heartlands of food history.

---

1   Super 2002, Ferguson 2005, Spary 2005, Scholliers 2007.
2   Oddy, Atkins and Amilien 2009.

Given this institutional thickening and surge of popular and professional interest, what has food history achieved? This is not an easy question to answer but, first, one has to acknowledge the success of a number of broad brush histories, particularly the *Cambridge World History of Food* and Flandrin and Montanari's *Food: A Culinary History*.[3] Although the quality of contributions and the breadth of coverage is uneven in these and similar encyclopaedic tomes, they nevertheless provide convenient points of entry to an expanding literature. What is striking about both of these edited volumes is the lack of a common theoretical ground or epistemological consensus among the contributors. There is everything from bioarchaeology to semiotics, so it is probably best to conclude that modern food history is a multidisciplinary field of great variety but with no codification of purpose or method. This is not necessarily a negative conclusion because there are many who would wish to celebrate such variety.

In trying to make sense of the field of food history in my preparatory reading for the present volume, I have concluded that the following nine dimensions capture the state of the art reasonably well. These sub-themes do not necessarily have equal numbers of publications or intellectual weight behind them but they fairly represent the efforts of the various disciplinary perspectives in food history. Inevitably there are overlaps and, of course, there are temporal and spatial sub-divisions that could be made.

1. Food systems, chains, marketing, retailing, advertising, branded products.
2. Diets, nutrition, hunger, famine, standard of living, family budgets, anthropometry.
3. Feasting, fasting, foodways, food habits, meals, manners, taste, beliefs, taboos.
4. Food and health, food scares, food safety, disease, obesity, eating disorders.
5. Cooks, cooking, cuisine, cookery books, recipes, menus.
6. Catering, restaurants, eating out, communal feeding.
7. Technology in eating, transporting, manufacturing, processing, and preserving food.
8. Food politics, policy, legislation, regulation.
9. Food symbolism, food and identity, images of food.

Since my own interest in food material finds no voice in this historiography, the present book is an attempt to write a new kind of food history, different from and complementary to the above. There are no cookery books in my story, no restaurants, no diets or foodways, no standards of living, no farms, no markets or trade routes, no feasts or famines or food riots, and no nutrition. It is certainly not traditional food history and I accept that food historians may find it strange, maybe even threatening; but I hope to persuade them and other readers that food history can be about food. This last statement may sound flippant but I mean it. We have

---

3   Kiple and Ornelas 2000, Flandrin and Montanari 1999.

surprisingly little literature about the stuff in food*stuffs*. Yes, there are specialist histories of potatoes, sugar, chocolate, salt, pasta, cod, peanuts, tea, spices, beans, citrus and other individual commodities; and very good they are as expositions of a particular commodity's role in social, and economic, and sometimes even political, history.[4] But, with a few notable exceptions, the authors say little about the composition of these commodities, or even about their smell and taste. In other words, we are deprived of the *material* in food history.[5] I am not referring here to materiality in its common usage of a growing interest in objects in cultural studies.[6] No, material for me means matter, stuff, substance, things, objects, that have 'sensuous proximity'.[7] Material is my central focus, and this is what is new about the book. In this account, food will not be a passive bystander but an active, physical participant in its own history.

Material Histories and Geographies

In the 1980s and 1990s material histories and geographies were submerged in the hegemonic tide of representational analysis that stood for the analysis of culture. Recently re-emerging in a number of guises, material studies have developed some interesting features, one example of which is the flattening of the hierarchy of objects thought worthy of attention. In particular, anthropologists and archaeologists have been reassessing their attitude to mundane, ordinary items, no longer privileging the high value grave goods and fetish objects previously studied. There is a recognition that the tools of everyday life – Bergson's 'immediate givens' – help to structure life and consciousness and are therefore deserving of attention.[8] Appadurai's account of the 'social life of things' and 'tournaments of value' adds a cultural dimension to exchange and, in the same tradition, other authors are now showing an interest in the 'biographies' of even the most basic material objects.[9] Miller is a pioneer at this frontier, with his work on objects as diverse as pots, mobile phones and saris, and Molotch has intriguingly introduced the study of 'stuff', at a stroke assuaging any guilt that researchers might have had that their work on toasters, microwaves, or pot noodles was too trivial for 'proper'

---

4 Salaman 1949, Mintz 1985, Adshead 1992, Kurlansky 1997, 2002, Serventi and Sabban 2002, Smith 2002, Coe and Coe 2003, MacFarlane 2004, Turner 2004, Albala 2007, Laszlo 2007.

5 Food historians have also neglected quality in their assumption of homogeneity about food products. This hides the uncertainty in many consumers' minds when they bought 'meat' or 'bread' and it neglects a vital element in the history of living standards.

6 I will be using the term materiality in the sense of physical properties that have social consequences. This is a subtly different emphasis from some recent writers. See Miller 2005.

7 Kearnes 2003: 140.

8 Gronow and Warde 2001, Hilton 2008, Bergson 1889/2001.

9 Appadurai 1986. For an anthropological update, see Hoskins 2006.

academic study.[10] My recent favourite is a study of reusable glass milk bottles, which comments on their role in reinforcing a sense of place, in discourses about the freshness of food, and also as symbols of resistance against the disposable plastic and cardboard packaging found in supermarkets.[11] The authors of the paper seem to suggest that the bottles used by their interviewees are filled as much with cultural meaning as with milk.

Geographers are now also engaged with the mundane, as shown by a recent collection edited by Jon Binnie and his colleagues.[12] A key point is that the use of words such as banal and everyday does not equate to 'lack' or inauthenticity. Places or events associated with these terms may well be sites of alienation or surveillance, but they may well also be sources of routine reassurance in the grooves of habitual behaviour. The geographical perspective on 'material encounters' often involves places, such as gardens, home and work; but bodies and minds are not far away.[13] Latham and McCormack explore even wider limits by including the materiality of 'automobile urbanities' and psychoactive substances such as alcohol.[14] In material culture studies, much of the writing on the everyday has been at the scale of household objects but there is no reason why such reasoning should not be extended to the study of, say, ordinary buildings or other aspects of place-making. Gareth Millington's account of the melancholia associated with the Palace Hotel, Southend, is an example, in which he stresses 'the placing, arranging and naming spatial ordering of materials and the system of difference they perform'.[15] Another is Peter Merriman's discussion of the material manifestations – road signs, sleeping policemen, speed cameras – of governmental attempts to shape the behaviour of motorists.[16]

The literature on mundane materials is growing but it is still uneven in terms of the choice of objects for attention. Most studies so far have been on objects of consumption in the 'global north'; a certain level of disposable income is taken for granted and there are hidden assumptions about lifestyle. Outside the realm of material symbolism, less has been published on the everyday objects of poor consumers in poor countries. Also, the implication of the act of consumption is usually a positive one of voluntarism. But what about what we might call the non-objects and the ex-objects that are forgotten, discarded, waste, ruined or rotten? These may have little residual value or power to excite. They may be hazardous, and they are very likely to be marginalized and stigmatized. Here we may particularly

---

10   Miller 2007, Molotch 2003. Miller says that he is working in the tradition of Pierre Bourdieu, which means that he is concerned just 'as much with how materiality makes people, as with what people make'. Miller 2005: 4.

11   Vaughan et al. 2007.

12   Binnie et al. 2007.

13   Lorimer 2005.

14   Latham and McCormack 2004.

15   Millington 2005.

16   Merriman 2005.

mention Tim Edensor's exposition of ruins, with its tremendous potential for the reinterpretation of place, but there is clearly also work to be done on food that is past its use-by date; overripe fruit; redundant processing technologies; unused recipes; corner shops closed in competition with supermarkets; abandoned farms; and maybe (why not?) even the human waste that is the inevitable consequence of food consumption.[17] Dominique Laporte has made a start on this last mission in a *History of Shit*.[18]

Fresh or processed, food is gaining attention in history and historical geography, and I am hoping that I can show in this book that its material substance and form have had a profound impact upon how it is produced, transported, sold, consumed, regulated and policed.

Material Can be Lively

It was Noel Castree who, some time ago, reminded us of the liveliness of material. David Goodman then imported the idea into food studies in his discussion entitled 'ontology matters'.[19] The point that both were making is that it has become increasingly obvious that an appreciation of the active capacity of material is crucial to an understanding of the present crisis of confidence about food safety, particularly with respect to animal foods. Michel Callon has suggested this bloody-mindedness of material is a form of counter-performativity or 'overflowing'.[20]

Consider the following. Of the infectious organisms that cause some 1,620 human diseases, 60 per cent are zoonotic, many of them food-borne.[21] Of the 175 emerging pathogenic species, 75 per cent are zoonotic, including possibly the most dangerous of all, avian influenza, which again has possible links with the food system.[22] Most of these infections are unknown to the general public but several have been among the biggest news stories of recent years. I am thinking here, of course, of salmonella in eggs, BSE, and E. coli O157. So 'lively' have these particular zoonoses been that writers such as Ulrich Beck have suggested that they are representative of a new historical era, the 'risk society'. First, the scale of catastrophe is said to be beyond anything previously witnessed, and who could gainsay this when considering the cattle slaughter figures in the BSE and foot and mouth outbreaks in the last 20 years, or the gloomy predictions about the number of people who might have contracted CJD from consuming contaminated beef in the 1980s?[23] Second, the impact has been global, crossing borders, so that

---

17   Douny 2007, DeSilvey 2006, Edensor 2005.
18   Laporte 2000.
19   Castree 1995, Goodman 2001.
20   Callon 2006.
21   Bokma 2006.
22   Cleaveland et al. 2001.
23   Powell 2001, Powell and Leiss 1997.

consumers increasingly calculate the risk of imported foods. Third, the authority of science and of the political process is said to have been undermined, and there has been a loss of credibility in official public health messages.[24]

> The recent BSE crisis in Europe is an example of a new kind of risk in late-modern society. Conventional risk politics are no longer considered suitable for such new risks, leading to a new kind of risk politics.[25]

Goodman, Sorj and Wilkinson added a further dimension to this material dimension of food when they commented that its biological characteristics are difficult for capitalism to capture. There are problems with industry harnessing the rhythms of the seasons or the length of animal gestation in a way that guarantees timely profits, although the industrial strategies of appropriationism and subsititutionism are making inroads, most recently through laboratory-created GMOs.[26] Further, Ben Fine claims that there is differentiation between the various food systems of provision according to their organic characteristics and their potential for transformation into manipulable chains of value creation, although he denies smuggling biological determinism back into food studies. More recently, however, David Goodman has withdrawn his support for any notion of biological exceptionalism in food studies and has criticized Fine for a modernist ontology in which 'the "organic content" of agro-food systems of provision is viewed exclusively from one pole of the nature-society dichotomy'.[27] Goodman is searching for a 'middle kingdom' and settles for one discovered by the tools of actor network theory: translation, quasi-objects, and black-boxing.

None of these insights are particularly surprising for those of us who have lived through the horrors of food safety alerts in the last 20 years, but applying the theory of the risk society to more distant periods of food history has not yet happened to any degree. In fact, curious statements by risk society theorists indicate that this would not, in their view, be appropriate.

> In all traditional cultures, one could say, and in industrial society right up to the threshold of the present day, human beings worried about the risks coming from external nature – from bad harvests, floods, plagues or famines. At a certain point, however – very recently in historical terms – we started worrying less about what nature can do to us, and more about what we have done to nature. This marks the transition from the predominance of external risk to that of manufactured risk.[28]

---

24   Kitzinger and Reilly 1997, Dornbusch 1998, Washer 2006.
25   Oosterveer 2002: 215.
26   Goodman et al. 1987.
27   Goodman 1999: 22.
28   Giddens 1999: 26.

Contra Giddens, my own contention is that 'manufactured' risks have been with us for centuries and that food systems, particularly those based upon animal products, have provided foci of risk concentration.[29] It is therefore worth developing a series of arguments about the liveliness of milk and the problems this has posed for public health and intervention by what I shall call the food police.

An Outline of Part I

The book is divided into four parts: Part I (Chapters 1 and 2) is a theoretical introduction; Part II (Chapters 3 to 5) about the 'search' for the material of milk; Part III (Chapters 6 to 8) deals with the 'disciplining' of milk; and Part IV (Chapters 9 and 10) with dirt and disease.

Chapter 1 is divided into six subsections, each of which presents a perspective relevant to 'thing history'. The first of these introduces a number of philosophical insights, the origins of which can be said to stretch back to the Greek metaphysicians and, more recently, in the twentieth century, to Martin Heidegger. My point here is briefly to introduce the debate on historical ontology which recently led to the publication of Ursula Klein's fine book on material histories.

Second, I will make a connexion between material and quality, arguing that there is a significant mutuality. Recent writings about food quality have braided into three streams of unequal volume and vigour. I will endeavour to reflect these in the chapter, starting with quality as 'goodness' in a number of ways that in short are about ethics and the social norms that flow from that. Next, we will introduce the notion of quality as a network concept, emerging from inter-actor relations of one kind or another. This is a rapidly growing approach that deserves a longer treatment than we will have space for in this introductory chapter. Finally, there is quality as the innate property of material. Since this is the vision of quality that is of greatest interest to me, I will explore its detail as a further introduction to the thinking behind the book.

Third, there is the claim from non-representational theory that material qualities are emergent, in our case the act of producing, processing, preparing, or consuming food. This is a more radical assertion than it sounds and it will be necessary to extend the commentary into an area sometimes called ontological politics, where the fragile nature of material is foregrounded, along with our daily performances.

Fourth, we will discuss relational materiality in assemblages. The research of Becky Mansfield is an inspiration here; her papers on catfish and surimi (fish paste) show how the biophysicality of food is fundamental to the systems which can be set up to construct its marketable quality. This section of Chapter 1 will also encompass the more traditional aspects of material and economic relationality in the shape of food systems and chains, and will mention the fruitful research on

---

29   Atkins 2008.

food in conventions theory. This in turn will lead on to a more intensive discussion in Chapter 6 of orders of worth and their relation to goodness and value.

Fifth, hybridity is crucial to material in the metaphysics of Bruno Latour. While his Actor Network Theory appears only on the margins in this book, Latour's ethnographic observations on the constitutive role of materials in, for instance, scientific laboratories or judicial process, are valuable. Also, food may be seen as a hybridization of nature and society, as confidently and convincingly shown by Sarah Whatmore in her account of *Hybrid Geographies*.

Finally, another way of seeing the role of material in the contact between humans and their food is through the processual nature of material qualities. There are several varied literatures illustrating this, for instance on material ontogenesis. One of the most interesting authors is Gilbert Simondon, the French philosopher of technological change. Although it is not the place for a full critique of Simondon's work, Chapter 1 will show why he is important for an understanding of material change.

Chapter 1
A Material World

Material Quality

> In both popular and scientific writing (in history, economics, sociology), a timeless and objective definition of quality is commonly used, which appears to have been mobilised by an outside observer. But in real markets, the quality of a product may diverge through time and in different places.[1]

As Alessandro Stanziani points out in the quotation above, material quality in practical reality is rather different from our representations of it. We idealize, we simplify, we create object identities which help us to deal with a complex world, but which may also serve a purpose, such as reassurance or the assignment of status or value. Stanziani's work is important because he refuses to take quality for granted and insists on a genealogical perspective and a critical questioning of the very nature of a commodity. An example is his observation that new products often begin as variants or imitations of existing products, which in the eyes of some may be equated as plagiarism or adulteration. He argues that choice and changes in taste are bound up with the differentiation of quality, as is the diffusion of consumer behaviour among social classes and income groups.

It is in this spirit of questioning that the present chapter will be devoted to musings about material and material histories. It will introduce a number of perspectives that are relatively recent in their development. It does not claim to be comprehensive because a great deal in this general area has already been published, to the extent that Graham Harman claims that 'philosophy today is either materialist, or is intimidated by materialism'.[2]

Metaphysics: Do You Take Milk with That?

> Knowledge always refers to an object that is defined step by step through its history.[3]

The mission of this book is not philosophical, although its flavour is derived from a number of ingredients that have philosophical connexions. It concentrates on

---

1   Stanziani 2007b: 209.
2   Harman 2009.
3   Gaston Bachelard cited in Bolduc and Chazal 2005: 81.

emergent material qualities, which, as noted in the Preface, distinguishes it from the efforts of most food historians. In this section I will look at a history of relevant ideas at the metaphysical level. Subsequent sections will then explore selected recent approaches to the material.

Most sociologists of scientific knowledge have the very proper, mainstream desire to unravel the epistemological problems that place limits on our knowing the world. Because they study knowledge-generation in context, this leads to historically- and spatially-contingent epistemologies. A good recent example is John Pickstone's history of science, technology and medicine, entitled *Ways of Knowing*. His account is a summary of the intertwined histories of the gathering and making sense of knowledge in natural history, experimental science, and the technoscience of industrial complexes. Nikolas Rose espouses similar objectives in his 'epistemology of assemblage'. Rose's concern

> is to reconstruct the epistemological field that allows certain things to be considered true at particular historical moments, and the kinds of entities, concepts, explanations, presuppositions, assumptions and types of evidence and argument that are required if statements are to count as true ... This approach does not seek to deny as such the 'objectivity' of knowledge, but to describe the ways in which objectivity is produced, and the consequences of the production of objectivity.[4]

Along the same lines, part of the stimulation for the present project came from the historical epistemology of Lorraine Daston and the historical ontologies of both Ian Hacking and Ursula Klein. Taking historical epistemology first, this is the study of the organization of knowledge through concepts such as objectivity, wonder, and error.[5] Daston's own definition is 'the history of the categories that structure our thought, pattern our arguments and proofs, and certify our standards of explanation'.[6] I am mainly interested in the material properties of milk, but its history, like the objects that appear in Daston, is the result of competing forms of facticity: laboratory experiments, legal ontologies, and legislative/regulatory/ administrative ideas about what is 'natural' about milk. These will be discussed in Chapters 3 to 9.

A good exemplar of historical epistemology is Mary Poovey's wonderful book, *A History of the Modern Fact*, where she traces the genealogy of the epistemological unit. She argues that historical facticity, starting with the Baconian revolution of the seventeenth century, concerns the failure of universals to 'coalesce out of the common experience of particulars'.[7] Although there are traces here of Foucault's genealogy, Poovey develops her own vision of a

---

4   Rose 1999: xiii–xv.
5   Daston and Galison 2007, Schickore 2002.
6   Daston 1994: 282.
7   Poovey 1998: 8.

contingent historical understanding of the particulate nature of knowledge that is prior to the arrangement and deployment of facts in discursive contexts, and she rejects the Foucaultian focus on the identification of practices or events that constitute temporal ruptures. She uses double-entry book-keeping as an example of the creation of modern facts, as the disinterested 'nuggets of knowledge' that lubricated trade and allowed the construction of a rule-governed, systematic mercantile knowledge base. Peter Miller summed this up in his observation that it is 'calculative practices that make the economy visible'.[8]

Daston's own version of historical epistemology is about emergence. In her Aristotelian metaphysical accounts, material objects are only seen as such by her historical actors when certain conditions are met, most of all a solidification in the light of their perceptual and knowledge-making resources. In these terms, artists have changed their revelation of the human body from the flat, stylistic paintings of Cimabue in the thirteenth century to, say, a Bill Viola video installation today. Likewise, food scientists in the early nineteenth century experienced a sense of emergence when entirely new aspects of the structure of foods came to light under the gaze of the microscope. In this way, new worlds and new possibilities materialize, where they have not existed before, although such objectivities are never fixed for more than a few years in the constant churning of new 'facts' being born.

Daston and Galison explore this phenomenon of epistemic contextualization most fully in their 500-page exegesis of scientific atlases. These books were made, they say, in the contexts of distinct codes of epistemic virtue.[9] What began as truth-to-nature in the seventeenth century, became objectivity in the 1860s, and then turned into expertise in the twentieth century. Each of these was part of the same spectrum, refracted differently according to the epistemological circumstances of each era and each laboratory. Observation, then, is relational and using the same nomenclature at different periods of history does not conjure the same object. The bread of two hundred years ago, artisan-made or domestic, was very different from the mass-produced, factory article of the same name today. But then the consumers are also different, as is the flour, the yeast, and the even the butter that is spread on it. A history of knowledge-making is needed to make sense of this, and material histories have no meaning otherwise.

Daston and Galison show that objectivity was never quite achieved. It was always just beyond reach, due to the technological constraints of laboratory equipment, of the chemicals used in experiments, or of the accuracy possible with various forms of mensuration. The recording of results and their interpretation were further uncertainties, as were attempts to find a neutral means of conveying them to other scientists.

In an autobiographical moment, Daston recently recalled that Ian Hacking has always disliked her historical *meta*-epistemology, as he calls it, preferring instead

---

8   Miller 2001: 379.
9   Daston and Galison 2007: 18.

the label 'historical ontology'.[10] What he has in mind for this is the study of 'general and organizing concepts and the institutions and practices in which they are materialized'.[11] Taking up Hacking's challenge, Ursula Klein and Wolfgang Lefèvre state that their purpose for historical ontology as nothing less than a 'new history of material objects in general' and, in particular, a reconstruction of eighteenth century scientists' ontology of materials through a study of their practices of identification and classification.[12] Klein and Lefèvre's 'thing history', then, is the contextual history of material taxonomies.

Stuart Elden provides further clarification of the concept of historical ontology through a reading of Foucault. He traces the roots of Foucault's thinking to Nietzsche, where a version of the genealogical method is to be found. But this does not amount to a type of history. What Foucault is doing, rather like the later Heidegger, is searching for 'different possible ways of being' at a much deeper level, in other words '*how* what is *is*'?[13] The histories and geographies that arise from this are the result of a rejection by him of the need to seek metaphysical universals.

Joseph Pitt finds a different form of words in a discussion of writers he calls 'the New Technists'.[14] They share, he says, an interest in the knowledge embedded in technological artefacts. This materialized version of historical ontology is close to my own intentions, although my material ontogenesis makes no claims about metaphysics at the deepest level. Rather, my search for substance histories involves: (a) the coming into being of milk as a scientific object and commercial commodity; (b) its recognition as a site of power; (c) its disciplining and harnessing as a lever of biopolitics; and (d) what it can tell us about changing practices of knowledge. At the risk of sounding pretentious, this amounts to 'a history of the modern object'.

Material has Qualities; Quality is 'Material'

The linguistic turn of post-structural social theory in the 1980s and 1990s meant a privileging of the semiotic over the material, with texts, images and discourses coming to dominate thinking.[15] As a result, we were entertained by Roland Barthes' accounts of mythologies such as 'steak et frites', awed by Michel Foucault's ability to find evidence of power everywhere, mesmerized by Deleuze's cinematic vision of metaphysics, dazzled by Baudrillard's reading of signs, and, like as not, puzzled

10   Daston 2007: 806.
11   Hacking 2002: 12.
12   Klein and Lefèvre 2007: 1.
13   Elden 2001: 60, quoting Heidegger; Elden 2005: 356.
14   These include Bachelard, Galison and Hacking. See Pitt 2003.
15   Shove et al. 2007: 6.

by Derrida's claim that 'there is nothing outside of the text'.[16] This pseudo monopoly of French theory was complemented in the 1980s and 1990s by a worldwide surge of social constructionism, which has also had a profound effect upon interpretations of human impacts on nature (including food), as we will see later.

From the 1990s onwards the theoretical basis of critical opinion has widened, with some scholars returning to philosophical basics in the work of twentieth century giants such as Heidegger, who argued that the being of things is never fully present before us. An object is therefore more than its appearance and more than its usefulness.[17] There have also been moves towards a rematerialization of the social sciences, on the premise that matter does matter despite it presently being under-represented in our research.[18] Scholars have been particularly keen to stress materiality as things-in-lives but there has also been an encouraging post-human prioritization of the material thingness of objects in their own right.

This new theoretical high ground was first occupied by ethnographers, notably again in France but increasingly also in anglophone publications. Anthropologist Daniel Miller has played a pivotal part in this emerging debate about materiality in a sub-field known as material culture studies, with contributions, among many others, from geographers Chris Philo and Peter Jackson, sociologist Tim Dant, and, in English literature, Bill Brown.[19] There is also a large and multivocal sociological and ethnographic literature on things in Science and Technology Studies (STS)/Sociology of Scientific Knowledge (SSK), which has been hugely influential in ontological reflections upon objects.

The most thing-orientated of the social sciences, archaeology, has added its own contribution to material studies. In a recent bout of soul-searching, attempts have been made by archaeologists to shed both the vulgar materialism of their past, 'in which human behaviour was more or less shaped by non-human constraints,' and also some of the more extreme aspects of their 'post-processual' stress upon the experiential and the phenomenological.[20] Although Meskell indicates that archaeologists have some way to go before they can be said to be pursuing the study of 'object worlds' in the full sense of the material turn, they have at least made a start in what they call 'symmetrical archaeology'.[21]

It has not all been plain sailing for these various reborn materialists. A commonly-heard complaint is that much of the new literature reduces its analysis of material to the social. Matthew Kearnes, for instance, is particularly critical of the tendency to imply 'that matter only matters in so far as it objectifies "social

---

16   Barthes 1972, Derrida 1976: 158.

17   Harman 2007: 1.

18   For a critical discussion of the rematerialization trend in geography, see Anderson and Wylie 2009.

19   Miller 1987, 1998, 2005, Philo 2000, Jackson 2000, Dant 1999, 2005, 2008, Brown 2001, 2003, 2004.

20   Oestigaard 2004: 79.

21   Meskell 2004, Witmore 2007, Webmore 2007, Olsen 2003, 2007, Shanks 2007.

relationships" ... rather than as fundamentally independent and active'.[22] Bjørnar Olsen has similar doubts:

> To the extent that things are allowed to speak it is largely to bear witness to those human intentions and action from which they themselves are believed to originate. Things may be social, even actors, but are rarely assigned more challenging roles than to provide society with a substantial medium where it can inscribe, embody and mirror itself.[23]

There are suspicions, then, that the material turn may involve some rebranding of existing ideas rather than the radical break with the past that many were hoping for. An example of the friction generated as a result of this disappointment is to be found in the pages of the journal, *Archaeological Dialogue*, where a lively debate between Ingold, Tilley, Knappett, Miller and Nilsson was published in 2007.

From our point of view, one solution to these uncertainties is to focus on material qualities, prompting a big question: what is food quality? Although this might seem to be a relatively straightforward, basic kind of question, soluble perhaps by asking producers and consumers for their own criteria, the literature suggests a greater degree of complexity and difficulty than is apparent at first sight. The idea of giving weight to consumer views, for instance, seems to be relatively new, perhaps arising from the sheer range of choice that is available in modern supermarkets, coupled with the fading of concerns about organic deterioration that used to dominate thinking before the spread of refrigeration technologies.

A number of recent papers identify the key dimensions of food's material quality. These are said to be: the naturalness of ingredients; qualities conferred by the method or place of production; food safety and traceability; nutritional value; organoleptic qualities and functionality; food's biological quality; and quality in terms of certification.[24] Although intended as commentaries on today's food scene, these dimensions also have some relevance to the historical period that is the subject of the present book, 1800 to the present. I will endeavour to show this as the argument progresses. In addition, I will set out to identify definitions of quality as they change through time, subject to a number of influences. This is a key part of the book's argument because insufficient attention has been paid to the historical roots of our modern debate about food quality. Also, like Harvey et al., I will argue that food quality is part of the more general question of how people choose what to consume.[25] Marie-Christine Renard puts this point succinctly: 'Quality may ... be defined as a product's capacity to satisfy explicit or potential consumer needs.'[26]

---

22 Kearnes 2003: 149.
23 Olsen 2007: 580.
24 Ilbery and Kneafsey 1998, Morris and Young 2000. See also Atkins and Bowler 2001.
25 Harvey et al. 2004.
26 Renard 2005, 421.

The notion of material quality has been undervalued in economics, to the extent that Michel Callon calls it an 'under-conceptualised and fragile notion'.[27] In an attempt to start a new debate, Musselin, Paradeise and their contributors have published an extended deliberation on quality, in which they explore the disadvantages that its present neglect brings for the study of markets and their mechanisms.[28] Supply and demand are rarely exactly balanced in real world markets, which are full of imperfections that result from the asymmetric distribution of information. Attempts to solve such mismatches often involve negotiations, not just about price, the usual consideration, but also about quality.[29] It is therefore essential for us to know more about understandings of quality and their origins.[30]

Callon et al. propose an 'economy of qualities' in which objects have shape, as a moment in a never-ending process of production, and also a life, a career. This is potentially applicable to food as it makes its way through a supply system, until it is transformed into an edible dish and then starts a new journey as it is metabolically absorbed or excreted as waste. Callon et al. then elaborate the definition of a good as a bundle of qualities that establish its singularity. There is no essence, as such, just hybrid characteristics, which, in a different combination, would make up a different good or a variant of the same good.

Callon et al. see the qualities of goods as emergent rather than designed, or revealed through 'trials which involve interactions between agents (teams) and the goods to be qualified'.[31] Thus the taste and texture of a new dairy product is tested using codified procedures and approved instruments, and, in the case of a well-established product, there will also be routines to check that quality guidelines are maintained.[32] Quality, in other words, cannot be taken for granted; it is forever threatening instability.

> The characteristics of a good are not properties which already exist and on which information simply has to be produced so that everyone can be aware of them. Their definition or, in other words, their objectification, implies specific metrological work and heavy investments in measuring equipment. The consequence is that agreement on the characteristics is sometimes, in fact often, difficult to achieve. Not only may the list of characteristics be controversial (which characteristics ought to be taken into consideration?) but so also, above all, is the value to be given to each of them.[33]

---

27 Callon 2005a, S94. See also Lévy 2002, Allaire 2004.
28 Musselin and Paradeise, 2005.
29 Karpik 1989.
30 For more on the astonishing neglect of quality until recently, see Parrott et al., 2002.
31 Callon et al. 2002, 198.
32 For a Norwegian example, see Straete 2008.
33 Callon et al. 2002, 198–9.

In Callon's economy of qualities the measurement of properties is important.[34] Andrew Barry has elaborated the theme by talking of a government of qualities in which various manifestations of the state invest in metrological regimes, such as the monitoring and analysis of food quality.[35] But these metrological regimes are frail. Any system of monitoring that depends on point sampling has a high degree of uncertainty built in from the outset, and then there are numerous means of cheating the process of sampling available to the retailer. The results are therefore always difficult to interpret and may have only a passing resemblance to the quality of the product reaching the consumer. In addition, tests for compositional quality say nothing about the other dangers, such as the spread of diseases like bovine tuberculosis.

Sensibly, Callon recognizes the temporal dimension of product characteristics. These change, with the material playing a role in the process of becoming, as does the customer. These metamorphoses require further investment for what he calls qualification-requalification, in order to stabilize and standardize goods for the market. A new model of car, for instance, will undergo many design alterations before reaching the market and then adjustments and redesigns before its career ends and it is replaced by a new model. The customer will buy a variant suitable to her circumstances but, Callon argues, part of the package will be intangibles such as the reputation of the manufacturer. A parallel with the food economy is where a shopper smells and feels a mango before purchase, but also takes on trust those material attributes, maybe its cultivar variety or organic certification, which she cannot judge for herself on the spot.

Take milk, for instance. In most mammalian species it is white but close inspection reveals a remarkable variation in its composition (Table 1.1). Even among specialist dairy breeds of cattle, the term 'milk' in effect is a homonym for liquids containing more or less butterfat and more or less solids-not-fat. So, we are entitled to ask: 'what is natural milk?'

One answer is that milk is such an astonishingly complex liquid that even now, at the beginning of the twenty first century, it has still not given up the full story of its organic chemistry.[36] In the simplest of terms, it contains water, lipids, protein, carbohydrate in the form of milk sugar (lactose), and also gases and minerals. It is an emulsion of fat globules, a fine dispersion of casein micelles, a colloidal solution of globular proteins and a colloidal dispersion of lipoprotein particles.[37] The lipids comprise neutral glycerides, free fatty acids, phospholipids, cerebrosides, gangliosides, sterols and carotenoids.[38] The most important protein

---

34   Callon et al. 2002. For a debate about Callon's recent work, see Miller 2002, Fine 2003, and Callon 2005b.

35   Barry and Slater 2002.

36   For a review of progress in dairy chemistry in the first half of the twentieth century, see Jenness 1956.

37   Walstra et al. 1999: 6.

38   Walstra et al. 1999: 50, Fox 1995.

is casein but also to be found are lactalbumin, lactoglobulin, and fibrin, along with the enzymes (which act as organic catalysts) peroxidase, reductases, lipase, phosphatase, catalyse, galactase and amylase.[39] The minerals (sometimes called ash in analyses of milk) are various salts of potassium, sodium, calcium, magnesium, iron, sulphur, phosphorus and chlorine. There is also a spectrum of vitamins and trace elements.

Further dimensions of milk composition have been revealed by dairy scientists. For instance, in the early twentieth century it was understood that feeding regimes influenced the quantity and quality of milk, but it was not dreamt that fodder type could be adjusted according to a desired breakdown into saturated, polyunsaturated and monounsaturated fatty acids, as dictated by human health concerns.[40] At the beginning of the twenty-first century, microbiologists now think that milk contains roughly 100,000 types of organic molecules, most of which have yet to be identified and studied.[41] It remains much more of a mystery than we are perhaps willing to admit.

Although rules and standards encourage threshold notions of quality, where foods either qualify or not for a special title or label, many of us nevertheless think of quality as relative. All milk may be better in quality than was the case 100 years ago, perhaps across every test one can think of in terms of cleanliness and freedom from disease or contamination, but we still have 'ordinary' milk and 'quality milk' that can be purchased in many forms. Once such premium milks become the market leaders, then, in their turn, they will eventually be consigned to the category of ordinary or 'regular'. Meanwhile, of course, we are all suspicious whether the 'new, improved' product is really any better than before.

Performed Materiality

One of the most important trends in materiality studies has been a scrutiny of the dynamic realm of practice in everyday life.[42] As a result, the hardware of material cultures is being reassessed in terms of its physical use. A particularly valuable contribution has come from Elizabeth Shove and her collaborators.[43] One of their projects has been about interactive design in DIY and digital photography – the co-evolutionary relationship between consumers and their objects that facilitates 'distributed competence' in the use of hammers and cameras, or indeed kitchen utensils, plastic cups and laptop computers. This type of work overcomes a blind spot in research hitherto that assumes the material of artefacts to be given.

---

39   Fox 1992.

40   Grummer 1991.

41   Singh and Bennett 2002: 1.

42   There is no space here to explore the important literature, mainly feminist, on bodily materialization. See, for instance, the discussion of 'bodies that eat' in Probyn 2000.

43   Van Vliet et al. 2005, Shove et al. 2007, Hand and Shove 2007.

Following Heidegger's tool analysis, Shove argues that a hammer has no essential characteristics that can be said to define it and its kind; and when it is picked up and used for hammering, it is the practice that tells us more than the implement.

In a similar vein, Emma Roe has worked on the processual emergence of the material qualities of foods, in other words 'the process of some *thing* becoming food' and 'the *doing* of eating'.[44] She does so by observing the embodied practices of consumers who have a material attunement to eating organic food and her interest is in the 'meaning-making event when a thing, a foodstuff, becomes food, becomes eaten'.[45] Her focus groups discussed understandings of the food that they had bought and prepared, and individually they then made video diaries in the kitchen. Unlike most consumption studies, Roe's methodology enabled an account to be written that 'does take the "guts" of eating practices seriously in terms of the diversity of individual eating practice, the consequential multiple meanings of edibility that arise, and the potential to make ethical and political material connections'.[46]

The work of Kevin Hetherington also has appeal. His book *Capitalism's Eye* deals with the experience of commodities in department stores, exhibitions and in other forums of spectacle, and his article on touch and textures is also interesting for its radical take on materiality.[47] For the latter he interviewed a visually impaired lady whose touch knowledge of objects in museums he takes to be performative rather than representational. Her claim is that 'when I am touching something there is no "me" and the object I am touching. It is just the object. So the me disappears ... for me it's just touching, identifying with the actual thing there'.[48] Hetherington concludes that one way in which place is generated is from this kind of unmediated experience, especially through routine, and this is confirmed, as he puts it, 'by the praesentia found in the dirt of time'.[49]

Although Hetherington does not discuss food as such, we could see taste in the same terms as a form of proximal knowledge. After all, the mouth is a major site of the performance of our intimate connexion with the world, hungrily incorporating it and sensually experiencing it. Each mouthful is an experimental test of organoleptic quality and a way of monitoring whether our expectations have been met. Taste, then, can be seen as an embodied quality of the material. It is both a physiological process with a distinct set of purposes and culturally specific.[50] Thus, similar foods may be valued differently by various cultural groups and the ability to distinguish subtly different flavours, for example in wine, can function as

---

44   Roe 2006a: 465 and 470, emphasis as in original.
45   Roe 2006b: 105.
46   Roe 2006a: 479.
47   Hetherington 2003, 2007.
48   Hetherington 2003: 1934.
49   Hetherington 2003: 1943.
50   Stassart and Whatmore, 2003.

Table 1.1   The average percentage composition of different milks

Selected species

	Fat	Protein	Sugar	Minerals
Bison	3.5	4.5	5.1	0.8
Cow	3.7	3.4	4.8	0.7
Donkey	1.4	2.0	7.4	0.5
Elephant	11.6	4.9	4.7	0.7
Goat	4.5	2.9	4.1	0.8
Grey seal	53.1	11.2	0.7	–
Horse	1.9	2.5	6.2	0.5
Human	3.8	1.0	7.0	0.2
Polar bear	33.1	10.9	0.3	1.4
Rabbit	18.3	11.9	2.1	1.8
Reindeer	16.9	11.5	2.8	–
Sheep	7.4	4.5	4.8	1.0

Dairy cattle breeds

	Fat	Protein	Sugar	Minerals
Ayrshire	4.14	3.58	4.69	0.68
Brown Swiss	4.01	3.61	5.04	0.73
Dexter	4.09	3.45	–	–
Guernsey	5.19	4.02	4.91	0.74
Holstein	3.55	3.42	4.86	0.68
Jersey	5.18	3.86	4.94	0.70
Red Poll	4.24	3.70	4.77	0.72
Shorthorn	3.63	3.32	4.89	0.73
Simmental	4.03	3.38	–	–
South Devon	4.19	3.58	–	–
Sussex	4.87	9.31	–	–
Welsh Black	3.96	3.49	–	–

Sources: Davis 1947, Fox and McSweeney 1998: 67, Jenness 1974: 91, Jenness 1999: 24, National Milk Records 1980/1.

a class marker.[51] Matters are complicated further when taste information is used to evaluate the degree of excellence of food. Doing so is predicated on a hierarchical evaluation of food characteristics that reflects power.[52] Such hierarchies are culturally defined, just as the characteristics that underpin them are culturally mediated.

One of my arguments in this book will be that the material of foodstuffs may be understood in ways that are emergent in practice. Tasting, consuming and metabolizing are elements of this but maybe the strongest observation for our purpose is that food is a site for ontological politics. This terminology is derived from the ethnography of objects and their associated knowledges undertaken by Annemarie Mol, who asserts that 'ontology is not given in the order of things … instead, ontologies are brought into being, sustained, or allowed to wither away in common, day-to-day, sociomaterial practices'.[53] In her world of hospitals, there are diseases and instruments and knowledges associated with their diagnosis and treatment, and these knowledges are enacted through practical, material events. She advocates a radical empirical philosophy in which the focus is on what she calls 'praxiographies' of the multiplicity of material reality. She pointedly criticizes the constructionism of STS/SSK, arguing that it ignores the fragility of object identity because 'matter isn't as solid and durable as it sometimes appears'.[54] Her principal contribution is in helping us to understand the slipperiness of making diagnoses of medical conditions such as atherosclerosis but her radical, empirical style is equally as applicable to food studies as to disease. [55] The present book is pitched at a different scale and is historical in intent, so a reproduction of Mol's method is impossible, but it seems to me that the underlying aim of her ontological politics is relevant. In what follows, I will attempt to demonstrate the elusiveness of knowing milk, which in the period under study was only partially overcome in the practice of science, commerce and the law.

Relational Materiality

There has been neglect in the theory of food studies of the material of production, a surprising omission. Becky Mansfield is prominent among those attempting to redress this imbalance, in her case by stressing biophysicality in the food geographies of fish (catfish and surimi).[56] The task she sets herself is to understand how definitions of quality emerge in specific economic contexts. Rather than making a general distinction between nature and society or arguing for socially

---

51 Guthman 2002: 295 and 300.
52 Gronow 1997.
53 Mol 2002: 6.
54 Law and Mol 1995: 291.
55 Mol 1999, Mol and Mesman 1996, Harbers et al. 2002.
56 Mansfield 2003a, 2003b, 2003c, 2003d.

constructed quality, she looks at individual production networks in order 'to understand how specific aspects of what we call "the natural world" participate in particular interactions'.[57] Her idea of quality is different because she argues that it arises from assemblages of practices within commodity chains.[58] These chains have particular histories that have established sets of relations and the commodities are differentiated as a result.

Mansfield uses the spatial metaphors of distancing and entanglement to understand the trajectories of fish products in their commodity chains. She looks in particular at surimi, a Japanese fish paste which is a mixture of fish protein and starch. Because of its physical characteristics, surimi has a wide range of functionality, including products such as 'krab' sticks (artificial crab). It can be made from a number of different fish species and so can be produced in many countries. Quality is based on fish biology, processing technologies, mouthfeel (texture, chewyness), whiteness, and the nature of the final product. Such issues can be dealt with at the local level and overseen by transnational food corporations, making this a food with a global reach.

Mansfield's conclusion is interesting, that 'economic activities materialize meanings'.[59] She insists that quality is relational. It is 'neither a subjective judgment (what different people like), nor an objective measure (the characteristics of a commodity), but instead it is produced within relations of commodity production and consumption'.[60] Jane Bennett's interest in the major electrical outage in North America's major power grid in August 2003 is similar in many ways to Mansfield's foregrounding of assemblages, known by Deleuze as *agencements*. These are arrangements of actants with a dynamic interconnexion. Bennett is willing to concede agency – 'thing-power' – not only to individuals, such as artefacts and technologies, but also to collectives of humans and nonhumans that 'pulse with energies, only some of which are actualized at any given time and place'.[61]

Support for the idea of that material quality is an emergent feature of interconnectedness may be found in Bill Cronon's book *Nature's Metropolis*, which presents a fascinating picture of early America.[62] Part urban history, part environmental history, and part a history of the frontier, it also makes a contribution to food history because Chicago was a focus for both the grain and livestock trades of the prairies and west. Cronon's breakthrough is in showing that there were strong links between the success and growth of Chicago and the exploitation of its wide hinterland. This took the form of the makeover of the landscape to such an extent that one could say that a field in western Illinois was just as much a part of Chicago as Madison Street or Michigan Avenue. Maintaining the fiction of a

57 Mansfield 2003a: 10.
58 For more on assemblages as a Deleuzian concept, see De Landa 2006.
59 Mansfield 2003c: 330.
60 Mansfield 2003a: 11–12.
61 Bennett 2005: 461, Bennett 2004.
62 Mansfield 2003a: 13, Cronon 1991.

'natural' landscape is not easy under such circumstances and so it is much easier to see the city and the countryside as fused in a relationship that was mutually constitutive. This was reinforced by the transport and telegraph networks and by a role that eventually made Chicago one of the world's great centres of commodity trade. The stock yards and the Board of Trade became international symbols of raw material capitalism. Two keys to success were the adoption of a system of grading wheat and the linked growth of a futures market. The development and stabilizing of quality was, then, a vital component in the success of Chicago's main agricultural industries.

London performed a similar gathering function in nineteenth-century Britain. Its docks, railway termini and vast food wholesale markets made it a nodal hub for the Empire. But, before we get carried away with Whiggish enthusiasm for growth and the wonders of complex system structure, it is important to point out that we are not assuming without challenge that the formation of long distance linkages is necessarily 'good'. One might be forgiven for taking this message from some of the published material, which evaluates, say, logistical efficiency or the formation of social capital. On the contrary, as we will see, there are networks of shared values that are at odds with those of mainstream society. In its early stages, the system of bringing railway milk to the big cities in the 1850s and 1860s, for instance, almost certainly facilitated large-scale cheating because no-one had a responsibility for standards and there was a reduction of traceability in a network that had a high index of anonymity. Extending this logic, we might go on to argue that the present-day 'crisis' of food identified by Beck in his *Risk Society* is really no more than a readjustment of consumers' perceptions away from the quality they attributed to their local butcher or supermarket and towards a networked idea of quality, where responsibility is lodged at the more diffuse level of the nation state or a supranational block such as the European Union. In other words, there has been a scale jump in the spatial texture of provision, which 'the system' has not yet been able to accommodate.

Peter Jackson finds the various accounts of commodity chains and systems to be 'too linear, too mechanistic and too focused on the simple metric of length as opposed to other issues such as complexity, transparency or regulation'.[63] By comparison, 'commodity circuits', exemplified in the work of geographers such as Ian Cook and Phil Crang, at least have the virtue of having 'no beginning and no end' and they acknowledge that 'origins are always constructed'.[64] Commodity circuits 'attend to cultural inflected relationships between production, circulation and consumption'.[65] They also pay attention to the knowledges and understandings of commodities held by farmers, traders and consumers, for instance in what is called by Kopytoff the 'biography of things'.[66] There is a link here with the

---

63   Jackson et al. 2006: 132.
64   Cook et al. 1998.
65   Braun 2006a: 645.
66   Kopytoff 1986.

materiality literature because of calls to take the circulating food commodities themselves seriously. Ian Cook, for instance, urges us to 'get with the fetish' and to 'follow the thing'.[67]

For the purposes of the present book, both chains and circuits are useful conceptual vehicles in as much as they consider the contractual arrangements and institutional governance of food, which helps to provide a basis for understanding the origins and driving mechanisms of food quality. Some authors suggest mediation between the different approaches: Leslie and Reimer, for instance, between systems of provision and commodity circuits. An important degree of flexibility in any chain study is also an openness to product-specificity. This is vital for work on food, where the characteristics of, say, wheat and horticultural products to a certain extent determine how they are handled.

Recent work in geography, economic sociology and institutional economics suggests that the fundamentals of food quality may be traced to rules and conventions, as set out, for instance, in the theory of conventions, a major thrust of action-oriented French pragmatism.[68] Starting in the 1980s, conventions theorists have been interested in the social relations associated with production and exchange.[69] As they remind us, there is much uncertainty between actors in food chains, but this can be coped with through conventions – formal or informal agreements that arise out of situations that vary according to the product in hand. Conventions emerge over time through the successful repetition of particular actions and may have their roots in tacit agreements, without written rules. They are shaped by the material environment of action and in turn become a guide for future action and a form of legitimation that can be discussed and modified.[70] Institutions may also arise in the long run, for instance to deal with disputes, and conventions may overlap with each other or be replaced, in time, by contractual obligations between parties. Even these formal arrangements are, nevertheless, social constructs, and, as a result, they are heterogeneous in their composition and deployment.

Early on in the conventions literature Storper and Salais looked at manufacturing industry. Their ambitious project was to reformulate the structure-agency debate in economic geography in terms of the pragmatics of action. They pursued the 'archaeology or genealogies of economic situations', showing a sensitivity to the emergence of 'worlds of production'. For them, the product is a central feature of their approach, because it 'embodies and thus realizes the potentialities of the resources of action', but their stress is not so much product input and manufacturing

---

67   Cook et al. 2004, Cook 2004.

68   For introductions in English, see Wagner 1994, 2001, Wilkinson 1997, Favereau and Lazega 2002. A comprehensive retrospective French language collection was published in two volumes by Eymard-Duvernay in 2006.

69   Storper and Salais 1997.

70   Thévenot 2006: 112, Lévy 2002: 263.

costs as upon the ordered practices that arise from long-run routines, agreements, assumptions and expectations.[71]

Eymard-Duvernay recognizes three types of coordination linking quality to the availability of information. First, where quality is unknown, unreliable, or unstable, the trust that is established over long periods between buyer and seller is a substitute. Second, questions of quality can be referred to a neutral adjudicator, such as a public analyst in the case of doubts about the chemical composition of or physical characteristics of foods. Third, concerns about quality may be judged in terms of the health of the consumer, as was frequently the case with the infectious diseases spread by milk.[72]

Conventions theory has been extensively used to understand the development of quality in food networks. According to Jonathan Murdoch, it has advantages when comparing networks, whereas actor network theory is more introspective because of its inductive approach.[73] Conventions are also of interest because they encapsulate the interpersonal world of contacts in the world of speciality and high quality foods. Conventions theory by now has a richness that encompasses several interpretations of the role of conventions.[74] The most appropriate for our purposes are conventions that were customary and, as such, were treated informally as social rules. It was not until the late nineteenth and early twentieth centuries that normative expectations among traders were replaced by legally binding regulations and, even then, there remained an element of negotiation with nature in the minds of most actors.

In their 'worlds of production' analysis, Storper and Salais speak of the 'personality' of products, by which they mean a commodity's profile of specialization, substitutability, perishability, quality differentiation, its potential for technological development, and so on. In Britain's dairy industry it is possible to identify three worlds of production in the late nineteenth century, reduced to two by the end of the twentieth. These were: (a) urban and suburban milk producers, called cowkeepers, who employed intensive systems, often with cattle confined permanently to their dark and poorly ventilated shed; (b) country producers who had the good fortune to be located close to the ever-expanding railway network, and who were able to find a city market for their output up to 200 miles away; (c) specialist dairy farmers who were primarily producers of butter and cheese, but who from time to time were called upon to supply 'accommodation milk' in times of general shortage.

The main feature of the conventions associated with the milk trade from the early nineteenth century until the 1960s was that the perishability of the product, and its vulnerability to infection with disease, meant that it had to be produced and transported in a timely and cleanly manner if it was to travel efficiently along

71   Storper and Salais 1997: 14–15.
72   Eymard-Duvernay 1989.
73   Murdoch 1998, Murdoch and Miele 1999.
74   Woolsey, Biggart and Beamish 2003.

networks to the customer's doorstep. In the very simplest of terms, we can identify some of the principles behind drinking-milk in our period that eventually came to be commonly accepted after decades of negotiation:[75]

- 'milk as it came from the cow';
- daily delivery, fresh not sour;
- stability in compositional quality over period of contract;
- freedom from disease and dirt;
- wholesaler responsible for supply chain management, i.e. balancing differences between supply and demand;
- 'invalid' and 'infant' milk from special cows.

Milk seems to fit the 'industrial' convention of Boltanski and Thevenot. It was produced intensively in several specialized areas, with the just-in-time flexibility of additional accommodation milk when the supply/demand balance was disrupted for one reason or another. Table 1.2 shows the conventional responses made by different actors to the problems that arose on a regular basis. These amount to a lived critique of the system, which could only be maintained as it was by the cynicism and fatalism of the consumer. By the early twentieth century there were economies of scale in production, processing and transport, making the industry attractive to investment.

Table 1.2   Typical reactions to milk problems by actors in the period c. 1850–1940

	Producer	Trader	Analyst	Medical Officer	Consumer
Insufficient or unreliable supply	Increase herd	'Accommodation milk' or adulteration	Test	N/A	Use condensed milk
Not rich enough	Stet	Colorants	Test	Sale of Food and Drugs Acts	Use whole rather than skim milk
Sour	Cool milk	Preservatives	Test	N/A	Complain to retailer
Diseased	Sell diseased cattle	Pasteurization	Test	Examine cows	Buy heat treated milk
Dirty	NIRD methods	Filter	Test	Examine cowshed	Buy certified milk

---

75   See Sylvander and Biencourt 2006.

Hybrid Materialities

The 1990s saw a gradual shift away from the chain metaphor of food systems and from political economy, both of which had served agri-food scholars well over the previous 20 years.[76] There had already been the work of Michel de Certeau on the texture of daily life and of James Scott on 'the weapons of the weak', which had suggested that the juggernaut of international capitalism and its associated cultural traits could not roll without resistance into every corner of daily life.[77] Granovetter's timely intervention had also stressed the power of the social in shaping market structures and processes. In addition, the post-structural turn in social theory was sending shock waves through the cosy certainties of the social sciences. When actor network theory came on to the scene in the 1980s, the time was right for its rapid acceptance and spread. Starting in science studies as an ethnographic means of understanding the working of scientists and their laboratories, it was soon applied in human geography and in sociology. Some agri-food writers have put it to work, if not to paint a coherent picture, then at least for its radical unsettling of many previously unchallenged assumptions.

Bruno Latour, the 'amodern' magician of objects takes a bow at this point. He has become an obligatory passage point for STS/SSK engagement with the material because actor network theory, as propounded by Latour, Callon and others, is the non-constructionist materialism that has attracted most attention overall. It asserts that objects have hybrid qualities and so avoids the classic dualism of nature and society, the Achilles heel of much constructionist writing. Actor network theory has attracted a great deal of attention: praise and criticism in almost equal measure. One the one hand, it has been called by one enthusiast 'the most promising philosophy of our time'.[78] On the other hand, one of Latour's most important recent books, *Politics of Nature*, was described by a broadly sympathetic reviewer as '300 pages of abstruse verbiage'.[79]

The foundational concept of actor network theory is that socio-natural hybridity is expressed in quasi-objects. In this spirit, Bakker and Bridge talk of 'the mongrel nature of the world', where 'we navigate a world made up of radically incommensurable things, suturing them together as we go'. Milk, for instance, can be seen as a blend of human and non-human agency ready to take its place in Latour's 'parliament of things'. This is a forum in which material is no longer mute, no longer reliant upon scientists, or anyone else, to speak on its behalf. Foodstuffs, which are simultaneously naturalized and socialized, deserve their place in the textbooks of food history on their own terms. Their stories will be biographies, but also, because they contribute control, partly autobiographies.

---

76 Busch and Juska 1997.
77 De Certeau 1984, de Certeau et al. 1998 , Scott 1985.
78 Harman 2009.
79 Castree 2006: 164.

This idea of hybrids – objects in motion blended with human action – promotes a relational and distributed view of materiality in which 'the competencies and capacities of "things" are not intrinsic but derive from association'.[80] The principal expression of this relational and dispersed agency is the network. Networks may include humans; milking machines; 'inscription devices' such as milk yield records and laboratory analysis results; cows; and materials like railway wagons and delivery carts. These are joined together in alliances that vary in strength and persistence. Some coordination and stability may be provided by 'centres of calculation' such as the official Government Laboratory in London, but this cannot be taken for granted and networks face constant challenge and change. For example, the long-distance milk supplies that came to London in winter in the late nineteenth century, which were only possible if every connexion and technology worked smoothly, in summer had to be reviewed because hot weather reduced the souring time and threatened to unsettle the whole network. But actor network theory is not really about spatially discrete networks in the sense of a sewage or a telephone system, nor is it structural as with a social network. A better metaphor is said to be the rhizomes of Deleuze and Guattari, which in my lighter moments I think of as couch grass theory.

Hybrids and networks are in a constant state of becoming, as a result of the process of translation, through which actants come to represent 'a multitude of others by defining and linking their identities in increasingly simplified and fixed forms'.[81] This may be by persuasion, seduction or force. One result of translation is the creation of 'mixtures between entirely new types of beings, hybrids of nature and culture'.[82] Another result is the creation of what we might call 'associational truth', which is the product of alliances in networks. According to Latour 'the strongest reason always yields to reasons of the strongest'.[83] This, by the way, is not a sell out to relativism or constructionism but a demonstration that the germ theory, which dominated medical science for decades from the 1870s onwards, derived its strength from an accumulation of alliances, including many weak ties, which were generated and exploited by Pasteur in his campaign to establish the truth of his science. In retrospect we know that he was ultimately successful, but at the time this was by no means guaranteed. He faced rivals for the explanation of the spread of anthrax and the souring of milk, and in each case he also faced resistance from the materials that he worked with.

Critics argue that actor network theory escapes socio-historical contingencies, and Latour responds that history is made of networks, for instance the much cited work by Thomas Hughes on the historical growth of electricity supply systems, or Braudel on the growth of the networks of capitalism.[84]

80   Bakker and Bridge 2006: 16.
81   Shiga 2006: 41.
82   Latour 1993: 10.
83   Latour 1988: 186.
84   Hughes 1983, Braudel 1982.

This is a very superficial presentation of actor network theory. While I am more than willing to concede the vital folding of the non-human world into agency, and that 'societies do not construct nature as they please', I cannot go further.[85] The strong symmetry between human and non-human actors that is proposed by Latour, and brilliantly taken up in her own way by Sarah Whatmore, seems to me in its extreme versions to be counter-intuitive.[86] I have more sympathy with the call of Pels for a 'weak asymmetry'.[87] Whatmore's hybrid geographies are balanced nature-culture-spaces, whereas mine emphasize the hybrid milk-in-society. My rather different type of narrative is structured around human knowing and intervening within the limits of material potentials and constraints. While actor network theory works for me as kind of agora for ideas about objects, it leaves too little room for cultural politics.

Material Qualities are Processual

> Matter comes to matter through the iterative intra-activity of the world in its becoming.[88]

We come now to the section which most closely reflects the approach to materiality adopted in this book. I say that guardedly because, as the book unfolds, the reader will see that this initial theoretical reflection provides a departure point but not a destination. Further commentaries will be necessary from time to time to provide support for a text that is largely about food. The special qualities of this vital part of our lifeworld have not been considered in depth by material philosophers, and the conceptual depth of food history, the discipline, is insufficient to carry us through the story that I want to tell.

Martin Heidegger's exceptionally powerful philosophy can help us, especially that to be found in his *Being and Time*. To begin with, he suggests that apprehending an object and its material substance is impossible in the sense of immediate phenomenology because the very being of that object is embedded in time and, of course, we cannot press a pause button in order to allow us to walk around it and see it from every angle. The properties of milk are always at least partially absent and only become 'present at hand' when someone encounters it with one of the commonly agreed needs for 'milk'. Objects to Heidegger are not bundles of properties, but are formed in a system of relations, which in turn is temporally contingent.[89]

---

85   Castree 2001: 17.
86   Whatmore 2002.
87   Pels 1996: 296.
88   Barad 2003: 823.
89   My interpretation of Heidegger relies heavily upon Harman 2007.

Milk's thingness, then, is not just a matter of its whiteness, or its nutritional value, or its usefulness in making my tea drinkable. Yes, under certain circumstances it does have these and other material and functional characteristics. But the full materiality of milk is always partially withdrawn. Heidegger did not have in mind histories in the sense of clock time but this should not discourage an interest in genealogy and ontogenesis. We can now see that a study of milk's facticity will be far more than a matter of observation, even under the controlled conditions of a laboratory. It will be multidimensional. Even at the end of this book, devoted as it is to only one material, we will be left thinking that there is much more to say. Indeed there is, but even the subsequent three volumes of the series I am intending will still not be enough to exhaust the possibilities.

Another of Heidegger's key concepts, that of dwelling, has been explored by Tim Ingold. Through this he devises an anti-constructionist account of objects and landscapes that is based on a mutually co-constitutive, dialectical process in which people are a part of the ecology and 'take part in generating their own form and the form of their surroundings'.[90] Ingold argues that society's features are a co-production with its surroundings, which presents opportunities but also restrictions according to time and place. In fact, Ingold's confident interventions make him a good person to start this last lap of Chapter 1. This is because he is uninhibited in his comments about the material turn. He is critical of social constructionism and deeply sceptical of work on 'materiality', which, according to him, is 'expounded in a language of grotesque impenetrability on the relations between materiality and a host of other, similarly unfathomable qualities, including agency, intentionality, functionality, sociality, spatiality, semiosis, spirituality and embodiment'. Instead, he is interested in the processual properties of materials, which to him need to be contextualized in temporal and spatial terms.[91] This is because the

> very property is a condensed story. To describe these properties means telling the stories of what happens to them as they flow, mix and mutate … The properties of materials … are not attributes but histories.

Ingold refuses to concede agency to materials but sees them as active 'because of ways in which they are caught up in these currents of the lifeworld'. His materials are not bystanders: 'far from being the inanimate stuff typically envisioned by modern thought, materials … are the active constituents of a world-in-formation.'

Some of Ingold's readers may interpret his statements in a realist sense. This is not his intention but one can see why his thoughts on processual properties might appeal, for instance, to historical ontologists working in the philosophy of medicine. An example is the assertion by Smith and Brogaard that 'each substance is a bearer of change'. This provides a philosophical basis in logic for a dynamic vision of biomedical ontologies, and they illustrate its potential in their article

90   Berglund 1998: 70.
91   The quotations are from Ingold 2007.

'Sixteen Days', which seeks to identify the point when a growing embryo can be called human.[92] Predictably, this is controversial, but Barry Smith's work generally seeks to demonstrate the value of rigorous ontologies in medicine and philosophy. Rooted in Spinoza's *Ethics*, it is helping biologists to think through the significance of spatial boundaries. Smith has worked extensively on social and biomedical ontologies, and also on the formal ontology of Edmund Husserl, which, he suggests, amounts to a general theory of objects.[93] Smith's epistemology has much in common with Geographical Information Science, which is widely used in geography, history and other disciplines where dynamic cartography is important.

Despite the power of Smith's biological ontogenesis, a genetic concept of change is problematic for our purposes because it is pre-programmed with certain limits. The story I wish to tell about the history of food quality has fewer constraints and might have turned out very differently. This prompts questions about the nature of material and technological change that have answers in two very different literatures, both depending to an extent upon systems thinking.

The first is the so-called 'path dependence' theme of economic history, which has been energized principally by the efforts of Paul David.[94] Although backed by a statistical vocabulary familiar to econometricians, David's argument is about the nature of historical narrative and his message is that 'history matters' in explaining present economic circumstances. A path dependent system is one that evolves as a result of its own history, and so events bear the traces, often distant memories, of the past. The purpose is to counter the assumptions of classical economics that human behaviour is subject to timeless laws, which have tended to lead to a low priority in explanation for spatio-temporal contingency and for events that lead to what David calls a 'forking in the road'. None of this will shock historians, even those reluctant to consider counterfactual versions of their stories. So what is special about path dependence?

David's formulation is interesting, not just because it posits alternative stable states of an economic (or social, cultural, political) system. He also deals with 'regrettable' branchings, which, because of irreversibility, may mean the adoption of sub-optimal technologies, institutions, and possibly lead to market failure. The example given, repeated now so often that is has become a cliché, is that of the QWERTY keyboard dating from the 1870s.[95] I am typing on one right now (very slowly with one finger) although I know that it is not ergonomically the best solution to minimize my effort. Design inertia is not an adequate explanation of this curious survival. We must turn instead to the concept of lock-in, through which a system state is stabilized, without immediate prospect of moving forward

92   Smith and Brogaard 2003: 47.
93   Smith and Smith 1995. It is important to note that Husserl was Heidegger's senior colleague but the eventual intellectual break between them was irreparable.
94   David 2007.
95   But see Liebowitz and Margolis 1990.

and seeking alternatives.[96] This may be because of institutional self-reinforcement and/or dynamic increasing returns through learning mechanisms. Other well-known examples of this phenomenon include Edison's direct current versus Westinghouse's alternating current, and the triumph of the VHS video system over Betamax. It seems that decisions made early in a technological cycle may have a lasting impact despite their sub-optimal nature. Chance may play a part in this but system features are also crucial.

Building on this, Martin and Sunley claim that there has been what we may call a 'precedent turn' in economic geography, economic sociology and evolutionary economics in recent years.[97] This has amounted to taking history seriously and has arisen from a general and profound disquiet with the assumptions of rationality implicit in the general equilibrium model of economic behaviour. Path dependence might help to explain the origin and persistence of regional clusters of firms, and maybe also various forms of negative regional lock-in that manifest themselves as aspects of technological and institutional rigidity that are difficult to break when a new competitor region appears. Martin and Sunley point to existing ideas that have some resemblance to this literature, for instance those on palimpsest economic landscapes in the work of Doreen Massey and to a certain extent David Harvey. But they are also critical of the many unresolved aspects of path dependence. It is not clear, for instance, how new paths are established and old ones maintained, and what their relationship is with economic evolution.

The second literature of technological change is based on the work of the French philosopher, Gilbert Simondon. Simondon was interested in inventions and technical developments, for instance engines used in the motor car industry. Design changes and practical adjustments mean the gradual emergence of improved technologies and it is sometimes difficult to identify when a new object has replaced the old one. What is the character of the object and how does it become stabilized? These are the sorts of questions asked by Simondon, for instance in his 1958 book *Du Mode d'Existence des Objets Techniques*.[98] As Isabelle Stengers has pointed out, Simondon was seeking answers to the chicken and egg problem because his objects rarely have a point of origin, nor even clear cut stages of development.[99] To use a simile from Simondon himself, they are rather like a coral reef that grows slowly, building upon its own previous achievements. But, to be recognisable objects, they must have self-referential independence, as in the ignition of a diesel engine or the whiteness of milk.[100]

Although very little of Simondon is in translation, we are fortunate to have summaries of his arguments provided by Massumi, Mackenzie, Toscano and

---

96 Arthur 1989.
97 Martin and Sunley 2006. See also Boschma and Martin 2009; Grabher 2009.
98 This has been translated by Ninian Mellamphy in a thesis for the University of Western Ontario (1980). See also Dumouchel 1992.
99 Stengers 2004.
100 Dumouchel 1992: 414.

Stiegler, and a number of French interpretations and special journal issues have also been published.[101] Simondon's key concept is that of 'individuation', as expounded in books such as *L'Individu et sa Genèse Physico-Biologique* (1964) and *L'Individuation Psychique et Collective* (1989). Simondon's ontological vision is one of becoming, in which the material and the forces that shape it are seen as a pooled presence.[102] He sought a philosophy capable of grasping developments, of genesis, not of identities or substances, which meant returning to a pre-Socratic view of existence. He seems to have been influenced by Bergson and Whitehead, and in turn Deleuze was influenced by Simondon.[103]

Like Simondon, Whitehead saw objects as historical events. His often quoted example is Cleopatra's Needle, the Egyptian obelisk sited by the River Thames in London, which he claimed is a happening made up of 'event-particles'.[104] To him, to Simondon, and to Deleuze, 'all "beings" are just relatively stable moments in a flow of becoming-life'.[105] Deleuze later developed Simondon's individuation further, for instance in the concept of 'machinic assemblages' that he formulated with Guattari.[106]

Simondon understood technical objects 'as evolving composites of relations rather than in terms of function, use, material or form'.[107] This means that they acquire a concretized character as the result of an historical process in which human intervention is both facilitated and constrained by the potentiality of the object. An indicator of this is internal coherence.

Like Whitehead, Simondon sees an individual, not as an entity, but as a process.[108] Henning Schmidgen calls this 'serial being' and for Ajit Nayak it is the 'indivisible continuity of reality'.[109] Following this lead, for us 'becoming milk' is a way of seeing the means by which humans and nature have formed a hybrid with a degree of stability. Simondon's 'ultimate phenomenon' is an assemblage of individuation and individual in which the material object we observe is only one of its phases, the most recent.[110] The successive stages of individuation are accomplished by a process of transduction: the ontogenetic repetition or modification of form.[111]

Simondon's ideas are different from those of the social constructionists. For him, change emerges in the process of individuation, not as a result of the

---

    101   Stiegler 1998, Massumi 2002, Mackenzie 2002, 2005, 2006, Toscano 2006, Combes 1999, Barthélémy 2005, 2006.
    102   Chabot 2003: 77, Chabot 2005.
    103   Stengers 2002, Halewood 2005.
    104   Whitehead 1920.
    105   Colebrook 2002: 125.
    106   Deleuze and Guattari 1987.
    107   Mackenzie 2006: 200.
    108   Shaviro 2007.
    109   Schmidgen 2005: 17, Nayak 2008: 177.
    110   De Beistegui 2005: 118.
    111   Mackenzie 2002. For more on transduction, see Dodge and Kitchin 2005.

interaction between subject and object. This is because individuation is preceded by what he calls a 'pre-individual', which has the potential for change.[112] An example is the DNA of the fertilized egg that produced each of us. This provided a set of genetic potentials but our actual form was a matter of unfolding. This continues now and will continue in the future as we individuate further in the process of ageing. The individual is never final.

One aspect of our bodily individuation is what we eat. We take on the form of that food through the process of nutrition. The scholastics saw this but, rather than embracing it, they developed a theory of substantial change, in which food loses its shape in the body due to the digestive process.[113] But bioarchaeologists can infer much about past diets because certain groups of food leave their signatures in the chemistry of bone collagen. The skeleton, then, is a record of food consumption that can be recovered by stable isotope and mineral analyses.[114]

Conclusion

There are, of course, many other avenues for exploring material and materiality. A favourite of mine is Henri Focillon's *Vie des Formes*, which was first published in 1934 and is still in print. Tom Conley, in his translator's preface to Deleuze's book, *The Fold*, points to the relevance of Focillon's work.[115] Focillon's essay was a theoretical justification of his major work, *The Art of the West*, which rejected the tradition of closed genealogies and grid-like classification of styles of painting or architecture. On the contrary, rather than watertight periodizations, Focillon found artistic hyperlinks across the centuries that produced eclectic hybrids and non sequiturs where they might not have been expected. Focillon's approach was influenced by Bergson, and, later, Focillon's student, George Kubler, continued the interest in the flow of history in his *The Shape of Time*.[116] While making parallels between art history and food history is stretching explanation to its limits, particularly since Focillon and Kubler were both interested in form rather than material, we can nevertheless admire their conclusion that 'things are materialized attempts to solve problems'.[117] Also Kubler offers us a sharp and timely reminder about the role of the scholar in material histories:

> The 'shape of time' is not immediately given in the things themselves, but results from the work of the historian. The series and sequences into which he or she

---

112 Mackenzie 2006.
113 Chabot 2003: 87.
114 Unfortunately, the technology as it stands does not allow us to distinguish between the consumption of meat and dairy produce. Müldner and Richards 2006.
115 Deleuze 2006: ix.
116 Molotiu 2000, Kubler 1962.
117 Max Planck Institute for the History of Science 2006: 6.

groups the forms of things and problems retrospectively alter the arrangements of
things that were hitherto accepted. As a consequence, the historian changes even
the forms themselves. Against this background, the history of experimentation
might be read as a succession of shapes, the production of which sets in motion
a cascade of retroactive re-shapings.[118]

So, sources of inspiration for the present volume are potentially many and
varied. I suggest that so far we have formulated four conceptual seeds that can
be grown on and planted out. First, no matter what the labelling on their work –
historical epistemology or historical ontology – I take heart from those authors,
all heavily influenced by Foucault, who have taken a genealogical stance on
knowledge. Starting with Chapters 2 and 3, we will learn how intellectual context
was crucial for the way in which farmers, experimental scientists, politicians, and
the attentive public saw and understood milk. This tells us a great deal about ways
of thinking but also sheds light on the material itself. What was milk after all?

Second, readers will not be surprised that social and material constructionism
has influenced my thinking to a certain extent. It has been a powerful movement
in the last twenty years and its logic remains deeply embedded in much post-
structuralist writing. But constructionism has passed its apogee and post-human
theory and the various forms of materialism have chipped away at its credibility.
Rather than reproducing the social-natural dichotomy yet again, scholars are
trying to find ways to let the material speak for itself. Also, some new writing is
stressing, in retrospect, how foolish it was to ascribe god-like, nature-changing
powers to individuals, institutions, and whole societies who patently lack the
resources, expertise, technologies, and political willpower to intervene, for
instance, in the major socio-environmental problems of the day, such as climate
change. The inherent resistance and messiness of socio-natures is now a focus of
study, and in science studies Andrew Pickering's mangle of practice looks much
closer to a practical framework of analysis than much of the earlier, frankly naïve
constructionism.

Third, our account of non-representational theories in general has thrown up
a number of ideas to take forward. Work on practice, especially Mol's discussion
of the fragility of object identity, will be of help, although obviously my historical
interpretations along these lines are in a different register to her praxiographies.
The hybridities of actor network theory will also provide some energy to my story,
although I must repeat that borrowing some of Latour's concepts does not mean
that we have to swallow his metaphysics.

Fourth, and finally, there are several processual literatures that strike a chord.
Ingold's reading of Heidegger provides a helpful insight into *thing*ness, as does
Smith's Husserlian ontogenesis of bodily form. But it is Simondon's philosophy
that, for me, is the most exciting. It is potentially relevant to the present book if we
count milk as form of technology. As we will see, it was a fluid that was 'made'

---

118   Max Planck Institute for the History of Science 2006: 6.

by producers. They bred cattle that gave large volumes of low fat milk and they designed feeding regimes for maximum output, a development so profound that Manuel De Landa sees it as a new form of biological history.[119] There were then issues about the legitimacy of manipulating the constituents of milk as it made its way through the supply chain. Much of our discussion will be about the means of detecting fraud and imposing sanctions. Finally, public policy with regard to a legal definition of milk will be another of our themes. Although a Simondian analysis is possible for items of food and drink such as bread, meat, wine, it seems to me that milk is the most appropriate commodity.

In summary, then, this book is about a product which, though it retained the undifferentiated and unqualified name of 'milk' throughout the period under study, was highly variable. As a technological object, its material form went through bursts of rapid change and long periods of stability. It has existed in a state of what Simondon called 'metastability', exhibiting tension, never static. To treat milk in the nineteenth and twentieth centuries as a homogeneous, natural foodstuff would be inappropriate. Its individuation, its change, is the subject of this book.

---

119   De Landa 2000: 163–4.

Chapter 2

Daniel Schrumpf: A Trial by Instruments

I would say that there are no two samples that have the same composition, nor is there any such thing as an average composition of milk.

Testimony of Charles F. Chandler,
New York Court of General Sessions, 20 December 1876.[1]

'I never have tried a more important question.' So Judge Sutherland of the New York Court of Sessions summed up a case that he presided over in December 1876 – *The People v. Daniel Schrumpf: Misdemeanour, Adulteration of Milk.*[2] But what was the question? Was it the means of establishing the genuineness of a common foodstuff or was it the more fundamental, material question of 'what is milk'?

In the last three decades of the nineteenth century there were thousands of milk adulteration cases each year on both sides of the Atlantic but what set this one apart was a scientific controversy about methods of detecting deviations from the expected characteristics of pure, natural milk. It was a test case, in the form of a disagreement between experts appearing on behalf of the New York Board of Health on the one hand, and for the local Milk Dealers' Association on the other.

Since its inception in 1874, the Board had been trying to clean up the city's milk trade and the instrument of choice for its inspectors was the lactometer, a simple device that compared the specific gravity of a sample against that of 'genuine' milk. By this date lactometers had been in regular use for about eighty years in Europe, but the technology was controversial and became the central issue in *The People v. Schrumpf*. There were 30 milk adulteration cases awaiting trial in New York at the end of 1876, and it seems that both sides (the city authorities and the representatives of the milk trade) had been waiting for one that was suitable for intensive scrutiny. This was it. The hearing began on 15 December with a jury being sworn in and Schrumpf was charged under Section 186 of the New York Sanitary Code, which stated that:

> no milk which has been watered, adulterated, reduced, or changed in any respect by the addition of water or other substance, or by the removal of cream, shall be brought into, held, kept or offered for sale at any place in the city of New York, nor shall any one keep, have or offer for sale in the said city any such milk.

---

1   Court of Sessions 1881: 64.
2   Court of Sessions 1881: 269.

The defence case was built upon the claim that the use of lactometers in policing the city's milk supply was inappropriate given their well-known deficiencies in distinguishing between adulterated and genuine milk. This argument was made out in great detail, over eight days of testimony, using 25 witnesses. The official transcript runs to 330 pages and the very fact of its publication suggests that the Board of Health thought that it would serve an exemplary function. The unofficial account in one of the city's newspapers, the *New York Times*, recounting what they called this 'now celebrated milk case', was also exhaustive. It totalled 8,400 words in 10 articles.

The case was unusual in that the dry ingredients of legal argument were supplemented by 'performance', with some of the experts making illustrative experiments in real time. The Court Room was temporarily turned into a laboratory as both sides attempted to recruit science to their ontological cause.

> When the jurors … entered the Court of General Quarter Sessions yesterday a curious array of exhibits met their gaze. Over the judge's bench was hung an immense linen placard resembling an election bulletin board, on which was painted in black letters the result of a number of scientific experiments on the milk of dairy cows. Similar placards giving the result of experiments by celebrated French chemists, were displayed on the walls in front of the jury, and attracted general attention. The tables used by counsel for the defence were covered with chemical books and papers, and on the table in front of the bar was an imposing array of glass tubes, bottles, measures, and other vessels, a miniature drying oven, and two steaming water baths, heated by gas drawn through a rubber pipe.[3]

Daniel Schrumpf was a small-time milk retailer at 206 Avenue B in Manhattan. He had been visited on 26 August 1876 by Dr John Blake White, who worked as an occasional milk inspector for the Board. The lactometer reading taken by Dr White on this occasion indicated an adulteration of about 15 per cent and it was this measurement that was taken to be sufficient proof to justify a prosecution. The measurement represented the right of society to comment on food quality and to enforce standards. The assertion made by the Board of a deviation from natural milk was one of the earliest and highest profile claims anywhere in the world about the knowability of natural milk in the normative sense.

All of the 'experts' called in this case claimed some scientific background in order to establish their credentials. When they took the stand, a short biography was requested by counsel, including a travelogue of the European laboratories where they had trained or just visited to observe the great experimenters at work. All of these institutions were in Germany, which in the middle of the nineteenth century was a centre of expertise on organic chemistry. Three witnesses had studied at Göttingen, and two at Heidelberg; and there had also been visits to

3 *New York Times* 27 December 1876: 3.

Berlin, Stuttgart and Leipzig. The expert witnesses were also required to list the published authorities upon which they based their views. This appeared to take several off guard and they had to return to court later with lists of references. Of the 26 works cited overall, 11 were British, seven German and six French. No papers or books of American origin were discussed.

Lactometers had been used in the City of New York since 1874 when the first two milk inspectors started work. The simplicity of this type of equipment was part of its attraction, as was the cost (about $1 each). Lactometers also had a democratic quality since the milk dealers themselves could also use their own to check their supplies. In cross examination by counsel for the defence, the positive and negative technical aspects of lactometers were laid bare. It emerged that the Board of Health regarded a specific gravity of 1.029 at 60°F as the lowest that could be associated with genuine milk, basing their instrument upon the design of the 'galactometer' of the French chemist Dinocourt.[4] This was a somewhat arbitrary threshold and was deliberately set low in order to avoid capturing natural milk of a poor quality. The Board's pragmatism was matched by their preference for maintaining the use of the traditional organoleptic skills of milk inspectors: taste, smell and the visual test of whether the milk adhered to the glass, as genuine whole milk would do. In short, it seems that by this date the New York authorities had not plunged into the highly skilled and expensive systems of chemical analysis that were now appearing in Europe. Their solution instead was cheap, semi-skilled and practical, but it was also scientifically rather crude and certainly open to question. Even the defence witnesses were at times unsure of the technology. Professor Chandler, for instance, was challenged, no doubt because of his vested interest in the Board of Health's lactometers, which were tested and calibrated in his laboratory. He was reminded by counsel for the defence of an article that he had published in the *American Chemist* in 1871 in which he stated that 'the lactometer is a very unreliable guide, as skimming causes the milk to appear better, while watering exerts the opposite effect'. The *New York Times* reported that this exchange between counsel and witness drew 'derisive and prolonged laughter' in the court room.[5]

Some of the witnesses were asked by the defence to say whether a white liquid given to them was milk. The 'is this milk?' question was delicately sidestepped by those who guessed that this was a game of ontological politics and that they were being set up in order to demonstrate the difficulty of confirming whole milk visually. One, Professor Henry Morton, a physicist and President of the Stevens Institute of Technology in New Jersey, instead offered a contextual definition. He claimed that if a substance was sold to him as milk, then it was reasonable for him to assume that it was the genuine article.

---

4 See Chapter 3, 'The First Story: The Physical Properties of Milk', pp. 61–7.
5 Berry 1993: 80.

I may illustrate what I mean perhaps by another example; if I went into a grocery store and asked for some eggs, and articles were given to me which looked like eggs, and I bought them and paid for them, I should feel myself justified in swearing that they were eggs, that I knew them to be eggs; if I went into Heller's establishment up Broadway, the gentleman who sells material for legerdemain, and saw an article which looked exactly like an egg so that I could not tell them apart, I should not feel myself justified in swearing, nor would I feel sure that that was an egg. I consider that my knowledge of anything is not derived from one, two or three circumstances involved, but from all of them, and it is only upon the basis of all the circumstances involved that that knowledge can be weighed and established.[6]

Professor Robert Ogden Doremus, an analytical chemist, was called for the defence. Although famous in New York mainly for his testimony in poison cases, he had also taken a special interest in milk for nearly 30 years, writing a report on milk supply for the city's Common Council and claiming expertise in the equipment used in testing milk quality. He came with a hard-line reputation, having previously called the lactometer: 'that perverse and mendacious instrument.'[7] His evidence on this occasion was similar:

Knowledge gained by the lactometer is of no value; the lactometer will test the gravity of milk or any liquid approximating the gravity of milk; the lactometer tests the gravity of milk, which knowledge is of no value at all in determining the goodness of milk.[8]

Doremus objected to lactometers on principle because a 'genuine' reading of specific gravity could be simulated by a combination of watering and extraction of cream. As a technique, their use was therefore indiscriminate. But he also objected on the grounds that lactometers were crude instruments that were rarely properly calibrated:

If you get an instrument costing a dollar, I defy any man ordinarily to expect that you will have accurate gradations any more than you will find thermometers hanging on the wall costing fifty cents to agree … If we are to depend upon the fraction of an inch to decide whether a man is guilty of a crime or not we ought to have the instrument exact.[9]

---

6   Court of Sessions 1881: 100.
7   *New York Times* 29 January 1876: 4. See also 'Professor Doremus on Milk', *New York Times* 27 January 1876; and *Chemical News* 14 January 1876: 20.
8   Court of Sessions 1881: 186.
9   Court of Sessions 1881: 187.

Doremus' tactics in the witness box were to show that milk is a variable substance. He and his two sons, Thomas and Charles, had visited farms in Orange County, a dairying area close to New York, which supplied the city. He gave long lists of detailed laboratory observations that they had made using a pycnometer, showing, cow by cow, how milk samples taken in their presence had had a wide range of specific gravities.[10] Doremus' testimony was long-winded and occasionally he had to be reminded by counsel or the judge to answer the question put. But his *coup de théâtre* was a demonstration of how a chemical test could be performed in the laboratory. The court seems to have been taken with this and even allowed defence counsel to proceed with questioning him while a Bunsen burner was heating a sample of milk. Doremus was also indulged in being permitted to read out long translations of passages from French and German authors about the status of the lactometer. This performance dominated the trial but did not provide the knock-out blow desired by the defence. Doremus could not resist giving answers that showed his (and other scientists') lack of control over the material substance of milk and the resulting on-going uncertainty about its characteristics.

Cross-examination by counsel for the prosecution:

Q. Is there any normal standard of milk as to the percentage of water it should contain?

A. I give it up; I leave that, your Honour, to the Board of Health to regulate; I have no standard.

Q. You have no standard?

A. I have no standard.

Q. If there is no standard how can you tell if water has been added to it?

A. I can tell that, but as to the standard of what the specific gravity of milk shall be, is another question.

Q. I did not ask that. A. I understood you so.

Q. Can you tell what is the standard of water in milk?

(Objection)

Q. Can you tell me the percentage of water in milk?

A. It varies with the different authors; and different chemists have different standards; and the Board of Health have adopted a different standard; I do not profess to say what shall be the standard.

By the Court Q. The question is, what is the average percentage of water found in milk?

A. I can state in answer to this what I have found, but I won't state that as a standard, by any means.

Counsel for the prosecution Q. Is there any such thing as an average percentage of water in milk? A. There is an average percentage differing with different chemists; some claim one and some another.

---

10 A pycnometer is a special bottle used for determining the specific gravity of liquids. It is weighed first when filled with water and then with milk.

Q. Is there any agreement upon it at all?

A. There is none.

Q. Within what limits does it range?

A. Well, sir, from perhaps 80 to 90 per cent., or a little over; I could refer your Honour to various authorities; some put the lowest gravity down to 1.026 and others again to 1.028, others at 1.030, others at an average of 1.029 and a half, others 1.030, others 1.028; I could give you half a dozen of authorities; there is no fixed standard.

Q. I understood you to say, that you could determine the amount of water added to the milk in adulteration by a process of your own?

A. Yes, sir.

Q. If there is no fixed standard of the percentage of water in milk, how can you distinguish between that which is added and that which belongs to the milk?

A. There is in the serum of the milk an average standard; it ranges from a certain point to a certain point.[11]

The prosecution was also able to counter Doremus with experts of their own, who were willing to speak favourably of the lactometer; it was then up to the jury to balance the credentials of witnesses on both sides. The defence exposure of the variability of milk, by quoting the composition of milks on farms in Orange County, seems not to have helped. It was successfully ridiculed in his closing address by W.P. Prentice, counsel for the prosecution, whose witty and forceful speech convinced the jury. They took only 45 minutes to convict Schrumpf and the Board's use of the lactometer was vindicated.[12] In his final address to the accused, the judge seems to have accepted some mitigation.

> I have not any feeling against you. I suspect you are as honest a man as there may be in your line of life; that is my own opinion. The sentence of the Court is that you be confined in the City Prison for the term of ten days, and that you pay a fine of two hundred and fifty dollars.[13]

This was a very heavy punishment for what everyone knew was routine cheating in the milk trade. Schrumpf's reputation lay in tatters. He found the legal and media spotlight stressful and, when his appeal to the New York Supreme Court later failed, he suffered a mental collapse.[14] The lower court had already accepted that he had been the unwitting victim of his suppliers but commented that it had nevertheless been his responsibility to keep and sell only milk that was genuine.

---

11   Court of Sessions 1881: 201.
12   *New York Times* 29 December 1876: 2.
13   Court of Sessions 1881: 270.
14   *New York Times* 12 October 1877: 3.

It would seem at first, however, it was rather a harsh law that you should be convicted of keeping watered milk when you did not know it was watered, but I can see plainly that unless that law can be enforced I do not see how the Board of Health is going to protect the public against watered milk. The idea, considering the quantity of milk that comes to this city, that they must resort to analysis and have a skillful chemist to analyze the milk would be very impracticable it appears to me.[15]

As a non-English-speaking immigrant, Schrumpf was unfamiliar with trade practices in New York and was almost certainly cheated by his supplier. During the trial, counsel called him the 'vicarious sacrifice' of the New York Milk Dealers' Association and he certainly seems to have been a convenient lightening rod for the controversy. Before the trial there was uncertainty and bitterness; afterwards both sides at least knew where they stood and were able to adjust accordingly.

We will never know whether Schrumpf was complacent or complicit, but his alleged crime, his trial and conviction stand at an important intersection between nature, science and the law of the kind that we will explore in this and following chapters. The questions are easy to formulate but exceptionally difficult to answer. Why was it that milk was so difficult to define and to measure? Why did dairy physics and chemistry develop so late that in 1876 there were still fundamental debates in public about the best measuring instrument? We have seen that the New York Board of Health was searching for a means to stop milk adulteration and to send a message to the milk trade that, no matter how minor their offence, the will was there to pursue it through the courts. The lactometer continued to be used in New York until 1895, when at last the Board had to acknowledge that dealers had become so skilful in beating the instrument that a chemical standard was set and the regular collection and analysis of milk samples then began.[16]

---

15 Judge Sutherland, Court of Sessions 1881: 269.
16 Parker 1917: 373.

PART II
In Search of Milk

Introduction to Part II

As the story of only one man, Chapter 2 is little more than an *amuse bouche* in the vast banquet of food history. Nevertheless it provides an important preliminary view of the problems of knowing milk. Early methods and equipment, in the everyday practical world at least, faced difficulties in perceiving and measuring. Frustrated observers on all sides of the milk industry were reduced to looking for traces of what otherwise remained hidden. Steve Hinchliffe has written on this in the rather different context of biogeography. His urban water voles are elusive creatures and often fresh footprints and active latrines are the only evidence of their continued presence. He suggests that knowing them is not a matter of revealing, but rather one of making them present.[1] So it was with milk.

Chapter 3 will look in greater detail at the history of making knowledge about the properties of milk through dairy physics and chemistry. First, we will reflect upon some methodological concerns about measurement and the various approaches adopted in the sociology of scientific knowledge. Next, we will try to understand why milk was so important in the early history of organic chemistry, and why animal chemistry generally was such a focus for great chemists of the early nineteenth century, such as Berzelius and Liebig. After that, the chapter resolves itself into 10 stories. These are, in effect, the successive individuations in the technological and theoretical development in dairy science, from the 1790s onwards. The unfolding of these stories, their intertwining and overlapping, is the result of what Andrew Pickering calls a dialectic of resistance and accommodation. My argument throughout this book is that looking at material paths and entanglements of food is more fruitful than attempting an historical meta-narrative.

Chapter 4 then turns to expertise. The Schrumpf case demonstrated the problem of uncertainty in scientific knowledge, an issue that continues today in animal biochemistry, as the recent BSE Enquiry showed in great depth.[2] My emphasis is upon two major themes. First, there is the intriguing prospect of struggles between rival groups within the circles of analytical chemistry at the end of the nineteenth century. This was partly a matter of personalities but methods of working were also at stake, and, ultimately, so was scientific authority under the law. Second, we will discuss whether expertise should be seen as concentrated and hierarchical, or distributed. There is a strong case for the latter, and the logic of knowledge and power being everywhere is that we may need fundamentally to reassess our understanding of material quality. If the adulteration of foods can be seen as the

---

1   Hinchliffe 2007: 127. See also Hinchliffe et al. 2005, Hinchliffe and Whatmore 2006.

2   Hinchliffe 2001.

application of skill and entrepreneurial initiative, where does our moral judgement stand? It also reopens the debate as to whether the human modification of nature can ever successfully be subject to legal sanctions, because one person's fraud is another person's new product.[3] To put this argument into a recent context, the British consumer now buys semi-skimmed milk with barely a thought about its naturalness; but one hundred years ago selling such a product under the designation of 'milk' would have contravened the Sale of Milk Regulations and attracted a punishment.

Nature, Material Production and Metabolism

> Henceforth, nature will be the title given to those things that have been solidified and made transcendant and society will be the label given to things that are malleable and in the making.[4]

At this point I wish to pause for a theoretical deliberation that expands a little upon the reflections in Chapter 1. This will assist with what is to follow. In particular, I wish to ponder the relevance of constructionism and its variants for the tale of milk.

According the Kearnes, there are times in the cultural studies literature when 'the material operates as a sign for the natural'.[5] We put this to the test whenever we eat raw foods such as fruit, nuts and seeds in the belief that, in their unprocessed state, they are somehow good for us. Raw, unpasteurized milk throughout most of the twentieth century, and right up to the present day, has been sought out for the same reason, although the public health authorities have made it increasingly difficult to find, to the extent that in America it has now acquired contraband status in some states. Milk, anyway, has multiple additional signs of 'naturalness' derived from its 'essential' white colour and its association with lactating motherhood. The latter has such an extraordinarily iconic status, seemingly beyond criticism, that it deserves a full study in its own right.[6]

There are links here to the common Aristotelian understanding of nature identified by Raymond Williams as that portion of the material world that is not of our own making.[7] It is 'out there', an inherent force which directs the world or human beings or both. Thus, histories of the essentials of human life, such as food and drink, may, in their association with that pristine myth of purity, the 'organic', and with the romance of a simpler past untouched by chemicals in either

---

3   Stanziani 2007b.
4   Ward 1996: 127.
5   Kearnes 2003: 144.
6   In the meantime see Hayes 2008.
7   Williams 1980.

agriculture or the processing industries, lose sight of the complex constructions of quality and trust that facilitate the growth of all commodity chains.

A number of other perspectives on nature have proved to be popular in recent decades. One of the most important, in the Marxist tradition, is that of the 'production' of nature. Henri Lefebvre, David Harvey and Neil Smith are particularly associated with this idea, which, for those of us interested in food, has an obvious relevance in an era of technologies so advanced that the DNA of crops can be modified and animal clones produced.[8] Another writer in this area is Erik Swyngedouw, who uses Marx's concept of 'metabolic relations' to understand the urbanization of nature.[9] He traces the metaphor back to seventeenth century discussions about the circulation of blood, adapted in about 1750 to the circulation of money. In the nineteenth century the idea was then the underpinning of the sanitary movement's replanning of cities. Sewers, water mains, and other improvements in physical infrastructure such as food markets, were conceived in terms of the circulation of both the basic needs and wastes of the human body. The city was to be nourished and cleansed as if it was the collective body of its inhabitants.

Swyngedouw's cities are complex, layered palimpsests of inputs (commodities, energy, sounds, bodily performances) that operate at spatial scales varying from the local to the global.[10] He introduces the notion of technological networks, by which he means that the city's metabolism is powered by underground water and gas pipes, electricity and telephone cables, and sewers. He argues that 'technological networks are the material mediators between nature and the city; they carry the flow and the process of transformation of one into the other'.[11] Writers such as Steve Graham have expanded on this idea and pointed out that increasing reliance upon these technologies introduces vulnerabilities as well as efficiencies.[12]

The production of socio-nature has a great deal to tell us about the material world. Swyngedouw's work on water is a good example. He argues that Spanish modernization has been about historical spatial-ecological transformations, particularly with regard to the use of water. He has also written at length about the water supply of Guayaquil in Ecuador and concludes that such urban environments are 'combined socio-physical constructions' reflecting the materials and power that circulate within them. He does not see anything unnatural about such produced environments and insists on a political ecology approach to understanding their physical conditions and social qualities. By this he means that he can analyse 'who gains from and who pays for, who benefits from and who suffers (and in what ways)'.[13]

---

8   Lefebvre 1991, Smith 1984, Harvey 1996, Braun 2006b.
9   Swyngedouw 2006.
10 Swyngedouw 2006. See also Kaika 2005.
11 Kaika and Swyngedouw 2000: 120.
12 Graham and Marvin 2001.
13 Swyngedouw 2004: 23–4.

Bakker and Bridge are somewhat sceptical of the historical materialism of Smith, Swyngedouw and those who talk of the production of nature.[14] This is, first, because there is a common emphasis upon use value generated by the labour process rather than any discussion of the agential capacity of biophysical processes. Second, they point to the neglect of 'the empirical stuff of nature' in commodity chains and systems of provision. But one suspects that their real, underlying criticism is unease that this production of nature literature is a subset of the vast, complex and controversial strand of social science known as 'social constructionism'.

In its simplest terms the argument of constructionists is that the ontological 'reality' and stability of the world cannot be taken as self-evident. A famous example is the concept of wilderness and how, in our search for a nature external to human control, we have lost sight of the conservation strategies that have created a particular version of the wild in national parks and other remote areas.[15] Similarly, a great deal of work has gone into making 'organic food' seem the most 'natural' that can be obtained, for instance, with no chemical fertilizers or pesticides used in its production. The Soil Association is the current arbiter in Britain but the origins of this particular effort at nature-making go back more than 80 years to when food was political in a rather different way from today.[16]

Extending the argument still further, it would seem reasonable to a material constructionist to claim that a 'natural' product such as milk is, on close inspection, fully in the human realm. They would use similar evidence to that presented in this volume: (a) breeding has affected volume and composition; (b) trade practices, including fraud, have made a milk that is more congenial to the structures of profit taking in a capitalized dairy industry; (c) the state has intervened to standardize composition and reduce the risk of disease. Constructionism has also invaded the laboratory with claims that science is a domain of human action from which nature is excluded, because laboratories are artificial environments in which everything is controlled as far as possible.[17]

As is well known from the friction generated by the so-called Science Wars of the late 1980s and 1990s, constructionist arguments are deeply controversial. It is not necessary for us to dwell on the enfilade fire that they have received from the natural scientists, on the one hand, who felt affronted that their methods of making truth claims were being questioned, and, on the other hand, from social realists, whose attacks were because constructionists were said to be undermining the possibilities of political engagement.[18]

---

14   Bakker and Bridge 2006.
15   Cronon 1995.
16   Conford 2001.
17   Knorr Cetina 1981, 1983: 119.
18   For more on the problems of social constructionism, see Hacking 1999, Latour 2003, Sismondo 2004.

It is worth noting Andrew Pickering's *Mangle of Practice* as a mature variant of constructionism which has much to commend it. In this we learn about the performances of scientists and their materials and what Pickering calls the 'dance of agency' back and forth. He has developed a 'dialectics of resistance and accommodation', and, in so doing, he moved away from the stronger versions of social constructionism, which had gone too far, in his view, towards a kind of sociological fundamentalism, by making social interests the key to scientific knowledge. Pickering, instead, sees science as muddling along towards understandings that it often fails to predict in advance, and adapting to the messy situations that arise, again to the surprise of all of the participants.

An example of Pickering's mangle from my own research on water quality is the problem of the arsenic contamination of the groundwater in Bangladesh.[19] In the 1970s many infants and children there were dying from diarrhoeal diseases caught from drinking unclean water from ponds and rivers. This prompted an admirable response from the international community encouraging the drilling of tube wells down to aquifers in the sediment where the water was of a much higher quality microbially. What no-one anticipated was that this water carried a small but deadly load of arsenic, which over the subsequent decades has been responsible for internal cancers and skin conditions, putting tens of millions of people at risk. Geochemists have responded to this emergency, which incidentally is said to be the world's largest environmental health catastrophe to date, by publishing thousands of scientific papers on the chemical processes behind the release of arsenic, and there is also a growing social science literature of this 'release' of mortality and morbidity by human intervention. The disaster has led to advances in field measurement and in technologies of mitigation on the back of research funding that would not have been justified otherwise, and here we have a mangle of scientific and administrative practice that was a surprise to everyone concerned, with outcomes that are still difficult to anticipate in the immediate future. The situation in Bangladesh gives the impression of a transgressive underground world that is out of control, which cannot be dominated, and which has shaken the self-confidence of all of the experts involved, including those who recommend models of governance to cope with the vast scale of the hazard. Modernity in all of its guises has been put in its place, and it is easy to see that, under these circumstances, an ontology of becoming makes sense.

I will refer again to Pickering's mangle in Chapter 3, which will be our bridge into a history of the knowability of milk through the laboratory protocols of modern science and technologies of measurement.

---

19   Atkins et al. 2006, 2007a, 2007b.

Chapter 3
Seeking the Natural:
Laboratories and the 'Knowability' of Milk

Materials and Measurement: *À Peu Près*

> The dance of agency ... takes the form of a dialectic of resistance and accommodation ... The contours of material and social agency are mangled in practice, meaning emergently transformed and delineated in the dialectic of resistance and accommodation.[1]

One task of this chapter is to illustrate the obstinate reluctance of the material – the milk, the laboratory equipment, the equations – to cooperate in the process of analytical exploration. For liquids in general, Bernadette Bensaude-Vincent finds that hydrometers are the perfect illustration of this phenomenon.[2] Whereas, by the late eighteenth century, the precision balance had already allowed a simplification of the interrogation and analysis of solids, the specific gravity of liquids remained difficult to apprehend and difficulties with designing, manufacturing and using the hydrometer confirmed instead nature's 'complexity and the limits of our rationalization of the real world'. Hydrometers were, in her words, a materialization of Archimedes' law, but they were, throughout our period, capable of little more than approximate measurement: 'à peu près' [almost]. Even the great Lavoisier was unable to make one that could be used in all liquids, and eventually the world of hydrometry fragmented into many different sub-fields, each geared to the measurement of one commercially significant fluid. By 1800 there were lactometers for milk, saccharometers for sugar in the brewing process, acidimeters for laboratory reagents, salinometers for measuring salt in solution, and so on. These were not designed to be interchangeable, although the underlying principle is universal and local adaptations do seem to have been common. For instance, the London Hospital in the 1840s was using a urinometer (suitably sterilized one hopes) to measure the specific gravity of milk and in 1865 Payen claimed that Baumé's areometer, designed to test the strength of alcohol, was regularly pressed into service for milk analysis in French wine-producing districts.[3]

As we will see, temporary alliances between humans and materials were necessary to overcome such complexity. These were contingent, especially in the

---

1   Pickering 1995: 22–3.
2   Bensaude-Vincent 2000.
3   Pereira 1843: 254, Payen 1865: 155.

nineteenth century, when dairy chemistry had not yet advanced to the point of reliable experimental replicability, so results were variable, as were the opinions based upon them. New equipment, better chemicals, and revised theories led to new alliances, also temporary, and the beast gradually edged forward to something approaching a consensus at the turn of the twentieth century. Even then there was plenty of controversy about measuring and knowing milk, right through to the 1950s.

James Sumner has explored the by now well-established idea that measurement and the quantities that arise are social constructs.[4] His object of study was the eighteenth century brewing industry and particularly the origins of a measure that came to be called 'pounds-per-barrel extract'. The pioneer of this was John Richardson and the story is about his struggle to justify the concept and secure its adoption. Sumner finds that a number of criteria that must be applied to innovations of this type if they are to be successful. First, an already well-known material property must be used and enhanced in the minds of those in the trade. This can be what Sarah Whatmore calls an 'everyday familiar'. Second, the new measurement must be 'reliable' and in line with the 'common sense', naturalized properties of the material – in other words, not too radical a new perspective. Third, it must convey practical benefits in order to encourage adoption. It so happens that Sumner was writing about a saccharometer used to measure the properties of wort, but he might equally well have been discussing the lactometer used in Daniel Schrumpf's case. Both were used to measure the specific gravity of commercial liquids and both were controversial in their day.

This is progress because Sumner's approach helps us to envision the way measurement became a central feature of the exploitation of commercial liquids. But it is worth taking a pause here in order to consider an epistemological tension.

Sumner is writing in the style of what Demeritt calls 'heterogeneous constructionism'.[5] This is ontologically realist about entities but epistemologically antirealist. Such a stance does not deny the objective reality of objects but it does assert the right to view those objects, and the science that studies them, as constructed 'through specific and negotiated articulation of heterogeneous social actors'. For example, the quality checking systems that emerged for milk were conceived from a network-shaping series of social practices, standardized measuring instruments, and analytical procedures. We will look at these in detail below.

From a rather different perspective, Andrew Pickering reminds us that constructionism draws its purpose from human agency. In other words it underplays non-human agency in what Pickering calls 'the machinic field of science'.[6] He prefers to stress temporal emergence in scientific practice, which for him is performative rather than representational or semiotic. His evocative metaphor is a

---

4   Sumner 2001.
5   Demeritt 1998, 2001, 2002. See also Sismondo 2004: ch. 6.
6   Pickering 1995: 15.

'mangle of practice' in which the goals of science emerge in real time, driven by chance in a matrix of 'brute contingency'.[7] In a recent paper, Pickering counters any notion of naive constructionism by using a 'de-centred' (i.e. symmetrical) account of industrial chemistry in which the properties of the materials that combined in the manufacture of the hugely successful and innovative artificial dye of the 1850s, mauve, were not predicted or planned. Here was the discovery of a new transit of matter, although its harnessing was in the social realm, particularly its successful translation from laboratory to factory production. Pickering argues for 'the irreducibility of material performances to the social'.[8] He sees the science associated with synthetic dyes as 'evolving under the sign of an impure, decentred, material-social dynamic, structured jointly by what was possible in the laboratory, by what happened in the laboratory (for example by whatever material transits were achieved there …), and also by social and extra-scientific concerns with specific materials (themselves evolving in relation to the science) …'. This was an early example of technoscience, the mixing of science and industrial technology.

Pickering's de-centred histories are an inspiration for the present book, as will shortly become clear. One reason is the sheer material force of milk. The complexity of its physics and chemistry made the instrumental apprehension of its qualities by no means straightforward. At the outset, and right through the nineteenth century, its particular characteristics determined what could be measured and so we can say, in the words of Gaston Bachelard, that its physicality 'transformed the method of knowing'.[9]

What to measure was less of a problem than how to measure. With regard to the composition of milk at least, there was consensus up to the 1980s that fat was the most important ingredient. The interest occasionally shown in total solids and in water content was merely a way, by subtraction, of calculating fat. This is certainly different from other measurement debates of the day. Graeme Gooday, for instance, entitles his book on Victorian electricity *The Morals of Measurement* because of the fierce contention between actors about the parameters of interest.[10] His instrument makers were confused by the sheer breadth of user needs and by their values or lack of them. Gooday's four moral questions would certainly have been of interest to science, but they were also poignant in a society where measurement was becoming a matter of legal contention: (a) were previous measurements reliable?; (b) were measurements conducted in a trustworthy manner?; (c) were published accounts reliable?; (d) what were the wider implications of a measurement? In a climate of uncertainty, accuracy came to be a marker of trust and the difficulty of achieving that trust was a motivation for further technical progress.

One of the most fascinating recent studies on measurement at first sight seems to be a long way from the sphere of food studies, but its general principles are

---

7   Pickering 1995: 24.
8   Pickering 2005: 369.
9   Bachelard 1949/1998: 56.
10   Gooday 2004.

certainly relevant. Donald McKenzie's account of the enhancement of accuracy in post-war missile systems is a triumph of careful scholarship. He teases out how and when crucial decisions were made about whether to refine existing technologies or to go with new ones. He finds key individuals responsible for those decisions and for persuading politicians to choose one funding option over another. He looks at controversies between scientists and delves into the most fundamental of all, whether to invest in strategic bombers or inter-continental ballistic missiles. In short, his book tells the story of the 'social' nature of technology and the constitutively 'political' character of deploying resources, even those in the humblest laboratory. What MacKenzie achieves is an understanding of the creation of accuracy that draws upon the sociology of scientific knowledge. This suggests to him that central to his story

> are matters of the interests, goals, traditions, and experiences of the social groups (technological and other) involved; of the convention surrounding technological testing; and of the relative prestige and credibility of different links in the network of knowledge.[11]

In what is to follow, we will put laboratory practice and instrumental measurement at centre stage and use them to demonstrate the difficulties of generating knowledge about milk and therefore also the knowing of nature.

So, What is Milk?

> The difficulty of fixing standards for natural products is universally admitted, Nature does not allow herself to be confined in narrow bounds, but delights in pleasant variations ... The very name of 'milk' has been used by different persons in quite a different sense. Some understand by it the unchanged product as drawn from the cow, while others do not trouble about a little more or less fat, and cannot see why 'skim milk' should not be called milk. 'Fresh milk' has been used as a word of distinction, but in our days of centrifugal cream separators, the poorest skim milk may be had as fresh and sweet as one can wish.[12]

Milk was an object of scientific curiosity in the eighteenth and early nineteenth centuries, for at least three different reasons. First, it was of economic significance and there was a desire to understand the everyday observation that the milk of different breeds of cow changed in butterfat content. It was known that from day to day the milk of the same animal might vary depending on the fodder it was given and the stage of its lactation.[13] A chemistry of the processes of manufacturing

11   McKenzie 1990: 10–11.
12   Paul Vieth, in Anon. 1886: 6 and 8.
13   Haller 1779.

butter, cheese and other dairy products was also emerging. Together, these contributed to the new 'animal chemistry' of laboratory experimentalists such as Jacob Berzelius, the renowned Swedish chemist, who wrote his first work on milk in 1808.[14] The study of bodily fluids such as blood, milk, saliva, bile and urine had at first been academic and medical in intent but then became an important sub-order of organic chemistry. However, when organic chemistry took off in the 1820s, as discoveries led to the industrial manufacture of products such as acetic acid and urea, it left behind those substances which lacked an invariable, definite quantitative composition.[15] In the 1830s and 1840s, in the hands of scientists such as Liebig, the analysis and synthesis of complex organic molecules acquired a status of global significance.[16] Animal physiology was now only a minor part of this agenda, although it still had controversial potential, for instance the bitter dispute between French and German workers about the processes of metabolism, which was settled in the 1840s by nutritional studies of cow and pig feeding.[17]

Second, food fraud was a live issue at this time and milk was the commodity most tampered with. But 'water added to the milk after it leaves the cow cannot be distinguished from the water that the cow gives in the milk',[18] so there was no direct trace of the adulteration, only inference from the balance of the constituents overall, which required a knowledge of its 'genuine' constituents. The two went together and the latter was a driving force behind dairy chemistry in the second half of the nineteenth century, when such falsification was at its peak. In Kantian terms it was first necessary to create the precondition for the possibility of a phenomenon by imagining the composition of a natural, whole milk in order then to understand its deviant forms. This was the driving force behind nineteenth- and early twentieth-century dairy science; not just scientific curiosity but the need to protect consumers from what were judged to be morally reprehensible practices by the milk trade. Milk had to be whole and unaltered in order to qualify as natural. And anything less than natural milk was thought to undermine one of the givens of European and North American food culture. In the sections that follow, we will narrate ten stories to make this point. Together they represent the foundation of our present-day knowledge of milk, or at least that which came before the emergence of advanced molecular biochemistry in the last few decades.

Third, human milk was included as an object of interest because there were concerns about the potential spread of infection from working class wet nurses to the infants of the wealthy. Wet nurses were so common in Paris that there were commercial agencies matching the supply of lactating women to the demand.[19] There was also a literature deploring the practice as unnatural for the infant and

---

14   For more on Berzelius's animal chemistry, see Rocke 1992, 2000.
15   Klein and Lefèvre 2007: 290.
16   For Liebig's animal chemistry, see Brock 1997: ch. 7.
17   Carpenter 1998.
18   Read Committee 1874: Q.2413.
19   Fildes 1988.

condemning the morals of the nurses and commenting on the communicability of syphilis and tuberculosis.[20] The scientific interest was in variations of milk composition in health and disease, and also the difference between human milk and possible animal substitutes, such as cow milk or goat milk.[21]

In his work on animal chemistry, Berzelius had been influenced by the writings of French chemists such as Fourcroy. From the 1780s to the middle of the nineteenth century, Paris was an important centre of scientific investigation of milk. In the 1790s the best known research on milk was by Parmentier and Deyeux, and in the early nineteenth century by Boussingault and Le Bel.[22] The French capital was also a clearing house of information on organic chemistry generally, with work from various European scholars appearing in francophone journals such as the *Annales de Chimie* (1789–1815), the *Annales de Chimie et de Physique* (1816–1913), the *Journal de Pharmacie et de Sciences Accessoires* (1815–1942), and the *Journal de Chimie Médicale* (1825–1870). German language journals were cited increasingly in the milk literature in the mid-nineteenth century, for instance Liebig's *Annalen der Chemie* (1832–1997), *Dingler's Polytechnisches Journal* (1820–1931), and the *Journal für Praktische Chemie* (1834–1873). British and American scholarship was less frequently cited until the last decades of the century.

At the outset of the nineteenth century there was a divergence between the in-depth chemical analysis of milk, on the one hand, which was both time-consuming and expensive in equipment, and, on the other hand, the need for a rapid assessment of milk quality that would deliver an instant verdict on the likelihood of adulteration. At this time the latter could only be achieved by assessing the physical character of fluids. This second perspective, on the physics of density and opacity, is the tradition in which the lactometer sits, and we will start our story here.

Milk was an 'epistemic thing' in the first half of the nineteenth century, when it was subjected to disparate modes of scientific curiosity. The empirical data generated did not enable better definitions of its 'true' nature, however, because the scale of enquiry was too small and therefore representation generally remained more in the realm of organoleptic experience than observations by physicists or chemists. Techniques of organic chemistry were in their early stage of development and it was later found that the data generated in the nineteenth century about average composition was misleading and that there were wide variations around the proposed means. Certainly, until the First World War, milk remained what Rheinberger has called a 'question-generating machine' that inspired investigation, reflection and frequent redefinition.[23]

The natural variability of milk was its main defence against human control. Farmers, analysts, traders, consumers were all puzzled by its material fuzziness

20   Faÿs-Sallois 1980, Sussman 1982.
21   See especially Vernois and Becquerel 1853, but also L'Héritier 1842.
22   Parmentier and Deyeux 1790, 1799, Deyeux 1793, Boussingault and Le Bel 1839.
23   Rheinberger 1997: 32.

and were forced to surrender agency to laboratory equipment to acquire knowledge about it. But the mission of dairy science was dogged by the emergence of further uncertainties as that knowledge grew. A stable relationship and full sharing of agency was delayed until the twentieth century, when, at last, technology was able to deliver equipment and protocols that could cope at the basic analytical level required to enforce the compositional requirements of the law.

The First Story: The Physical Properties of Milk

> An experimental system is full of stories, of which the experimenter at any given moment is trying to tell only one. Experimental systems not only contain submerged narratives, the story of the repressions and displacements of their epistemic concerns; nor, as long as they remain research systems, have they played out their potential excess. Experimental systems contain remnants of older narratives as well as shreds and traces of narratives that have not yet been related.[24]

Milk has numerous physical properties, such as smell, taste, density, viscosity, colour, but only four were employed to any great extent for precise measurement in our period: specific gravity, microscopic structure, refractive properties and freezing point.[25] In addition, organoleptic testing by experienced dairy workers remained a common qualitative form of screening, at least until the recent development of electronic noses.[26] Davis pointed out, as late as 1950, that, from the commercial point of view, 'the checking of farm supplies remained, without question, the most important test which is applied to raw milk in the dairy industry'.[27] He meant sniffing and tasting, and his superlative refers to the risk involved for bulk supplies of introducing even a small amount of tainted milk, which could potentially ruin a whole batch. Davis also noted that, in terms of cost-effectiveness, 'there is no other test which can take the place of smell, apart from the fact that it is by far the quickest and cheapest test known'.[28]

Another simple test in the early years was to measure the quantity of cream that rose to the surface of a batch of milk. In 1817 Sir Joseph Banks, President of the Royal Society, put this on a scientific footing when he was the first to use a tall, graduated glass jar to measure cream.[29] This jar was made by Thomas Jones of Charing Cross.[30] Banks called it a 'lactometer', somewhat confusingly

---

24 Rheinberger 1997: 185–6.
25 For a comprehensive modern account, see Sherbon 1999.
26 Marsili 2002, 2006.
27 Davis 1950: 171.
28 Davis 1950: 172.
29 Anon 1817.
30 See Clifton 1995.

in view of the hydrometer that had gone under the same name for over twenty years. Eventually the new instrument was renamed the 'creamometer' to clarify its purpose.[31] Banks's assumption was of a relationship between the head of cream and the fat content of the milk – its 'richness'.[32] In genuine milk about 8 per cent of cream is to be expected.[33] In reality this is a loose relationship, mediated, amongst other things, by the size of the fat globules.

The glass of the creamometer was graduated from 0 to 100 and beyond, so that readings of 95 to 105 showed an approximately normal amount of cream, but 70 meant an adulteration of 30 per cent. Quantification and qualification, in the sense of judging a sample's genuineness, were therefore possible simultaneously. In Ludwik Fleck's terms we might call the notion of using cream as an index of quality a 'pre-idea', because it used a primitive characteristic. By comparison, the measurement of physical properties with a lactometer and other equipment was a 'proto idea', allowing greater depth of scientific investigation, beyond the visible.[34] Andrew Pickering sees such equipment as 'the balance point, liminal between the human and non-human worlds (and liminal, too, between the worlds of science, technology and society)'.[35] If we include laboratory equipment as machines in the broadest sense, then we can begin to see that a humble creamometer or lactometer, both frankly rather dull items in their own right, can become, in the hands of the right experimenter, a key to unlocking knowledge about the world.

This science of the 'knowability' of milk had already begun unfolding in the 1790s when the first lactometer was sold by John Dicas, a mathematical instrument maker from Liverpool.[36] This was a modification of the hydrometer or areometer principle described by Robert Boyle in 1675, and then exploited commercially from the 1720s onwards by John Clarke and others in the brewing industry to determine the specific gravity of alcohol (weight per volume compared to that of water) from the volume of displacement.[37] The Excise authorities adopted Clarke's hydrometer in 1762 to determine the duty payable, according to the strength of alcohol. The use of areometers for other liquids spread in the 1770s following the initiative of the Parisian pharmacist-chemist, Antoine Baumé. Dicas had patented an alcohol hydrometer in 1780, which had been so successful that it was adopted as the federal standard in the USA.[38] He now turned to milk, enabling a judgement

---

31   Banks's 'lactomètre anglais' was imported into France by de Valeur in 1830. Chevallier and Henry 1839: 206–12, 220–23, Garnier and Harel 1844: 313.

32   The cream was left to rise for 10–12 hours in summer and 15–16 in winter.

33   Normandy 1850.

34   Fleck 1935.

35   Pickering 1995: 7.

36   Dicas died in 1797 but his lactometer continued to be sold by his daughters. Clifton 1995: 83.

37   Filby 1934, Morrison-Low 1998, Bensaude-Vincent 2000, Sumner 2001, Ashworth 2001. Archimedes' Principle is put to work here.

38   Berry 1993: 78.

to be made as to whether the milk had been tampered with by watering, or was still whole and therefore natural.[39]

The particular application of the lactometer technique derives from one theoretical point and one practical. Milk, being a fatty liquid, has a specific gravity that deviates from water, and this facilitates measurement. If water is added to milk the specific gravity is lowered, because water is lighter than milk; and if cream is skimmed, the specific gravity is increased because fat is lighter than milk. The specific gravity increases about 0.001 for each per cent of fat removed, and decreases 0.003 for each 10 per cent, of water added. Each 10 per cent of water reduces the solids about 1.2 per cent, as shown in Table 3.1.

Table 3.1   The reduction of total solids with addition of water

Water added (%)	0	10	20	30	40	50
Total solids in milk (%)	12.0	10.8	9.6	8.4	7.2	6.0
Water in milk (%)	88.0	89.2	90.4	91.6	92.8	94.0

Source: Sommerfeld 1909: 290.

In the words of Bensaude-Vincent and Stengers, applied to a rather different experimental context, Dicas may be taken to be a 'spokesman for nature' by popularizing a simple instrument that made nature knowable, along with her variations – those 'other' natures.[40] But his lactometer was expensive and appears to have been rarely used, other than by public charities.[41]

Since the idea of the areometer was well-developed in France in the eighteenth century through the efforts of Baumé and others, it seems likely that the concept would have been applied at some point to milk, if only by way of experimentation. In 1787 Brisson, for instance, had calculated the specific gravity of cow's milk to be 1.0324, and his most important finding was that the milk of each species (human, cow, ass, mare, ewe, goat) had different densities.[42] The first published reference I have been able to find for a widely used lactometer is the specific design of Antoine-Alexis Cadet de Vaux in 1805, based upon a pèse-liqueur (hydrometer) and manufactured in Paris by the internationally-renowned instrument maker, Jean-Gabriel-Augustin Chevallier (1778–1848), who marketed it as a 'galactomètre' or 'galamètre'.[43] Chevallier's instrument was improved by Dinocourt in the 1830s to

---

39   Holt 1795: 160–63. Dicas provided a slide rule that allowed compensation for readings in different ambient temperatures.

40   Bensaude-Vincent and Stengers 1996: 45.

41   Johnson 1817: 304, Smithers 1825: 368, Loudon 1825: 982, Hartley 1842: 187.

42   Brisson 1787: 391.

43   Chevallier 1819: viii. See also Chevallier 1812. To add to the terminological uncertainty noted above, the instrument is referred to by Rozier et al. (1806: 487) as a

allow the inference of watering in cases of suspected adulteration.[44] But undoubtedly the most successful design of the mid-century was that of Quevenne, introduced in 1841, again with the assistance of Chevallier.[45] His 'lacto-densimeter' was suitable for daily use in the dairy industry and not just as a specialist tool for laboratory research. Even to this day Quevenne's design remains a classic. A scale of 30 divisions was used for readings of specific gravity, from 15 to 45, corresponding to gravities of 1.015 to 1.045.[46] Most normal milks fall in the middle of this range and the lower values indicated adulteration.[47] Tables were published so that the experimenter could allow for changes due to temperature variations.

Many other lactometers were designed in the nineteenth century and it seems that, for marketing purposes, each instrument had to have a recognizable name. Vernois and Becquerel called theirs a 'hydrolactometer'; and there was also a 'thermolactometer' that conveniently combined readings of temperature and specific gravity. Lactometers were available in many shapes and sizes. Many were large, to accommodate the lead shot or mercury counter weights that enabled them to float steadily. But the sophistication of the advanced models priced them out of the practical world of dairy commerce and reduced what in mid-century had been widespread use.

The expansion of the London milk trade from the 1860s stretched lines of supply and created opportunities for fraud. The wholesale trade in particular needed quick, simple and cheap methods to check quality, and evidence given in the sittings of the 1874 Select Committee into the Adulteration of Food Act indicates that both the creamometer and simple lactometer were still in favour at that time.[48] Some of the manuals published for the guidance of Medical Officers of Health recommended use of the lactometer without reservation, for instance in Dr Edward Smith's *Manual for Medical Officers of Health*, first edition 1873, which was said to have 'a sort of pseudo-government sanction'.[49] Using a lactometer and creamometer together was a means of triangulation, since they measured different characteristics, but the increasing technical complexity of frauds demanded the development of chemical tests that would be more robust under legal scrutiny.

Lactometers eventually came under critical analysis, as we saw in Chapter 2.[50] For instance, cream decreases the density of milk and a sample's specific gravity can therefore readily be manipulated by skimming part of the cream to raise the density and then adding water to reduce it back to the original reading. There is also 'Recknagel's phenomenon', the observation that the specific gravity continues

galactomètre. For Cadet de Vaux, see Berman 1966.
44   Chevallier and Henry 1839, Dinocourt 1846.
45   Foucaud 1841: 446.
46   Farrington and Woll 1901: 93, Snyder 1906: 35.
47   Pelouze and Frémy 1865: 629.
48   Read Committee 1874: QQ 2446, 2610.
49   Wanklyn 1874: 13.
50   Berry 1993.

to rise for about ten hours after milking. Recknagel suggested that this is due to the escape of dissolved gases but there are other processes involved, such as the solidification of fats.[51] Because of these observations, the analyst Alfred Wanklyn claimed in 1874 that 'there hardly ever was an instrument which has so utterly failed as the lactometer'.[52] In the same year Charles Cameron, analyst and medical officer, similarly stated to a parliamentary committee that 'the lactometer test is a perfect fallacy'.[53]

As a result of what became a relentless campaign of criticism, lactometers lost ground in the 1870s and early 1880s. Writing in 1893, Fleischmann reflected that for a period of about ten years the technology had fallen out of favour 'on account of the careless, unscientific method in which some early investigators carried it out', but it had then subsequently been re-established.[54] From the mid-1880s lactometers returned to popularity in combination with other methods of analysis because new mathematical formulae made it possible to estimate the fat content of milk by inference from such multiple readings. The physics of specific gravity had also developed further, with more accurate methods of weighment using either a gravity bottle of known weight and precise capacity, or a Sprengel tube, which was filled with milk and weighed. The net weight could then be compared directly to the weight of the same volume of distilled water, giving a reliable specific gravity. The Westphal balance was another alternative, where a float of known weight was immersed in milk and the displacement measured.[55]

Although French scientists continued to have an interest in methods of milk analysis, it is noticeable that the locus of dairy research shifted in the second half of the nineteenth century. The literature on milk analysis had an increasing proportion of German language publications.[56] Moreover, the economic opportunity presented by the great expansion of liquid milk, butter and cheese consumption in Britain drove an Anglo-Saxon investment in techniques and systems of analysis. Also, an increasing share of world trade in manufactured dairy products was sourced in North America and, as a result, more of the scientific research came to be located there. The creation of laboratories and teams of researchers at the land grant colleges, for instance, was a major factor in encouraging this work.

---

51   Recknagel 1883. A solution for the Recknagel phenomenon is the 'density hydrometer', which is used in milk that has been warmed to 40°C for five minutes and then cooled to 20°C.

52   Wanklyn 1874: 8. Nevertheless lactometers continued to be used by the wholesale trade for reasons of speed and economy. Read Committee 1874: Q. 2446, Barham.

53   Read Committee 1874: Q.4955.

54   Fleischmann, English translation 1896: 67.

55   The use of lactometers continued in the twentieth century but a decision was made in 1937 by the British Standards Institution to recommend the measurement of density rather than specific gravity. Density hydrometers then became the norm and measurements switched to grammes per millilitre at 20°C. See BSI 1937.

56   Rothschild 1901.

It was not until the development of the factory system of manufacturing cheese (1850) and butter (1870) that some means of easily determining the composition of milk, particularly as to fat content, became important to both producers and manufacturers.[57]

Besides the widely used lactometers, there were other, relatively minor physical tests. One was to exploit the optical characteristics of milk, its refractive index. The first instrument using this was developed in Paris by Alfred Donné in the 1840s and named the 'galactoscope'.[58] The light from a lamp or candle was viewed through a hand-held device in which a thin film of milk was trapped between two glass plates. The greater the opacity of the sample, the more fat globules were said to be present, and so the milk's richness was greater.[59] This type of test proved to be impracticable for commercial use, however. The equipment was expensive and needed to be dismantled each time it was used: 'If not thoroughly cleansed and dried in every part, the screw becomes clogged and its action embarrassed; in fact, if it gets into careless or unskilful hands, it will not fail to become, in a short time, unserviceable.'[60]

Other designs followed, such as the 'lactoscopes' of Vogel (1863), Rheinbeck (1871) and Heusner (1877); Feser's 'galactometer' (1878); and Heeren's 'pioscope' (1881).[61] Some of these were quite widely used but, due to their lack of precision, they tended to be screening devices for identifying samples that then needed further investigation.[62] Wollny's 'refractometer' was another that measured the refractive index, although it was used more for butter than milk.[63] The problem with all of the instruments in this category, which at first undermined their credibility, was that results varied with the size of the fat globules in the liquid, which was not directly related either to its richness or genuineness.[64] A work-around was to prepare a milk serum, for instance by the addition of acetic acid or copper sulphate. The 'immersion refractometer', introduced by Zeiss in 1900, was submerged in such a serum and it quickly became popular amongst analysts because of its ease of use.[65] In 1925 it was recognized by the American Association of Official Agricultural Chemists as a viable technique.[66]

---

57 Wing 1913: 78.
58 Donné 1841, Donné and Foucault 1844.
59 Payen 1865: 154–5.
60 Minchin 1860.
61 Millán-Verdú et al. 2003.
62 Snyder 1906: 116–17, Jensen 1909: 198, Wing 1913: 85–6, Heinemann 1921: 100.
63 Leach 1904: 103–4, Barthel 1910: 40–47.
64 Girard 1904: 353–8
65 Elsdon and Stubbs 1927.
66 Richmond 1942: 87.

The Second Story: Animal Chemistry

> It is now admitted by all persons who possess some knowledge of dairy matters
> that the composition of equally genuine samples of milk varies greatly, and that
> there is no such thing in existence as normal milk.[67]

Studies of the physics of milk had their basis in its liquid characteristics. But
milk is composed of about 87 per cent water and it is not surprising that chemical
investigators wished to eliminate this in order to study the solids. Evaporation was
therefore their first step and accurate weighing was also usually involved. The
impressive work of Parmentier and Deyeux in the 1790s was proof that this could
be undertaken carefully, even under relatively primitive laboratory conditions,
with interesting results.[68] It is probably fair to say that French scientists were the
first modern dairy chemists.

Besides the obvious interest of milk fats, solids-not-fat, and the minerals in
milk, chemists also wished to know about milk sugar (lactose) and lactic acid.
In the eighteenth century, Haller had worked on the former and Scheele on the
latter. In the early nineteenth century, Berzelius also spent a great deal of time on
lactose, as did Poggiale and others.[69] Another topic of speculation among animal
chemists was whether milk was formed from blood and therefore shared certain
of its features.[70] But the main concern of the nineteenth century was the fat (lipid)
content of milk and the ratio of total solids to water. Both were driven by the need
to understand milk fully if adulteration was to be stamped out.

In 1828 Anselme Payen was an early proponent of what came to be known
as the 'French method' of milk analysis.[71] After evaporation, he added alcohol
or a mixture of alcohol and ether to absorb the butterfat. The lactose was then
washed out with water. This gave an underestimate of the casein, however, which
is slightly soluble.[72] Péligot's method (1836) was similar, as were those of Simon
(1838) and Boussingault (1839).[73] Henry and Chevallier preferred to precipitate
the casein using a moderately strong acetic acid, and then to remove the fat from
the dry precipitate using ether.[74] Again there were problems because the casein is
chemically combined with calcium phosphate, which the acetic acid dissolves.[75]

---

67   Augustus Voelcker in Institute of Chemistry 1879: 28.
68   Parmentier and Deyeux 1790, 1799, Deyeux 1793: 323–4. For more on Parmentier,
see Balland 1902.
69   Berzelius 1814–15, Poggiale 1849.
70   Dumas, Boussingault and Payen 1843.
71   Payen 1828.
72   Simon 1845: 44. For more on early analysis of casein, see Braconnot 1830,
Pertzoff and Bell 1932.
73   Péligot 1836, Boussingault and Le Bel 1839. For more on Boussingault, see
Cowgill 1964.
74   Quevenne (1841: 74) also used acetic acid, following Donné 1839.
75   Le Canu 1839, Berzelius 1841: 332–3.

Variations on this method became standard procedure in the second half of the century.

In 1843 a member of Liebig's research group at Giessen, Paul Julius Haidlen, developed a method that avoided the error in the determination of the casein.[76] It was a modification of a suggestion by Müller, who had used a mixture of one part alcohol and three parts pure ether, which dissolved in the milk, something that ether would not do alone.[77] Haidlen added dry gypsum, thus coagulating the casein and rendering it less soluble in water. The residue was then dried and pulverized to a fine powder, after which the fat could be extracted by ether, and the lactose and soluble salts by hot alcohol. The casein and insoluble salts then remained. Lehman in 1854 claimed Haidlen's to be 'the best method of analyzing milk which has yet been proposed' because the individual components of milk could now be measured accurately.[78] Haidlen's method was extended by Vernois and Becquerel in 1853. They evaporated the milk and divided the residue into three portions. The first was treated with ether to measure the fat. The second was looked at by polarimeter to estimate the lactose. The third was incinerated, leaving the ash. By subtraction it was then possible to calculate the casein.

From the 1870s onwards, as chemistry replaced physics as the milk analyst's primary tool, two methodologies emerged. The first, the gravimetric, involved weighing, as the name implies. The accurate weighment of both the reagents and the products was vital because the results were often to be used as evidence in court cases where the accused was thought to have watered or skimmed a sample of milk. The second, the volumetric, relied upon readings from graduated glassware, particularly for the identification of fat content. This latter group required less skill and was not as accurate.[79] It was employed for rapid screening by public analysts and by commercial dairy companies to monitor the quality of milk delivered by their farm suppliers.

A key date in Britain was 1872, the passing of the second Sale of Food and Drugs Act, which tightened the existing legislation and made the appointment of public analysts possible. The founding in the same year of the Society of Public Analysts (SPA) gave scientific legitimacy to the new profession and a context in which debate about standards and methods could flourish. Before this date the laboratory analysis of milk was full and painstaking but it soon became obvious that the scale of the problem of milk adulteration required techniques that could guarantee the rapid processing of large numbers of samples at minimum cost.[80]

Alfred Wanklyn was among the first to react. He had already conducted well-publicized analyses for the government in 1871 on milk supplied to workhouses

---

76   Simon 1845: 46, Fruton 1988.
77   Pelouze and Frémy 1865: 625.
78   Lehman 1854: vol. 2, 383.
79   Richmond 1920: 115.
80   Richmond 1893b: 271.

in London and in 1871–1872 for the *Milk Journal*.[81] In 1874 he published a book on the subject – one of the first in English.[82] Wanklyn's method was a development of the continental literature. First he evaporated 20cc of milk and, when dry, added ether to remove the fat. The ethereal solution was in turn evaporated and the residue weighed. This was as far as many analysts needed to go, in order to dis/prove adulteration. The next step, if required, was to add strong alcohol, followed by boiling water, to extract the lactose and soluble salts. The residue was casein. This was weighed and then incinerated, leaving the insoluble salts. Wanklyn decided that a minimum standard for genuine milk was 9.3 per cent solids-not-fat and 3.2 per cent fat, which was high enough to condemn 90 per cent of the London samples he tested.[83] The SPA standard adopted in 1874 was only a little more benign at 9 per cent and 2.5 per cent.[84] His dry extraction technique remained the state-of-the-art until about 1885 when it was found to overestimate the solids non-fat and underestimate the butterfat.

One point to make here is the parallel with water analysis, also very important at the time.[85] Wanklyn operated in both worlds and was difficult and disputatious in both. He fell out with his mentor, Frankland, over the relative merits of Frankland's combustion method for water and Wanklyn's own ammonia process. The dispute lasted from 1868 to 1877 and became personal and bitter. Wanklyn, the less well networked of the two, was ostracized by much of the professional academic community.[86] Frankland went on to be an acknowledged water expert of international reputation.

One difficulty for the analyst in the 1870s was that the procedure of evaporating milk had still not been perfected. Usually the evaporation was undertaken in a water bath or an oven, at 100°C, and there were debates about the length of time needed to drive off the water. In 1876 a French scientist, Magnier de la Source, found that this apparently straightforward procedure usually led to an overestimate of the weight of the residue because a small quantity of water always remained. As a result, a series of suggestions were made for how a more complete evaporation could be achieved, mainly by spreading the milk out over a surface of absorbent material, such as sea sand, blotting paper, asbestos, cloth, plaster of Paris, pumice, copper sulphate, sponge, lint, powdered glass, wood fibre, or the rare earth kieselguhr.[87]

Bearing this in mind, it was Matthew Adams, analyst for Kent, who was responsible for questioning Wanklyn's method. He devised a more efficient, dry

---

81   *Report of the Superintendent of Contracts, Admiralty, Relative to the System of Supply of Provisions and Stores for the Workhouses of the Metropolis*, P.P. 1872 (275) li.599: Appendix E.

82   Wanklyn 1874.

83   Read Committee 1874: Q. 6446.

84   Wanklyn 1874.

85   Hamlin 1990: 184–90.

86   Russell 2003, Hamlin 2004.

87   Richmond 1889: 122.

extraction method.[88] He used a blotting paper coil to soak up the milk and this was then weighed, dried in an oven, and the fat separated from the solids-not-fat by ether. Because the milk spread out more than was possible in a basin or flask, the extraction of fat by ether was more complete.[89] Adams' new method, he claimed, solved the problem of a fat residue being 'imprisoned in the dried proteids in such a way that the solvent cannot get at it'.[90] But he in turn was not without his critics. Vieth, for instance, showed that the blotting paper contained matter soluble in ether, so that the estimates of fat were too high.[91]

In 1883 the Society of Public Analysts appointed a Milk Committee to investigate the various methods then in use and their accuracy. It reported two years later. Their recommendation to all analysts was to adopt the Adams method but to insist that his paper coils should be subjected to at least 12 syphonings in ether in order to eliminate error. Their new threshold standard was now to be 11.5 per cent of total solids, of which at least 3 per cent had to be fat. They reached this conclusion after analyzing 283 samples: 138 using Wanklyn's method; 60 with plaster of Paris; 31 by maceration; 11 by Carter Bell's technique; and 43 with Adam's coil. Plaster of Paris was found to yield 0.6 per cent more fat from skim milk than Wanklyn had managed. The committee decided that maceration was too complicated and that the Carter Bell method was inaccurate. Adams' coil gave 0.2 per cent more fat than using plaster of Paris.

Both the Wanklyn and Adams methods were subject to variations in results between the various laboratories. Table 3.2 shows a subset of the data that illustrates this point. These data are significant because they show a high margin of error, which amounts to an 11.9 per cent watering.[92] In other words, the average level of adulteration detected by public analysts was *within* the experimental error of equipment and methods in use in the mid-1880s. One prominent analyst commented that such variations were at least partly due to differences in protocols between laboratories, particularly the length of time taken to dry solids and at what temperature.[93] The government chemist, James Bell, was worried about this.

> Unfortunately, the history of food analysis shows that this difficulty in dealing with natural products has been increased to some extent by the adoption of different processes of analysis, which, in the hands of various chemists, have yielded results differing so materially as to lead to contrary opinions upon the

---

88   Adams 1885.
89   Richmond 1920: 115.
90   Adams 1885: 47.
91   Richmond 1920.
92   To calculate this I employed the method described in Blyth 1927: 248–9, using the solids-not-fat.
93   Hehner 1882a.

Table 3.2   The results (percentages) of 14 analysts asked to look at 15 samples of milk in 1884/5

	Total solids		Fat		Solids non-fat	
	Highest	*Lowest*	*Highest*	*Lowest*	*Highest*	*Lowest*
Observed extremes	15.09	6.16	9.65	0.08	10.11	5.32
Average difference		0.37		0.62		0.66

Source: Russell Committee 1896: 748 (evidence of Richard Bannister).

same sample. To my mind, it is therefore most important that whatever analytical process is used, it should yield absolute, and not comparative, results.[94]

As a result of this confusion, some genuine samples of milk are likely to have been identified wrongly as adulterated, and some of those outside the SPA's guideline threshold were not picked up. We don't know the proportions of each.

The Adams method was common for about 10 years (Table 3.3), with various modifications.[95] Already by 1888 two innovations were to prove more accurate. First, Werner-Schmid added strong hydrochloric acid to the milk and the mixture was heated over a flame.[96] With constant shaking, a clear layer appeared on the surface, and ether was then added. Further shaking encouraged an ethereal layer, which was drawn off and distilled, leaving a deposit of dried fat. The process was repeated until all of the fat was recovered.[97] Stokes's tube was used in some laboratories because it enabled the ethereal liquid to be drawn off more easily. Also in 1888 Röse described a method (refined by Gottlieb in 1892) using ammonia to loosen the membrane that surrounds the fat globules.[98] Alcohol was then added to dissolve the casein and, finally, a mixture of ether and benzene created a solution of the fat. This ethereal liquid was drawn off and evaporated, leaving the fat to be weighed. As for Werner-Schmid, the process was repeated to extract the fullest amount of fat, and was therefore very time-consuming. Fortunately, Soxhlet's invention (1879) of automatic equipment for extracting lipids eased the burden.[99]

---

94   Bell 1884: 137.

95   Allen and Chattaway 1886. Storch's method was an alternative popular in Scandinavia. Richmond 1920.

96   Schmid 1888, Stokes 1889, Hill 1891.

97   Richmond 1920.

98   Röse 1888, Vieth 1889, Gottlieb 1892, Weibull 1898, Popp 1904, Richmond 3e 1920. The Röse-Gottlieb method is still used as a reference method of fat determination in the form modified by Mojonnier.

99   The Soxhlet extractor is a classic design and is still used in some laboratories. It consists of a glass reservoir which fills and empties automatically, a flask holding the solvent, and a condenser. A small amount of solvent is passed many times over the sample, leaving a residue. Soxhlet was Professor of Animal Physiology and Dairying at the School of Agriculture in Munich.

Table 3.3   Most popular chemical methods of milk fat analysis in Britain up to the 1960s

Method	Technique	Date	Commercial/ analytical	Comments
Gravimetric	Wanklyn	1871	A	Fat extracted from dried milk by ether. Underestimated fat.
	Maceration, Somerset House (Bell)	1875	A	For analysis of sour milk.
	Adams coil	1885	C	Fat absorbed by blotting paper, and then extracted with ether. 0.5% more fat than Wanklyn.
Volumetric	Marchand	1854	C	Lactobutyrometer.
	Soxhlet areometric	1879	C	Chemical extraction in ether, then SG measured.
	Lactocrite	1885	C	De Laval
	Werner-Schmid	1888	A	Similar to Gottlieb/Röse but used hydrochloric acid to loosen casein from fat.
	Gottlieb/Röse	1888/1898	A	An accurate method for legal determination employing ammonia, alcohol and ether.
	Babcock/Lister	1890	C	Disadvantage that three centrifugings required. Standard method in USA.
	Gerber	1891	C	Quicker, less manipulation, used less material than Babcock. Most used method around Europe.
	Leffmann/Beam (revision of Babcock)	1892	A	Used Beimling centrifuge.

Sources: Wenlock Committee 1901: 401, Blyth and Cox 1927, Richmond 3e 1920.

Maceration was the method used in the Government Laboratory for reference samples under the various Food & Drugs Acts, from the 1870s to the 1970s.[100] It was a laborious technique that sought accuracy but also one that had to cope

---

100   Bell, 1883: 10, Thorpe 1905, Richmond and Miller 1906, Hammond and Egan 1992.

with the sour and often decomposing samples submitted to it.[101] The aim was to estimate the fat and solids-not-fat as they stood and then to apply a correction for the changes that took place during fermentation. The mode of chemical analysis remained the same for over thirty years but, under fierce criticism, changes were eventually made to the standard corrections.[102] The milk was evaporated to a paste, methylated ether was added and the mixture dried.[103] The residue was ground up and more ether added, and the process repeated at least six times. After the addition of some alcohol, the paste was evaporated. There was then a further series extractions with ether, after which the residue was dried and weighed. This process allowed the measurement of the solids-not-fat, and distilling the ether gave the weight of fat.[104] The maceration method was at the centre of disputes between government analysts and those employed by local authorities, as we will see in Chapter 4.

The Third Story: Molecular Materiality

> Bodies are composite entities, at once biological, technological and political.[105]

Apart from his galactoscope mentioned in the first story, in the 1830s and 1840s Alfred Donné was also a pioneer of microscopy and the associated use of photography. With his collaborator, Léon Foucault, he was responsible in 1844 for the first daguerreotype photomicrographs of milk, which, although of no immediate commercial value, were nevertheless a step-change in the visualization of the natural world and in understanding the structure of milk.[106] Later, Hassall deployed a microscope in London in the battle against food adulteration, but this technique did not become significant in milk screening until the 1890s when it was used increasingly to identify bacteriological pathogens such as the tubercle bacillus.[107]

Donné and Hassall were contributors to the revolutionary idea that the true impact of the environment upon health was operating at a microscopic level, beyond the apprehension of the human senses, and therefore not principally through dirt

---

101 Thorpe 1905.
102 Stokes 1887.
103 In 1899 Richmond and Rosier suggested using petroleum ether in order to reduce volatility.
104 Richmond 1920.
105 Braun 2007: 7.
106 For more on Donné and Foucault, see La Berge 1991 and Tobin 2006. For more on the history of microscopy, see Bradbury and Turner 1967, Schickore 2003, 2006.
107 Smith 2001.

or smells. This extraordinary discovery – the molecularization of life[108] – was so unexpected and radical that it took decades to become universally accepted, and had an impact similar to that in our own age of the revolution initiated by the computer microchip, with its miniaturization of connexions and pathways. It paved the way for other aspects of life also to be interrogated and perceived at a similar scale: food for instance. Although organic chemistry generally, and food chemistry more specifically, had flourished since the early nineteenth century, it was this technical development of the microscope that finally launched the politics of food composition.

The molecular has been a scale of analysis in the material turn for a number of reasons. One is the tidal wave of science about DNA that has been enabled by rapid advances in technology associated with the mapping of the human genome. Although there are hopes of medical breakthroughs in this area, researchers have also alerted us to the twenty-first century form of eugenics – genetic screening and counselling – that Nikolas Rose calls ethopolitics.[109] The latter can be seen as a contemporary means of dealing with individualized risk, particularly that of genetic stigma, but there are also debates at the societal level about the ethics and scientific practicability of a DNA database to provide evidence for criminal trials.[110] Bruce Braun cleverly extends this argument about the molecularization of materiality by looking at the extraordinarily flexible, and therefore potentially dangerous, molecular make-up of the avian influenza virus, especially the strain H5N1. He argues that the best apprehension of this phenomenon is in terms of 'precarious' bodies inhabiting 'virtual' biologies.[111]

Where genetic materialities are to the fore, race will not be far behind, as part of the 'molecular gaze'.[112] This is true of food studies, where it has been argued that taste and dietary needs are somehow hard wired into our genes. We know, for instance, of the existence of super-tasters, for whom the presence of chillies and some other vegetables can make a meal an unpleasant experience. There are also some Asian and African genetic groups who after weaning lack sufficient of the enzyme lactase to metabolize milk sugar, and for whom dairy products are therefore unattractive. This is due to a combination of genetic mutation and natural selection that explains why the introduction of a dairy economy in the neolithic was associated with changes in human DNA.[113] Frederick Simoons and Gary Nabhan have written on this, and other food/gene translations, as adaptations to the histories of agriculture and migration, but their conclusions are controversial and will play no further part in this book.[114]

---

108   Braun 2007.
109   Rose 2006.
110   Lynch and McNally 2003.
111   Braun 2007: 25.
112   Fullwiley 2007, Rabinow and Rose 2006.
113   Burger et al. 2007, Tishkoff et al. 2006.
114   Simoons 1978, 1994, Nabhan 2004.

A second example is public concern about the possible corporeal implications of consuming foods that have been genetically manipulated. This is mainly the much publicized GM foods but in future there is also potential for technological developments at the nano scale (10^{-9}m). Since the latter is still at the development stage, there is a whiff of science fiction about it, but, if the predictions of the futurologists are to be believed, there will be impacts upon all of our lives.[115]

We may also speculate about nano-scale medical treatments in the future but there is still work to be done on the corporeal implications of presently available drugs. Andrew Barry's discussion of drugs as 'informed materials' is especially relevant here.[116] He draws this concept from the philosophy of Whitehead and partly from the history of chemistry written by Bensaude-Vincent and Stengers.

> An environment of informational and material entities enters into the constitution of an entity such as a molecule. Nor can this environment be perceived from a viewpoint which is external to it. The perception of an entity (such as a molecule) is part of its informational material environment.[117]

The 'sociology of promises' has attempted to understand the nano imaginaries of both scientists and the public, in order to unravel their risk perceptions and the politics that will follow. This includes the role to be played by nano technologies in the switch that Deleuze identified from a disciplinary society to a control society.[118] In fact, Deleuze's molecular ontology, as discussed by Kearnes, has much to offer the study of nano-scale materiality. Being monist and vitalist, it opposes mechanistic interpretations of molecularization and therefore opens opportunities for different understandings.[119]

The Fourth Story: Volumetric Chemistry

Generally speaking, gravimetric chemistry (based on weighment) had held sway in the era of Berzelius and Liebig. Gay-Lussac developed titrimetry in the 1820s and 1830s but it was not until the 1860s that volumetric methods (measuring volumes) offered a challenge in analytical chemistry.[120]

All of the gravimetric methods described above were laborious. In 1854 Eugène Marchand, a pharmacist from Fécamp in Normandy, devised a method for the rapid measurement of milk fat using a special glass tube that he called a

---

115   Kearnes and Macnaghten 2006.
116   See also McCormack 2007 on the 'molecular affects' of drugs.
117   Barry 2005: 58.
118   Kearnes 2006.
119   Lash 2006.
120   Szabadváry 1966.

lacto-butyrometer.[121] The process could be completed in eight to ten minutes for each sample and was thus a great improvement in terms of timeliness. The tube was filled to one-third of its capacity with milk, to which were added a few drops of caustic soda and then equal volumes of ether and alcohol. After careful mixing, the tube was plunged in a water bath that had been pre-heated to 40°C and then left to cool until the layer of fat that rose to the surface had stabilized. Markings, in degrees, on the outside of the tube enabled a reading of the depth of the fat, and this was then entered into a formula to determine the percentage of butterfat in the milk. The results were fairly accurate when dealing with milk that was not too rich, every 0.1 cc indicating 0.2 per cent of fat; but it could not be relied upon for legal cases because not all of the fat rose to the top of the tube; about 1.2 per cent stayed in solution.[122]

In the late 1870s Marchand's method was scrutinized by Schmidt and Tollens, who proposed a formula that would give better correspondence with accurate gravimetric tests. They also modified the method by substituting acetic acid for caustic soda and by using a higher strength alcohol. The results were now within 0.05 per cent of those obtained gravimetrically.[123] Marchand was criticized by Soxhlet and eventually condemned by Adams because of large variations in results according to the proportions of ether, alcohol and caustic soda added, along with sensitivity to temperature. Nevertheless, Marchand's method was widely used because of its simplicity and results that were reasonably accurate. Soxhlet's technique, suggested in 1881, had a similar impact and is interesting for being a mixture of the two stories. Milk was mixed with a potash solution, and then shaken with ether. The ethereal fat solution was allowed to separate, and its specific gravity was determined. The principle here is that the difference between the specific gravities of fat and ether is greater than that between milk and water, so measurement is facilitated. A table gave the corresponding percentage of fat. The results were similar to those arrived at by extracting milk dried up on plaster of Paris.[124]

The Fifth Story: Centrifugal Force

One of the most fundamental discoveries in the history of dairying was the mechanical means of separating cream from whole milk. It provided the basis for the industrial-scale development of butter-making. C.J. Fuchs at Carlsruhe was the first, in 1859, to use centrifugal force to achieve this separation, a great advance on the creamometer or the setting of milk in open pans.[125] He was followed in

---

121   Marchand 1854.
122   Vieth 1889, Fleischmann 1896: 71.
123   Schmidt 1878, Tollens and Grote 1879.
124   Fleischmann 1896: 70, Snyder 1906: 41, Barthel 1910: 36–40.
125   Faber 1887.

1864 by another German inventor, Antonin Prandl, but neither was as successful as Gustaf De Laval, working in Sweden. In 1877 De Laval modified Prandl's design, using turbines that allowed for continuous operation, and so his centrifuge could be emptied and refilled while the machine was running.[126] It had a capacity of 130 litres per hour and a speed of 5,000 revolutions per minute.

De Laval's machine allowed the measurement of cream. A modification, patented in 1885 and called the Lactocrite, enabled the testing of butterfat.[127] It was obvious to him that there was a major commercial opportunity in manufacturing such equipment for butter factories and large dairies, where the quality-control of inputs was a factor in profitability. The Lactocrite required a steel disk to be added to the separator; this held 54 cylindrical tubes, each with a sample of milk.[128] A modified volumetric test was used in which the milk with first boiled with a mixture of acetic and sulphuric acids to dissolve the casein and, after whirling, the column of fat could be read off on the tube's graduations in tenths of a per cent by weight. In 1890, the mixture of acids was replaced by a quantity of ethylidene-lactic acid and solution of hydrochloric acid, giving an improvement in accuracy and convenience. As a result, the method became very popular but its spread beyond Scandinavia and Germany was restricted by the expense of the Lactocrite equipment.[129]

De Laval was not alone in this market. He had Danish and German competitors for the cream separator and alternatives to his lactocrite also soon emerged. In the United States a crucial development was the Hatch Act of 1887, which enabled the establishment of an agricultural experiment station in each of the states. Within a couple of years of this major institutional investment, no less than seven distinct methods had been invented for the determination of fat in milk. Stephen Moulton Babcock's was by far the most successful of these, having the advantage that its relative simplicity facilitated its use in creameries and cheese factories, and on some farms:[130]

> Its manipulations are few and readily learned, and it is cheap, both in first cost and as regards running expenses. The test is furthermore speedy, accurate, and easily applicable to practical conditions.[131]

---

126   Bondeson 1983, Wohlert 1983, Jackson 2005.
127   Faber 1887.
128   Barthel 1910: 58–60.
129   Faber 1887, Vieth 1889, Fleischmann 1896: 70–71, Snyder 1906: 116–17, Barthel 1910: 58–60, Wing 1913: 87.
130   Farrington and Woll 1901: 4.
131   Farrington and Woll 1901: 6.

Babcock developed his method at the Wisconsin Agricultural Experiment Station in 1890.[132] Sulphuric acid was used to dissolve everything other than the fat, which, after whirling, could be read off against the bottle's graduations. The need for three centrifugings was a disadvantage because of the time taken. At first his method was used in conjunction with Lister's hand-cranked centrifuge, a simple machine that achieved only 400–500 revolutions per minute by comparison with the 6,000 for the Lactocrite. Nevertheless, within ten years the method had become the dominant method of milk fat estimation in North America.[133] Also American in origin, the Leffmann and Beam method, first described in 1892, employed the relatively inexpensive Beimling centrifuge, which had been patented in 1889.[134] Instead of sulphuric acid, amyl alcohol was used as a means of liberating the fat, and the time necessary for whirling was therefore reduced.[135]

There are several factors that make or break tests such as these. The first is adding just the right strength of reagent, to dissolve the casein but not burn the fat. Second, the significant improvement of centrifuges in the 1890s, by firms such as Gaertner and Hugershoff, reduced the time for each sample, for instance from the ten minutes that had been needed for a Leffmann and Beam analysis down to three to four minutes.[136] Better gearing was employed and then electricity replaced hand cranking. With a lower overhead cost per sample, it was thus possible to revolutionize the milk-testing industry.

In the 1890s Gerber's 'acido-butyrometer' became the most widely used milk fat test across Europe. It was effective, cheap and could be performed relatively quickly by staff at a low level of training – all of the qualities that it needed to be commercially useful. Its standard of accuracy, ± 0.05 per cent of fat, was acceptable for most commercial requirements but other tests had to be used for legal purposes. Sulphuric acid was added, leaving the fat suspended, and amyl alcohol was used to assist its separation from the acid solution. Centrifugal force completed the process, which was similar to the Babcock test, although simpler. Gerber had patented the test in 1891 and published it the following year, but it was not until 1895 that it was tweaked into its mature form, which then lasted until the 1960s.[137] Close inspection shows it to have been an improved hybrid, combining a version of Marchand's lacto-butyrometer test bottle with a modification of the

---

132   Babcock 1890, Shutt 1892, Stokes 1892, Embrey 1893, Barthel 1910: 61–4, Wing 1913: 94–107.

133   American dairy analysis textbooks focus very much on Babcock: e.g. Farrington and Woll 1901, Snyder 1906, Ross 1910. For a literature review, see Herreid 1942.

134   Leffmann and Beam 1892, 1893, Hehner 1892, Richmond 1892, 1893a, Richmond and Boseley 1893a.

135   Stokes 1892.

136   Snyder 1906: 139–42.

137   Gerber 1892, Barthel 1910: 69–77. A British Standard was devised for the Gerber method. See B.S.S. No. 696 (1936).

chemistry of Leffmann and Beam.[138] Later, Gerber devised a non-acid version, called 'Sal', using a mixture of caustic soda, Rochelle salt, and sodium chloride to dissolve the casein, and isobutyl alcohol instead of amyl alcohol.

By 1900 the landscape of analytical testing had changed irrevocably. There were now only two types of chemical method in use. First, there was the accurate assay required for legal purposes, such as the methods of Gottlieb/Röse and Werner-Schmid. In addition, Bell's maceration technique was the ultimate arbiter for sour samples, as used in the government laboratory for appeal cases. Second, more rapid, less labour-intensive techniques were used in commercial laboratories and also to an extent by public analysts. Soxhlet's automatic areometric method of determining fat, which gave results as reliable as those obtained by most gravimetric methods, was a particular boon because it could be left to operate while other tasks were undertaken by the laboratory assistant. Even quicker and cheaper, the Babcock and Gerber methods were universally popular. According to Fleischmann, all of the older methods had been replaced: 'indeed they possess now only historic interest.'[139]

Wanklyn had realized in the 1870s that the precision of his analysis was limited to a degree by the quality of the laboratory glassware then available.[140] The design of glassware robust enough to withstand centrifuging at speeds of up to 1,000 revolutions per minute, was an additional requirement from the 1890s onwards. The popular Babcock and Gerber techniques required both accurate and hard-wearing glassware and the solution in each case was the design of special bottles, known as butyrometers (Figure 3.1).[141] The design and calibration of these became a major issue.[142] In 1900 the newly established National Physical Laboratory was called in to guarantee the bottles' accuracy and the NPL subsequently became pre-eminent in the standardization of equipment and techniques,[143] in Latour's sense of a 'centre of calculation'.[144] These bottles soon were an important element in the income stream of the NPL and therefore had a central role in its early years of development.[145] Gerber bottles were vital to the dairy industry, not only to monitor quality and reduce adulteration but also to reassure farmers who sold their milk to butter factories that they were being paid fairly for the fat content of their milk.

---

138   Richmond 1920.
139   Fleischmann 1896: 67.
140   Egan 1976.
141   Shaw 1917: 9.
142   In evidence to the Wenlock Committee, in 1900, there were complaints that many of the bottles used were 'most inaccurate': Q. 2759. For a comment on the complexity of calibration, see Day and Grimes 1918.
143   Day and Grimes 1918, Ling 1944, Magnello 2000.
144   Latour 1987.
145   In 1903 charges were 3d per pipette and 6d for a test bottle. Any test bottle which gave results differing by more than 0.05 per cent, from the results obtained in the standardized bottle was discarded.

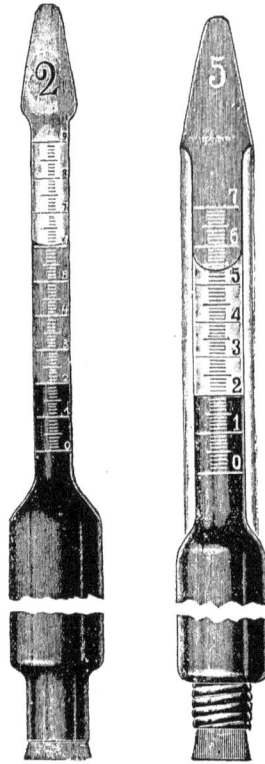

Figure 3.1   Gerber butyrometers

Source: Richmond 1920: 142.

Simon Schaffer has suggested that precision was a Victorian value and one
that was applied, to great effect, in British science and technology. There was
commercial value in precise mensuration, of course, but precision was also a way
of thinking. With regard to the NPL, and the diverse laboratories in which its
founders worked, Shaffer concludes that

> exact measures were the product of reorganizations of working practices and
> of the management of a range of different workplaces ... The label of precision
> attached to any measure hinged on cultures of communal trust and was the
> consequence of the strength of the social relations between these separate and
> complex institutions.[146]

146   Schaffer 1995: 164.

A similar history is evident in the United States, where there was initially confusion with the calibration of Babcock bottles as to whether the Mohr cubic centimetre (1 gram of water at 15°C) should be used or the 'true' centimetre at 4°C.[147] Eventually the US Bureau of Standards stepped in and formulated a specification, and the Babcock method of analysis itself was also standardized.

The Gerber and Babcock methods remained in use commercially into the second half of the twentieth century because of their convenience and relative cheapness. Meanwhile analytical dairy chemistry changed, as we will see in the ninth story.

The Sixth Story: Paper Tools

> All scientific instruments, from Galileo's telescope to contemporary physics' most bulky instrument, are ... materialized theory.[148]

Ursula Klein's evocation of organic chemistry in the early nineteenth century is especially interesting for its emphasis upon the formulae used by Berzelius. She points out that he was first, in about 1813/14, to use notations such as H_2O to represent the composition of water. British chemists were at first reluctant to adopt this new system but within twenty years major continental figures, such as Liebig, saw its advantage. Formulae eventually became 'enormously productive' and Klein claims that these 'paper tools [were] fully comparable to physical laboratory tools and instruments and that both kinds of tools contribute[d] to the creation of reference and meaning'.[149]

As dairy chemists became increasingly familiar with the key constituents of milk, they discovered a more or less constant relationship between them. As early as 1854 Cesaire Régnard had devised tables to read off the percentage of watering in a sample from its total solids. He saw 12.92 per cent of solids as normal; so a reading of 11.63 would mean 10 per cent added water, 10.34 indicated 20 per cent and so on. In the same year, as we have seen, Marchand used an empirical formula for the interpretation of his volumetric test. But it was the German researchers Behrend and Morgen who were the first, in 1879, to publish tables showing the covariation of the three major variables: fat, total solids and specific gravity.[150] Also in 1879, Clausnitzer and Mayer put these relationships into mathematical form, holding out the possibility of deriving an unknown component from the other two. Four years later in London, Hehner pointed out that a higher percentage of fat lowered the specific gravity, whereas the solids-not-fat raised it – a very

---

147   Herreid 1942: 342.
148   Gaston Bachelard in Bolduc and Chazal 2005: 81.
149   Klein 2003: 232, 245.
150   Behrend and Morgen 1879.

significant observation for the detection of adulteration.[151] Following this, the number of formulae multiplied and soon slide rules were being marketed to facilitate calculations.[152]

The following equations were the best known, in order of introduction. I have standardized the lettering and employ here the simplest algebraic formulae, as they would have been used in laboratories: D = density of water in grams/ml at 20°C; F = per cent fat; G = specific gravity (for instance 1.032); S = per cent solids-not-fat; T = per cent total solids; W = per cent added water.

1. Clausnitzer and Mayer 1879:
   $$F = ((G - 1) / 0.00475) - (0.789\ S)$$
2. Hehner 1882b:
   $$S = (0.725\ T \times G) / 4.33$$
3. Fleischmann & Morgen 1882; Fleischmann 1885:
   $$T = 1.2\ F + 2.665\ (100\ G - 100) / G$$
4. Hehner and Richmond 1888:
   $$T = 1.164\ F + 0.254\ G$$
   Often simplified to T + 1.2 F + 0.25 G
   Later revised to F = 0.859 T - 0.2186 G – 0.05 (G / T – 2.5)
   Where G = lactometer reading of specific gravity at 60°F, using the last two decimal places where water is 1.000.[153]
5. Richmond 1889:
   $$T = 1.17\ F + 0.263\ G / D$$
   Later modified to T = 1.2 F + 0.2625 G / D.
   Richmond's was a very important formula, especially in the commercial and analytical worlds. Sometimes it was simplified to T = 0.25 G + 1.2 F + 0.14. For legal purposes, the following formula was used:[154]
   $$W = 100\ (((((G \times 1000) - 1000)/36) + F) \times 100)$$
6. Babcock 1892:
   $$T = (L + 0.7\ F) / 3.8 + F$$
   Where L is the Quevenne lactometer reading
7. British Standard No. 809, 1938:
   $$T = 0.25\ D + 1.21\ F + 0.66$$
   Here D = (density of milk at 20°C – 1) x 1000

It is important to note that each of these formulae was relevant to only one method of analysis. Fleischmann and Morgen's was designed to go with Soxhlet's evaporation technique using plaster of Paris, Hehner's with Wanklyn's, and Hehner and Richmond's with Adam's coil. Only in the 1890s, with the coming

---

151   Hehner 1882b, Hehner and Richmond 1888: 26.
152   Richmond 1888, 1898, 1920.
153   Ling 1944: 54.
154   Richmond 1898.

pseudo monopoly of the Gerber and Babcock methods, did formulae have a wider currency.

Conceptual tools were one means by which pure chemistry was transformed into applied chemistry, for instance in the synthetic dye industry that eventually consolidated into a recognizable form of technoscience, from the late 1850s. The formulae of reactions, products, and compounds, which were such useful shorthand means of the storage and communication of knowledge, could be used as the bases of the routinization of work in the industrial laboratory.[155] They were not necessarily chemical formulae, *sensu stricto*, but could be the expressions of relationships in algebraic form. From 1879 onwards, these milk formulae acted in effect as summaries of empirical observations, which were at first based on the output of only a handful of cattle, but eventually used tens and hundreds of thousands of samples tested in industrial laboratories.

We are not saying here that these formulae captured, in a single expression, some 'essence' of milk composition. On the contrary, they generally included only fat, solids-not-fat, and specific gravity, which happened to be the observations that were considered important in an instrumental sense for the dairy economy. In addition, the data were contingent upon the practical conditions of milk production, including the cow's breed, how she had been fed, and so on. Thus the series through time a space was remarkably sensitive to prevailing conditions and could not represent the fullest potential of milk produced under 'ideal' circumstances. It was very much the chemistry of the everyday.

These dairy formulae were a significant representation of the workings of technoscience and a 'surface of emergence' of late nineteenth century organic chemistry. Through their substitutions, they reduced the amount of laboratory work necessary for dairy companies to monitor the milk supplied to them and the quality of the product that went to their retail customers.

The Seventh Story: Rapid, Commercial Methods

> The estimation of fat in milk, simple though it seems at first sight, has perhaps given more trouble to analysts than anything else, and on it, consequently, a very large amount of original work has been done.[156]

The demands of the dairy trade and the state for faster and cheaper analytical methods spawned a small industry in devising and marketing laboratory techniques for the mass production of milk analyses. Improvements in timeliness, however, often sacrificed precision. The time taken in whirling the samples tied up expensive equipment and therefore hampered a laboratory's throughput. As we have seen, from about the turn of the century the Babcock and Gerber tests held

---

155   Klein 2003, 2005.
156   Richmond 1889: 121.

sway and almost all of the competitors were ruled out for one reason or another. The Wanklyn and Marchand techniques were inaccurate, Adams' coil was too slow, and the lactocrite was too expensive.[157]

The Aylesbury Dairy Co. began trading in 1867 and soon had built up a good reputation for quality amongst its wealthy clientèle in London's West End.[158] In 1900 its 6,000 gallon daily turnover was drawn mainly from farms in Berkshire and Wiltshire and, unusually at this time, there was a contractual obligation on producers to cool their milk before despatch.[159] The methods used by the Aylesbury Dairy Co. show how a commercial laboratory worked. Paul Vieth, a German analyst, took over as chief scientist in 1880.[160] From then until July 1884 he used Marchand's lactobutyrometer for the direct extraction of fat.[161] This was a relatively straightforward and rapid procedure but not the most accurate. For a similar level of precision, but even simpler for unskilled laboratory technicians, he found it preferable to measure the milk's specific gravity and total solids. From August 1884 these were the only measurements taken but the estimation of the fat was now possible because of Fleischmann and Morgen's formula which modelled the relationship between these three characteristics of milk. Fleischmann's corrected formula was used from May 1885 to May 1891, and from then onwards Hehner and Richmond's formula. The total solids were extracted using Soxhlet's equipment (1884–1891), Adam's coil (1891–1894), and then the Leffmann-Beam method.[162]

Scaling up any method to deal with thousands of samples a year was an achievement. There were many hurdles to overcome (e.g. collection of samples without deterioration; a skilled analytical workforce; equipment and laboratory protocols because nothing on this scale had ever been attempted before). Then a dairy company had to transform its laboratory results back into the governance of supply: cancelling contracts, taking on new suppliers, perhaps changing price incentives.

One of the biggest changes in the early twentieth century came with new technologies of bulk transport. Until the 1920s, variations in composition were problematic because the average size of herd was small and the milk was delivered to market in cans, each of which contained the milk of one or only a small number of animals. The milk was then retailed at the doorstep through two or three deliveries per day, with the result that customers were drinking unmixed milk from the morning or evening milkings. This changed markedly from 1926/7 onwards

---

157   Stokes 1889.

158   Foster Committee 1894: Q.3592, Farmer.

159   I.A. Hattersley, Managing Director, in evidence, Maxwell Committee 1902: Q 5781.

160   Vieth stayed until 1892 and then returned to Germany to become Director of the Dairy Institute at Hameln.

161   Vieth 1892: 85.

162   Foster Committee 1894: 223.

when glass-lined road and rail tankers were introduced to bring milk to London. These tanks ensured the mixing of the produce of thousands of cows. Also, the number of small producer-retailers declined from the 1930s onwards in the face of competition from the large-scale dairy companies. As a result, there was a significant reduction in the variability of composition and of quality generally.

The Eighth Story: Freezing Point

Where adulteration continued, it became more subtle and more difficult to detect. One important analytical method, which during the twentieth century became important, was based upon a physical characteristic, the freezing point of milk. This is more of a constant than any of the other physical or chemical property of milk. Although it was controversial at first, the freezing point was used in Holland from the 1890s onwards.[163] The theory was that fraudulent additions of water or skimmed milk could be detected by the amount the freezing point was depressed away from that of normal milk (between –0.530 and –0.560°C) towards that of water (0°C).[164] In 1914 Monier-Williams reported to the British government on the science of the freezing point and Hortvet's classic paper, published in 1921, described how the test could be operationalized using his 'cryoscope'.[165] As some of the technicalities of test equipment were refined,[166] and as its price dropped, the idea of the freezing point became more popular amongst analysts in Britain from the early 1930s onwards (Table 3.4), particularly to test milks already considered suspicious because of solids-not-fat readings below 8.5 per cent.[167] By the 1950s the technology had been perfected and magistrates were accepting the results of this (unofficial) test as evidence in allegations of adulteration.[168] However, it was realized that using the freezing point test was valid only for milk pooled from many cows, usually in a bulk tank. Taking the freezing point of milk from individual cows risked ignoring their particular characteristics. The thin milk of high-yielding cows thus has a higher freezing point than lower-producing cows. The animal's diet, and time that it was fed relative to the taking of the sample, were other variables.[169]

The modern procedure is to assume that milk with a freezing point depression greater than –0.535°C is genuine. From –0.530 to –0.534°C is a zone of alert and a letter is sent to the producer advising them to check their equipment. Between

---

163   Raalte 1929, Andrew 1929.
164   Beckmann 1894. For modern accounts see Sherbon 1999, Novo et al. 2007.
165   Monier-Williams 1914, Hortvet 1921.
166   British Standards Institution 1959.
167   Elsdon and Stubbs 1930, 1933, 1934, Cook Committee 1960: 142.
168   Davis 1950: 33–52.
169   Department of Health for Scotland 1945, Cook Committee 1960: 60.

Table 3.4   Summary of milk tests standard in the mid-twentieth century

Constituent	Accurate tests	Routine tests
Fat	Röse-Gottlieb	Gerber
	Adams	Babcock
Solids-not-fat	Gravimetric	Lactometer
Added water	Freezing-point	Lactometer and conductivity-measurements
Lactose	Polarimetric	Polarimetric
Total protein	Total nitrogen (Kjedahl)	Sørensen titration (with formalin)
Casein	Precipitation at pH 4.6 and weighing	Precipitation at pH 4.6, redissolving and Sørensen titration
Salt (sodium chloride)	Silver titration after digestion with nitric acid	Conductivity
Ash	Heating to dull-red heat	–
Visible dirt	Official SPA method (centrifugal)	Filter-pad or 'sediment tester'

Source: Davis 1947: 1117.

–0.525 and –0.529°C there is 'appeal to the herd' testing of the cows.[170] Below –0.525°C milk is assumed to be adulterated unless the farmer can prove otherwise.[171]

The Ninth Story: Instrumental Methods

From the 1960s onwards the Gerber method was gradually replaced. Its use of strong acids made it a laboratory safety hazard and it was also becoming relatively expensive per sample because of the need for trained technicians. It was no surprise, therefore, that creamery managers looked carefully at automated techniques when they became available. Probably the best known early example was the Milko-tester manufactured by the Danish company, Foss Electric. This worked on the principle that, once it is homogenized and its casein solubalized, the turbidity of milk is directly related to its fat content. The instrument measures this turbidity by light scattering.[172]

A second generation of instruments included spectrometers that looked at milk's absorption of infrared radiation. They measured the direct absorption of infrared energy by the fat, protein and lactose at specific wavelengths. Fourier

170   This will be explained in Chapter 8.
171   Harding 1995: 63.
172   Harding 1995: 87.

Transform infrared milk analysers are a further development, using computer-based modelling to convert the data's interference pattern to a spectrum and then analyse the much stronger signal.[173] More recent developments include Near Infrared Reflectance Spectroscopy.

The Tenth Story: Vitamins, Enzymes and Micronutrients

The history of dairy chemistry is not only one of the revelation of greater complexity than ever initially imagined but also one of the anticipation of further mysteries to come. From the turn of the twentieth century to the 1930s there was a constant undercurrent in popular literature about milk. It reflected a wonder about unknown properties of milk that might be sacrificed by ill-informed scientific and technical interventions. The principal motor of this reaction was the gradual introduction of pasteurization. Its use of high temperatures was thought likely to have had unintended consequences in terms of chemical changes, and it was true that these had not been researched in-depth and were certainly not well understood. The discovery of vitamins in the teens and twenties of the century seemed to be conformation of the hypothesis of unseen complexity, and it did not take long for the destructive effect of heat treatment upon Vitamin C and other vitamins to be revealed.

Milk played an important role in the emerging newer knowledge of nutrition in this period.[174] Frederick Gowland Hopkins, who won the Nobel prize for his work on Accessory Food Factors (later renamed vitamins), was convinced of milk's centrality.[175] Although the scientific detail took time to unfold, vitamins had been an 'unrevealed presence' for many years through empirical work on deficiency diseases such as scurvy and beri beri.[176] Many people took vitamin research to be strong support for the raw milk case, along with the investigations of enzymes that were being published at about the same time. Again, it was shown that pasteurization affected the catalytic chemical properties of enzymes.[177]

These debates about the nutritional and other properties of raw milk rumbled on for the rest of the twentieth century, although in recent years the anti-pasteurization movement has had a lower profile than the positive affirmation of milk produced under environmentally sustainable conditions. The organic movement has made a variety of claims about its milk, including largely anecdotal ones about taste. But recent research has shown that cows fed solely on organic grass do indeed produce milk that is quantifiably different on certain micronutrients. One study of 25 farms found that 'good' fatty acids are higher than on conventional dairy farms. The

---

173   O'Sullivan et al. 1999.
174   Atkins 2000a.
175   Hopkins 1912, 1920.
176   Hughes 2000: 754.
177   Race 1918: 186.

premium for Omega-3 fatty acids is 39 per cent and for conjugated linoleic acid 60 per cent, while the less desirable Omega-6 decreased by 60 per cent. Vitamin E was up by 33 per cent and β-carotene by 30 per cent.[178]

Conclusion

> Because the conceptual separation between nature and society as categories is created in practice, and then affects subsequent practices, the result is new relationships between nature and society, even as both are the outcome of historical practice.[179]

The downfall of Daniel Schrumpf was a moral fable and a piece of theatrical tragedy played out with the scenery and props of a laboratory. He was condemned by a piece of equipment, the lactometer, that itself was widely criticized, not just in New York but around the world. Its enrolment as a scientific, administrative and legal instrument was a reflection of a technological genealogy that stretched back into the eighteenth century, and its continued use was proof of the ongoing difficulties faced in exploring and charting the mysteries that were milk. This was bad luck for Schrumpf you might say, and it is certainly likely that a trial ten years later based on the same evidence would have reached a different verdict. But is this not also true with many of the trials and tests of knowledge that face us every day? Different histories emerge according to the shape of what Latour calls 'sedimentary time', as we reassess 'facts' and outcomes. The delayed implementation of the pasteurization of milk between the wars, which at the time was for perfectly good reasons associated with non-interference with the natural, is now seen to have cost lives.

It was Schrumpf's misfortune to be indicative of one of the least edifying individuations of milk. He stood for the timeless uncertainty of nature, which always seemed to be just beyond visibility and knowability. The technology of hydrometers improved little during the eighteenth and nineteenth centuries because of the complexity of fluid mechanics. Nevertheless, lactometers were used throughout this period and into the twentieth century because of their portability, ease of use, and relative cheapness. There were some modifications in size, shape, and functionality but it remained the same piece of equipment. Taking a reading was straightforward and there was an aura of scientific data gathering, as Schrumpf found to his cost. The technological continuity was remarkable, but eventually the inspector-lactometer hybrid lost its credibility as stand-alone evidence, and was reduced to one component in the cluster of readings needed to produce reliable legal judgements or commercial decisions about quality.

---

178   Butler et al. 2008, Ellis et al. 2006, 2007.
179   Mansfield 2003c: 330.

The second individuation was that of laboratory science. Lactometers, and indeed other means of physical measurement, played little part, because this was the story of dairy chemistry. Starting with gravimetric techniques and moving on to the volumetric, the dozens of minor and major experimental improvements resembled a gradual process of ontogenesis in which 'objects' – test equipment, chemicals, equations – emerged from pre-individuals. We cannot necessarily call this 'progress' because Simondon's dialectical tension between humans and objects demonstrates resistance and also insists on the inevitability of old, concretized objects losing out in the process of new becomings.[180]

Third, the individuation of commercial milk analysis was particularly significant and not necessarily dependent upon the other two strands. As one might expect, the criteria employed were speed of analysis and minimum cost per sample processed. This meant that the technologies were never the most accurate or scientifically advanced and they ran in a parallel world to that of the public analyst as provider of evidence of a legal standing. We adduced 10 stories but these could have been framed differently and there is an argument for saying that there were no recognizable 'stages' of development, only redrawn hybrids of objects and people that were intertwined and intercalated between existing alliances.

Another exceptionally important aspect of the commercial laboratory, and one which trickled through to the more precise analytical world, was the identification of dimensions of milk for knowing and testing. This complex liquid has dozens, hundreds, thousands of components, most of which were difficult to detect and measure. Fat became a focus for six reasons. First it had been well-known for centuries in the making of butter, one of milk's most lucrative bye products. Second, milk fat had been compromised during the period of worst adulteration because of the ease of skimming cream. Third, it had been a target of academic and practical work in the laboratory throughout the nineteenth century, not as an object of easy knowability, but as one that had gradually revealed its secrets. Fourth, everyone knew that the fat content of cow's milk varied considerably by breed, giving the farmer the ability to produce thin or creamy milk at will. Fifth, by the end of the nineteenth century it had become clear that the fat content could be manipulated in the cow, before milking, by the adoption of certain feeding regimes. Finally, fat in milk was considered to be nutritionally important for the healthy growth of infants and its absence in much condensed milk was considered to be a scandal and a crisis at the turn of the twentieth century.

This combination of insights encouraged ever-more sophisticated means of measuring fat content, but there was also a change in ways of thinking about this natural product. Milk fat was the essence of milk for many people, with the remaining fluid being of less interest and importance. Because the means of analysis improved, it was possible to identify poor milks according to this criterion, and farmers and retailers put more effort into feeding and breeding

---

180   The 'perpetual perishings' of Locke as developed by Whitehead.

their cattle accordingly.[181] Their philanthropy knew limits, however, in a dance that remained just the right side of the law, which, from 1901 onwards, expressed a presumptive view of percentage fat content. In this way there was a reflexive dialectic between public expectations and the technology of milk analysis – a mini arms race of intention, investment, and judgement. Through practice, then, milk became known, but known as an object that was ever more 'natural', ever more disciplined and standardized than it had been at the beginning on the nineteenth century.

It is these stories of object resistance, character-emergence-in-practice, and the ontologies of socio-legal expectation that we will follow in the rest of the book. In this chapter we have begun our participation in what Graham Harman calls 'a poetics of objects'.[182] This has involved milk itself, of course, and its material constituents of butterfat, casein, milk sugar, mineral salts and water. It has also been a story of water baths, lactometers, Bunsen burners, chemicals, and paper formulae. There will be more objects as the book progresses and I hope that it will become clearer how it is alliances between people (farmers, milk retailers, dairy scientists, local authority analysts, and judges) and objects, sometimes in successful translations but also in ways that were at times unpredictable and dysfunctional, that produce the most convincing histories. Pickering's de-centred stories are a powerful exemplar in the history of industrial chemistry, as are the quasi-objects of actor network theory. Together, their object-orientation overcomes many of the weaknesses of those versions of social constructionism where material has no voice.

181   Eckles and Shaw 1913.
182   Harman 2009.

Chapter 4

Expertise

Analytical Expertise: Concentrated or Distributed?

> The subject of milk analysis was one of which every member thought he knew
> something more than anyone else ... .[1]

The vast literature on expertise is complex and riven by the assumptions and
disciplinary perspectives that have loaded so much meaning on to this one word.[2]
Recently Harry Collins has argued that attempting to understand expertise is the
foundation of a 'third wave' of science studies, which seeks answers to the question
'how do you make decisions based on scientific knowledge before there is an
absolute scientific consensus?'[3] He asserts that this is 'the pressing intellectual
problem of the age' because of the recent widespread questioning of scientific
authority.[4] He is referring here to the alleged undermining of the authoritative voice
of 'experts' in food scares such as BSE or genetically modified organisms.[5] Ulrich
Beck goes further; for him it is the challenge to the whole notion of expertise that
sets the tone of the latest phase of modernity – the 'risk society'.[6]

Collins's waves and the risk society of Beck and Giddens, while attractive
for the bold and far-reaching vision they portray of expertise, both lack historical
texture. For a sounder foundation in this regard, I prefer the insights on expertise
in two splendid collections edited respectively by MacLeod and by Rabier. On
society's attitudes to trust and risk, I also value the thoughts of several French
historians providing interesting interpretations.[7] Madeleine Ferrières is one,
writing on food scares and the lack of public trust in city retailers before the

---

1 Hehner 1891.

2 For the work of psychologists and others, see Ericsson et al. 2006, Crease and
Selinger 2006.

3 Collins and Evans 2002: 236. For critiques and a response, see Gorman 2002,
Jasanoff 2002, Rip 2003, Wynne 2003, Collins and Evans 2003.

4 Collins is a realist. To him, expertise is substantive, experts can be identified and
their expertise is a matter of a social process of acquisition. This is different from relational
interpretations, which are contextual. Collins and Evans 2007.

5 For more on the relationship between science and expertise, see Nowotny et al.
2001.

6 Beck 1992, 1999, Mythen 2004.

7 MacLeod 2003, Rabier 2007.

Second Empire.[8] Jean-Baptiste Fressoz is another.[9] He shows that Beck-type risks existed as far back as the nineteenth century in France. These were 'new' risks with incalculable consequences, mostly resulting from technological change in the industrial revolution, but including also environmental hazards such as the impact of deforestation on climate and the potential contamination of the groundwater around the capital by faecal matter. Expertise was involved, not only through scientists advising Paris's Conseil de Salubrité or giving evidence in court. There were also popular commentators articulating caution about technical progress that was very similar to today's 'precautionary principle'.

Timothy Mitchell's perspective is also historical but somewhat different in its theoretical context. He argues that experts were brokers of knowledge in the processes of modernization and development, and, as such, performed an important controlling function for the direction of both technoscience and the economy. For him, the quantity and quality of knowledge was not necessarily changed by their intervention; their achievement instead was one of redistribution: a process of removals and transfers to new sites.[10]

For food, the issues at hand in the nineteenth and early twentieth centuries, both in Britain and France, were, first, the indeterminacy of compositional expectations, second, the ferocious arguments that erupted around how standards (agreed or imposed) could be solidified into the form of regulations and then legally enforced, and third, the hazard that milk presented in terms of infectious disease. The present chapter will begin by recounting a major controversy about expertise and professional credibility in the eyes of the law. It was a long-running dispute between government scientists and local authority analysts in which expertise was seen as 'a relational attribution in which you acquire the status of an expert by virtue of your position in a network of social relations'.[11]

Michel Callon would have us call such debates 'hybrid forums', where laboratory expertise mixes with 'recherche de plein air'.[12] He argues that both knowledge and democracy benefit from the controversies that form here like storms at a meteorological front. In today's extensive debates about the quality of food and drink, which have increasingly taken on the guise of deliberative democracy, the interests of the consumer-citizen are at least represented, even if they are frequently overshadowed by the corporate power of the food industry; but in the late nineteenth and early twentieth centuries such voices were subdued.[13] In fact, more often that not, we must ask in whose real interests food laws and

8   Ferrières 2006: 302.
9   Fressoz 2007.
10   Mitchell 2002.
11   Evans 2008: 282.
12   Callon and Rip 1992, Callon et al. 2001.
13   For more on consumer-citizensip, see Chapter 7.

regulations were established: those of the public or those of certain sections of the food producers, processors, manufacturers and retailers?[14]

Expertise in the regulatory situations under scrutiny in this chapter and the next was a set of constructions of goal-orientated knowledge that was deployed in laboratories, in courts and in the corridors of power, in order to achieve the insertion of rational ordering and standardization into the realm of the food supply. But who were the experts: the government chemists or the public analysts? Or was chemical expertise distributed widely among actors, including those in the trade? There certainly seems to have been a sense in the nineteenth century that scientific authority in this area was fragile and open to challenge. This was true of the anonymous skill and considerable enterprise of the adulterators. The Scholefield Committee heard evidence along these lines in 1856:

> Hitherto the progress of legislation has not kept pace with the ingenuity of fraud, which has not scrupled to avail itself of every improvement in chemistry or the arts which could subserve its purpose.[15]

As we noted in Chapter 3, the chemical means of analysis in the mid-nineteenth century were time-consuming, expensive and uncertain. Both scientists and traders therefore employed the measurement of risen cream and the specific gravity of milk for the practical, everyday detection of adulteration. Both they and the adulterators knew very well that, with careful attention to detail, cream could be skimmed and water added to simulate the properties of genuine milk; so detection was not easy. This point about the ease of simulating genuine milk continued to be made for at least a further 50 years, until the coming of corporate laboratories in the early twentieth century led to a redefinition of expertise and of the skills required of an expert.

> The detection of adulteration is becoming more and more difficult, and is due, in the first place, to the astuteness of the vendors of the adulterated articles, and, in the second, to the more highly scientific means now practised ... The addition of separated milk to new milk has become almost a fine art with some milk purveyors, who, although they are known to receive large quantities of separated milk, which they do not sell, yet mix it so skilfully that it is impossible to bring them within the four corners of the law.[16]

The struggle for expert status in judging dairy products was not just between professional laboratory-based chemists. There were other actors in the various food trades who thought that their traditional knowledge – what Steven Shapin

---

14   Stanziani 2003a.
15   Scholefield Committee 1856: 8.
16   Bannington 1915.

calls prudential expertise – was superior.[17] Organoleptic skills were especially prized among dairymen, with taste and smell being important in detecting signs of the onset of souring. There were similarities here with, say, the wine industry in France, where use of the human senses continued to be at the forefront of expertise, despite the development of an organic chemistry as complex and insightful as that for milk.[18]

In Britain, product quality was increasingly linked in our period to a series of centrally defined rules that were negotiated between civil servants and representatives of the food industry. These were empowered by a combination of laws and official regulations, which were then tested and enforced by the courts, starting at the local level in the magistrate's courts and, in a small number of cases, appealed to the High Court. As a result, commercial and administrative rules and legal debate were inevitably bound together; but it is important to repeat and emphasize the contestation that was built into such a system. Because of disagreements between experts of the same background and between the expertises of traders, scientists, administrators and lawyers, our period has a rich literature and series of case law precedents to draw upon for research.

Since I have brought up the differences between the French and British experience, this is a convenient point to press further the notion of locally-specific meanings of expertise. We saw in Chapter 2 that scientists were often aware of key literature in foreign languages, and the expert witnesses in the Schrumpf case had even gone to the expense of transatlantic travel in order to broaden their experience. There was also, from the 1880s onwards, a circuit of international congresses at which papers were read and the experiences of laboratory and administrative successes were shared. And yet there remained cultural, intellectual and political barriers that were responsible for divergent paths in the evolution of the science and technology relevant to our story. The purchase of locally marketed laboratory equipment, especially glassware, was one limiting factor, and another was the variation of local legal frameworks. As a result, we need to consider contingency in our histories of commodity expertise.

The second part of this chapter will look at how chemical expertise and commercial interests fed into a political debate about milk standards. Although the British regulations formulated in 1901 were long-lasting in their impact, nevertheless deliberation about market milk continued throughout the twentieth century. This was partly philosophical, about the relationship between food and nature, and partly about the degree to which the practical methods used by the food industry to make profit were socially and commercially acceptable.

---

17   Shapin 2003: 293.
18   Atkins and Stanziani 2008.

Somerset House

> The Excise laboratory does not enjoy a high repute amongst chemists ... I do not
> think that any good could arise from having official referees in those disputed
> cases. I have had some experience of the kind myself, and I can promise you that
> if you were to elect any such body, and give them powers of this description, you
> would be giving them a license to blunder and to disregard the general views of
> chemists.[19]

The negative view of government science displayed in the quotation above was
already established among private analysts as early as the 1850s, as a result of
occasional disagreements about the composition of dutiable goods. By the 1870s,
when analysts were increasingly being employed under the Sale of Food and
Drugs Acts, a clash of wills had emerged with the state chemists.[20] This was partly
institutional in origin and partly the result of a disagreement about the nature of
expertise and the consequence of decisions about the 'natural' composition of
foods.

By far the most volatile body of chemists in the 1870s and 1880s were the
public analysts, no doubt because they were a newly constituted body that had yet
to achieve acceptance and develop a self-confidence in their role. These 'practising
chemists' felt they were under attack from the outset, and were especially angry
about the evidence of Augustus Voelcker to the Read Committee in June 1874,
when he said that he doubted there were more than a dozen chemists in the whole
country capable of carrying out the duties of public analyst properly. He went on
to claim that

> a good deal of mischief has been done by the so-called analyses, and the food
> analysts have been the greatest enemies to the Food Act ... Many are very
> incompetent analysts; they have not had sufficient chemical training, nor any
> experience in analysis, and therefore their statements are sometimes of a very
> flippant and unwarranted character, which has done a great deal of harm in
> rendering the Act contemptible in the eyes of practical men.[21]

This pointed condemnation was uppermost in the minds of those who shortly
afterwards, on 7 August 1874, met to create the Society of Public Analysts and
give credibility to a new profession.[22] Their agenda was decidedly defensive: 'the

---

19 Read Committee 1874: Q.6523, Wanklyn, referring to plans for a chemical court
of appeal at Somerset House.

20 Scholefield Committee 1854–55: Q.346, Hassall, Q.2212, Wakley.

21 Read Committee 1874: QQ.5589, 5861, 5600.

22 In 1953 it was renamed the Society for Analytical Chemistry and in 1980 it was
incorporated into the Royal Society of Chemistry.

refutation of unjust imputations' and 'the repudiation of proposed measures of interference with our professional position and independence'.[23]

The bitterest controversy to erupt was between the government's chemists and analysts employed by local authorities. It lasted for 30 years.[24] All aspects of the quality of food and drugs were at stake but the main flash point was milk. In essence, the debate was about the composition of 'genuine' milk and who had the moral and legal duty to declare that a deviation from 'natural' milk constituted a fraud. It was also a technical argument about methods of analysis and the identification of thresholds of acceptable composition. These are themes that flow naturally from the discussion in the previous chapter and they will now be considered in the sharper light of legal enforcement.

The 'Excise laboratory' referred to in the opening quotation was the state's official laboratory, located in Somerset House, London.[25] It was founded as the Board of Inland Revenue Chemical Laboratory (1842–1894), and was later renamed the Government Laboratory (1894–1911), and then the Department of the Government Chemist (1911–1959), before being absorbed by the Department of Scientific and Industrial Research.[26] As the result of a recommendation by the Read Committee, the 1875 Sale of Food and Drugs Act granted it powers as a chemical Court of Appeal, sitting in judgment upon disputed samples referred to it by the magistrates' courts.[27] Although the number of these 'reference samples' submitted was small, the proportion of disagreement between the government chemists and the local public analysts was substantial, leading to friction. In the case of milk, 33.1 per cent of those certificates issued by council analysts to indicate adulteration were challenged by Somerset House in the period 1876–1890 (Table 4.1).[28] Professional opinions were being contradicted on a regular basis and here we have an instance of expertise being undermined. Relations between the two sides were inclined 'more towards hostility than towards co-operation', and disputes frequently spilled over into the trade press and sometimes even into the popular prints.[29] Enmity developed at a personal level, particularly in the government laboratory, where Civil Service rules prevented the mounting of a

---

23 Chirnside and Hamence 1974: 8.

24 For a flavour of the vituperation, see the correspondence of July and August 1884 between Richard Bannister of Somerset House and Alfred Allen, analyst of Sheffield, reproduced in the *Analyst* 19, 1884: 231–40.

25 The laboratory was established here in 1852 and in 1897 moved to Clement's Inn Passage. Pilcher 1919, Hammond and Egan 1992, Hammond 1992.

26 The Inland Revenue laboratory and the Board of Customs and Excise laboratories merged in 1894. Hammond and Egan 1992.

27 Read Committee 1874: 248. Strictly speaking, the sample was sent to Somerset House by the magistrate and not on appeal by the defendant.

28 Foster Committee 1894: QQ.552, 555, Bannister.

29 B.D. 1908: 158.

public defence against critics.[30] This bitterness persisted for a generation until the protagonists from the 1870s to the 1890s had retired.

Table 4.1   **Analytical disagreements (per cent of reference samples) between the government laboratory and local authority analysts**

Years	Samples	%	Years	Samples	%
1876–1880	108	26.9	1901–1905	498	17.3
1881–1885	199	41.7	1906–1910	478	15.5
1886–1890	212	28.3	1911–1914	328	11.0
1891–1895	233	11.6	1876–1914	2315	20.4
1896–1900	259	29.7			

Sources: Reports of the Commissioners of Inland Revenue, Report of the Principal Chemist, Government Laboratory, NA: DSIR 26.

One problem was that, until 1901, there was no legislation, nor any official regulation legally defining milk, whether in words or in terms of scientific measurement. As a result, the government chemists were reluctant to label any sample 'adulterated' if there was any chance that it might have achieved its abnormal composition through natural processes or accidentally. As the wide variability of the constituents of honest milk became better known, their early caution seems to have been justified.

> For a purchaser to be prejudiced within the meaning of this clause, it is necessary that the article sold should contain some admixture of a foreign substance not specified at the time of sale; and therefore that the purchaser is not legally prejudiced when the article sold is of low quality but genuine … It has been urged that samples should be judged by those of average quality, which the purchaser might reasonably expect to get: but this was evidently not the view of our legislators, for Parliament deliberately abstained from fixing limits of quality for natural products, whether in a raw or prepared state.[31]

One reason for Somerset House's confidence in its own position was that they had undertaken empirical research into the natural variations of milk composition (Table 4.2), and this provided the basis for an evidence-based methodology.[32] As a result, they rejected the normative approach of the Society of Public Analysts and refused to engage in academic debate about appropriate laboratory methods of analysis.

---

30 Dyer and Mitchell 1932: 16.
31 Bell 1884: 135.
32 NA: DSIR 26/134.

We have felt ourselves unable to adopt the 'definitions' and 'limits' for genuineness laid down by the Society of Public Analysts, for the simple but all-sufficient reason that they are not borne out by our own analyses of hundreds of samples known to be genuine. In this view we are supported by many eminent analysts, including some of the members of the society in question, although that society still nominally adheres to the 'limits' laid down.[33]

Table 4.2   Somerset House analyses of milk composition, 1876

	Specific gravity	Cream (%)	Solids-non-fat (%)	Fat (%)
1	1031.02	9.17	9.46	4.30
2	1032.95	7.50	9.31	2.50
3	1030.64	12.10	9.51	4.96
4	1031.96	6.69	9.30	2.80
5	1030.61	9.89	9.39	4.39
6	1031.62	11.35	9.63	4.11
7	1031.61	9.90	8.98	3.58
Average	1031.30	9.33	9.41	3.99

Key: [1] whole milkings of 40 cows in the counties of Derby, Devon, Oxford, Surrey and Kent; [2] first part and [3] middle part of milking respectively of each of 14 cows in the London parishes of Brompton, Chelsea and St Pancras; [4] first part and [5] remainder of milking respectively of each of 19 cows in the counties of Devon, Stafford, Oxford, Surrey and Kent, and the parishes of Lewisham, Chelsea and St Pancras; [6] eight dairies each containing from three to 35 cows in the counties of Stafford and Oxford, and the parishes of Lewisham, Brompton, Chelsea and St Pancras; [7] six instances from churns from different localities immediately on their arrival in London, the sample being drawn in each case from three different depths of the churn.

Source: National Archives: DSIR 26/134.

Another factor in the official laboratory's perceived arrogance was its solid establishment and long-standing reputation. In addition to dealing with samples referred by local magistrates, the Inland Revenue Laboratory was also responsible for analysing samples from various government departments, such as Customs and the India Office, and under other legislation they undertook the sampling that went with excise duty, for tobacco, beer and spirits. The total throughput of the Government Laboratory in the year 1894, for example, was 48,255 samples, making it one of the busiest enterprises on the Strand. This was with a staff of

---

33   *Local Government Board: Twelfth Report, 1882–83*, P.P. 1883 (C.3778), xxviii.602.

57.[34] Only 71 of the samples were the so-called 'reference samples' sent in by magistrates under the Sale of Food and Drugs Act, so this food adulteration work was a relatively small part of the overall Somerset House mission.

Among the public analysts there were many outbursts against their protagonists. In 1874, for instance, while the Read Committee was still sitting, an editorial in the *Chemical News* declared that

> When we heard the Inland Revenue Laboratory at Somerset House suggested as the ultimate court of appeal we were disposed to think that the proposal was a joke intended for the columns of one of our comic contemporaries.[35]

Later, M. Henry, editor of the journal *Food and Sanitation*, was highly critical.[36] He spoke of 'the existing wretched, ignorant, and utterly untrustworthy system of food analysis at Somerset House'. It was a 'poor, bungling department struggling to perform work for which it has not got the skill or knowledge'. In his opinion, 'scientifically the Somerset House chemists are dead, and there exists no shadow of an excuse for their remaining unburied'.[37] Pearmain and Moor were more specifically condemnatory of what they regarded as a lax attitude to milk quality: 'It is not too much to say that the disgraceful state of the milk industry in this country is fostered, if not actually caused by this ridiculously low standard.'[38]

Among those who preferred more measured language, six types of complaint were aired. First, the analysts working for the government were accused of being underqualified and their knowledge of food chemistry was said to be limited to testing the strength of liquor.[39] Second, while it was recognized that the milk samples they worked with were always sour, nevertheless their unique 'maceration' method of analysis was heavily criticized. Third, they were said to be far too generous to the perpetrators of fraud and to err of the side of caution when declaring their results. Fourth, Somerset House officials were said to be somewhat aloof and secretive, being unwilling, for instance, to discuss their methods with other scientists, apparently on Civil Service legal advice.[40] Fifth, despite the available staff and resources, the government laboratory was censured for having made 'but few fundamental contributions towards the rapid and brilliant advancement of food analysis'.[41] The unspoken thought here was the widespread suspicion of centralized government services in the mid-nineteenth century; that they were both inefficient and intrusive. Finally, there was some debate about

---

34 Foster Committee 1894: Q.772.
35 *Chemical News* 30, 1874: 11.
36 *Food and Sanitation* 10 February 1894: 47.
37 *Food and Sanitation* 27 January 1894: 25.
38 Pearmain and Moor 1897: 17–18.
39 Dyer and Mitchell 1932: 15, Egan 1976: 110.
40 Dyer and Mitchell 1932: 16.
41 B.D. 1908: 158.

whether the legislation had ever intended the reference laboratory to have the final say.[42] Charles Heisch did not think so and pointed out that:

> Most magistrates act as if the certificates of the Somerset House officials were not only evidence, but final evidence. Now those of us who followed the stages of the Sale of Food and Drugs Act, will all remember that when Sir H. Peek proposed to insert after Somerset House the words, 'whose decision shall be final', Mr Sclater-Booth, who had charge of the bill, refused to insert them, and when the matter was pressed to a division they were rejected by a large majority. Not only this, but Mr Booth declared in his place in Parliament, that he did not intend the Somerset House decision to be final, but that the analysts should both be subject to examination on oath in case they differed, and should each have the opportunity of justifying their decisions if they could.[43]

It is true that Dr James Bell, Principal of the Laboratory, set a low threshold to judge the genuineness of milk. For fat this was 2.5 per cent, well below the 3 per cent adopted by most other analysts and the 3.6 per cent that was acknowledged by most commentators as being a good average of the actual output of healthy cows. Bell's concern, he said, was for 'justice'. The spirit of the Sale of Food and Drugs Acts, in his opinion, and presumably that of his legal advisers, was not against milk that was poor in quality but not adulterated. The thin milk produced by malnourished cows could be sold legally, as could the milk of cows fed on watery material such as brewers' grains, or from breeds that at certain times of year yielded milk that was exceptionally low in fat and other solids. As a result, a debate opened on whose rights were principally protected by the legislation: those of the consumer to purchase rich, whole milk, or those of the farmer not to be unjustly prosecuted.[44]

Table 4.1 above shows the level of disagreement between the government and local authority experts. In the case of milk, it was close to half of reference samples in the early 1880s, remaining at between a quarter and a third up until the late 1890s. After that, the number of disputes fell as techniques of analysis converged and the relevant skills became better known and practised. The proportion of disagreements about the degree to which cream had been abstracted was more variable than that about the percentage of added water, partly because the accusation of something having been extracted was in a sense hypothetical. Throughout the period, watering remained the most common charge to press.

In 1892 a letter was sent to Bell signed by 'a very large number of public analysts' in which they objected to the wording of the certificates issued by his laboratory. These had sometimes stated that the analysts in Somerset House 'were unable to affirm that water has been added', thereby giving the benefit of the doubt

---

42   Jago 1909: 231.
43   Heisch 1882: 14.
44   Hill 1876: 46.

to the accused.[45] Faced with such a statement, magistrates not unnaturally felt obliged to dismiss the case immediately. This procedural problem went to the heart of the argument between the two sides because it seems that often there was no scientific disagreement about the results of their analyses; only their interpretation differed, as expressed on the certificates. What the public analysts wanted to be made clear to the courts was that Somerset House's wording was consistent with the possibility of adulteration, giving the magistrate a clearer view of the balance of probabilities.[46]

After 20 years of hostilities, in the 1896 hearings of the Russell Committee, the Society of Public Analysts argued for the abolition of the Inland Revenue Laboratory and its replacement by a Board of Reference.[47] This new institution would have been given the wider responsibility of defining food standards by publishing the allowable limits of chemical composition. The suggestion was scrutinized in depth by the Select Committee and it appeared in their final report in 1896 as a firm recommendation to government.[48] Evidence on an earlier version of this idea, from Richard Bannister, Deputy Principal of the Revenue laboratory, however, had been somewhat negative at the 1894 hearings of the Foster Committee, emphasizing that the proposed voting structure of the Board meant the possibility of the public analysts gaining control.[49] The Board was never established but the new Sale of Food and Drugs Act (1899) did give powers to the Board of Agriculture to frame presumptive regulations, and this was a very important step for our history of milk.

Bell retired in January 1894 and was replaced by Thomas Thorpe, who was Professor of Chemistry in the Royal College of Science and, like Bell, a Fellow of the Royal Society.[50] He was brought in to oversee the amalgamation of the Inland Revenue and Customs laboratories and to introduce modern methods of investigative chemistry.[51] Defusing the dispute with analysts was also an objective. Thorpe immediately proved to be more pragmatic than Bell. In 1894, pre-dating the commencement of the Foster Committee, he raised the milk fat criterion from 2.5 to 2.75 per cent, and then further to 3 per cent in 1899.[52] Nevertheless, despite his best efforts, both he and his team continued to be vilified. An example is the statement to the 1896 Russell Committee by Frederick Lloyd, consultant chemist

45 By 1894 this form of words had been abandoned. Foster Committee 1894: Q.1697, Bannister.
46 Foster Committee 1894: Q.743.
47 Russell Committee 1896: 501–2.
48 Russell Committee 1896: 524.
49 Foster Committee 1894: Q.887.
50 Bell had become Principal in 1875. He was President of the Institute of Chemistry, 1888-91. See Pilcher 1914: 90–100.
51 Thorpe was appointed over the head of Richard Bannister, who had been with the laboratory for 31 years. Foster Committee 1894: Q.542.
52 Foster Committee 1894: QQ. 555, 563, 708, 1665, Wenlock Committee 1901: QQ. 10, 546–54.

to both the British Dairy Farmers' Association and the Metropolitan Dairymen's Society, neither of them organizations with any interest in exaggerating the situation:

> I am sorry to say that in my opinion the action of Somerset House in taking low standards for milk has been the cause of the enormous amount of adulteration at present in the country.[53]

Thorpe's principalship at the Government Laboratory was a period of progress. He established better contacts with the public analysts and they collaborated together in trying to influence the legislation in 1899.[54] He also put the laboratory on a firmer footing scientifically with regard to milk adulteration by recognizing that it was necessary to measure the products of the decomposition that inevitably took place in the reference samples that they dealt with, and so achieve a more reliable 'reconstruction of the original solids'.[55] This was more accurate than the sliding scale adjustment that had previously been made, based on the length of time since the sample had been collected.[56]

What can we learn from this long-running dispute and how does it inform this chapter and the next? First, it suggests that we need to reflect on the nature of expertise in the British analytical system. This will require a commentary on the role of the public analyst and the resources available. It will quickly become apparent that analysts were heavily dependent upon the political will of their local authorities, which varied quite markedly according to local circumstances.

Second, the steady evolution of the Sale of Food and Drugs legislation was partly a function of technical considerations. An obituary of Dr Bell in the *Analyst*, the journal published by his adversaries, admitted in 1908, with obvious reluctance, that his position had had a 'substratum of truth' and that the real problem had been 'ineffective drafting' of the 1875 act.[57] We will look at attempts to capture 'adulteration' as a concept in the various Acts in Chapter 7. In that chapter we will also look at the various Bills and Acts as a political process in which a balance was struck between opposing interests.

Third, there will be a long section in Chapter 7 on the various types of adulteration, fraud and bad practice with regard to milk. In addition to watering and the extraction of cream, there was much contemporary discussion of the addition of chemicals, preservatives, to prolong its shelf life and colorants to simulate a rich milk. We will look at their relative significance nationally and also comment on differences between town and country, between the different cities, and change

---

53   Russell Committee 1896: Q. 353.
54   Dyer and Mitchell 1932: 17.
55   Dyer and Mitchell 1932: 16.
56   Foster Committee 1894: QQ 1667–72, 1688, 1692. For a debate on the analysis of decomposed milk, see Estcourt 1887, Bell 1887, Stokes 1887.
57   B.D. 1908.

through time. There are some surprises here, for instance 'toning', which initially was seen as a crime but eventually became standard practice, throwing into doubt early conceptualizations of both naturalness and honesty.

Fourth, the enforcement of food standards required what I shall call 'the quality police'. Our discussion of them in Chapter 7 will not be a dry account of regulations but rather a theoretically-informed discussion of court cases, especially those that acquired notoriety in the appeals system though some point of law that bears upon our interest in the definition of quality. Then I will argue in Chapter 8 that such legal disputes have been undervalued by historians and that their study provides a way forward in understanding a key aspect of food quality governance.

Analysts and the Quality Police

The best historiography of the literature on the emergence of experts and expertise in nineteenth and twentieth century Britain has been written by Roy MacCleod in the introduction to his celebrated edited collection. His co-authors complement this with the results of their own empirical work on the history of administration.[58] Together they claim that the direction in which modern governance has evolved was at least in part dependent upon the specialist knowledge and judgement associated with expertise, and in this assertion they have some support from the work of Harold Perkin on the professionalization of leadership in administrative governance.[59]

Our public analysts seem to have approximated the first stages of MacDonagh's model of Victorian bureaucratic control.[60] Their numbers increased during the acceleration in the size and scope of the civil service in the last twenty years of the century and food quality monitoring then evolved in phases: first an ineffective piece of legislation was drafted, followed by the appointment of experts to enforce the law, and a central body (the Local Government Board) then acted as coordinator and had ultimate authority. The analyst profession had a vigorous pre-history as servants of industry and sanitary science,[61] as gas examiners, alkali inspectors and as water analysts, and they were part of the wave of public professional men and women of science appointed in the middle and second half of the nineteenth century, along with Medical Officers of Health. Frank Turner sees the 1870s as a hinge-point, with science after that being taken more seriously and the state feeling an increasing need to employ scientists rather than consulting them occasionally, as had been the previous pattern.[62] From the 1840s analysts had a proliferation of

58 MacLeod 2003.
59 Perkin 1989.
60 MacDonagh 1958.
61 Donnelly 1994.
62 Turner 1980.

professional organizations and journals to champion their interests.[63] At the outset, few had any experience of food analysis, however, and in 1856 Scholefield rightly predicted the mismatch between supply and demand of the relevant skills that was to occur 20 years later:

> At first, no doubt, some difficulty would be experienced in finding persons qualified to conduct the required chemical and microscopical examinations; but the want will soon give rise to the needful supply. This want has been already felt by the Board of Inland Revenue, which has been compelled, for the purposes of the analyses requisite in cases of adulterated articles of Excise, to educate persons for this special duty.[64]

Analysts could be appointed under the 1860 Sale of Food and Drugs Act, but by 1873 they had been appointed in only 17 counties and nine boroughs. The 1875 Act was a compulsory measure, with a duty to appoint a competent analyst being imposed upon each county, boroughs with quarter sessions or a police authority, and, in London, the City and all of the Vestries and District Boards.[65] For analysts generally, there was frustration at poor funding and weak access to the policy-making process, and the fact that sampling was not compulsory until 1899.[66] Some analysts were full-time, others were shared between authorities on a part-time basis, and a few were paid a small retainer against the day when funding and political priorities would enable them to take up their duties.

A majority of the first analysts appointed were not chemists at all but local Medical Officers of Health. They were instructed by their authorities, for reasons of economy, to accept this additional portfolio, but generally they had 'very little education in chemistry ...'.[67] As men of science it was simply assumed that they could perform the extra duties, in most cases without any training or additional resources. The required qualifications were modest. The 1872 Act mentioned medical, microscopical and chemical knowledge, the medical element being a consequence of the concern with poisoning as the main problem of adulteration at that time, not fraud. In 1875 the medical requirement was removed but the monitoring of appointments by the Local Government Board remained lenient. In 1874 it had been observed that the Board had 'hardly objected to anyone'

---

63   Dyer and Mitchell 1932, Moore and Philip 1947, Chirnside and Hamence 1974, Russell et al. 1977, Roberts 1979.

64   Scholefield Committee 1856: 8.

65   *Return of Number of Analysts appointed under Adulteration of Food Act, 1872, in each County and Borough of United Kingdom*, P.P. 1873 (280). In 1893 the powers of the quarter sessions were transferred to county councils.

66   Macleod 1967, 1968, 1976.

67   Read Committee 1874: Q.2448, Barham.

put forward.[68] Of those in post at that date, it was said that 'many of them are incompetent chemists'.[69]

Gradually, better qualified analysts came forward and the Board tightened its procedures, so that applicants by the 1890s had to 'produce evidence of competent microscopical knowledge, chemical knowledge, and also, to a certain extent, medical knowledge; that is to say, as to the effect of adulteration on health'.[70] Testimonials were required but still no formal academic qualification or professional membership.[71] This was despite the advice of the Society of Public Analysts that applicants should have a science degree, a Fellowship of the Institute of Chemistry, and a year's experience in the laboratory of a public analyst.[72] It was not until 1900 that the Local Government Board required evidence of a pass in the final examination of the Institute for Chemistry's Branch E: 'the Analysis of Water, Foods and Drugs.'[73]

Experts, in order to perform their expertise, need a clientèle and an audience. At first, the local authority public analysts in Britain were not quite sure who had a stake in their results. As we have seen, many local authorities simply went through the motions of appointing an expert analyst and resisted expenditure on the collection of samples or laboratory facilities for processing. But what of private clients? Under the 1860 Act, samples could in theory be submitted by members of the public but the cost was up to half a guinea per item and so few citizens did so.[74] Some public analysts refused approaches from the trade on the grounds that only 'purchasers' had access to their skills.[75] Compare this with the situation in Paris, where, as a result of pressure from the public, in 1882 the municipal laboratory processed almost as many samples of wine submitted by citizens and traders (5,188) as collected by inspectors (5,238).[76] For milk, the situation was more like that in London, with fewer complaints from civil society.

Although commercial traders were also entitled to submit samples, many analysts did not see their role as a support service for industry. This was almost certainly because many of them had private practices to supplement their meagre state stipend and they did not see why prosperous dairy companies should get their analyses done 'on the cheap'.

---

68   Read Committee 1874: Q.4696, Cameron.
69   Read Committee 1874: Q.5665, Voelcker, Dyer and Mitchell 1932: 5.
70   Foster Committee 1894: QQ.6, 194, 381, Preston-Thomas.
71   Foster Committee 1894: QQ.279–90, Preston-Thomas.
72   Dyer and Mitchell 1932: 34.
73   Local Government Board Order, 1900. Chirnside and Hamence 1974: 37.
74   Some local authorities charged less but were not able to elicit further private samples. Foster Committee 1894: Q.249, Preston-Thomas.
75   Read Committee 1874: Q. 2579, Barham.
76   Atkins and Stanziani 2008: 322.

> At first some of the dairy men tried to take samples to the analysts, but they said
> they could not take them unless they came from a purchaser … in point of fact,
> they shirked the question.[77]

Towards the end of the nineteenth century, expertise took on a procedural
nature. Many of the analysts working for local authorities were required to
process large numbers of samples. Efficiency was therefore a concern and time
and motion data were collected. At the turn of the century we learn that each
adulterated sample took about 100 minutes of an inspector's time and Table 4.3
also gives a breakdown of the expenses on 2,777 prosecutions in England and
Wales, indicating that chemical analysis represented about 75 per cent of the total
cost. The 'rituals of verification', then, were expensive and, in order to conform
with the law, they began to take on an institutional guise. More and more local
authorities brought analysts into their civil service establishment and provided
facilities in order to guarantee high standards of work.[78]

Table 4.3   Expenses associated with analyses

Expenses	£
Cost of samples, including packages	1,333
Cost of inspectors' time purchasing samples	3,070
Incidental expenses of inspectors	830
Cost of analyses	29,285
Cost of inspectors' time spent in court	359
Cost of Clerks' and Medical Officers' time in court in three-quarters of the adulteration cases	685
Solicitor or barrister for the other quarter of cases	1,457
Cost of appeal cases, about 10 a year	1,000
Fees and expenses of analyst if called by prosecution, stationery and postage, cost of summonses	1,000
Total	39,019
Less amount collected in fines	3,683
Net total	35,336

Source: Cribb and Moor 1899: 225.

Analysts were required by many local authorities to produce quarterly reports
of their activities and the law obliged them to fill in a certificate in any case of
suspected adulteration, as the pseudo-legal form of their professional opinion.

---

77   Foster Committee 1874: Q.2579.
78   Power 1997.

Most, from time to time, had to attend court to be cross-examined on their evidence. The articulacy skills of the witness are very different from those of a laboratory scientist, so this required a hybrid expertise that was performative. In the nineteenth and twentieth centuries disputes about food were dealt with under the British common law, which was based upon local justice and, in criminal cases, upon trial before a jury of lay peers. As Hildebrandt points out, this was not predisposed to participation by experts.[79] The adversarial setting and the sometimes hostile questioning by counsel made it an uncomfortable ride for many scientists, who were often unable to encapsulate their complex, specialist knowledge in simple yes or no answers. Hildebrandt calls this *épreuve* or coordinate justice and traces its historical origin to grassroots, popular justice, with its scepticism about elite knowledge. By comparison, the inquisitorial process – *preuve* – of continental civil law was more respectful of scientific specialists and less dependent upon the theatre of courtroom procedure.

Whatever their level of zeal in detecting fraud, analysts could only analyse what was sent to them, and in many districts few samples were collected in the early decades. There was a plain reluctance by many local authorities to use the permissive powers granted to them in the Act. The Local Government Board issued a circular in 1884 encouraging local authorities to use these powers but the reaction was mixed. The Board's own annual report for 1891–1892 noted that:

> in many large boroughs ... such as Devonport, Sunderland, South Shields, Preston, Ashton under Lyne, Burnley, Great Grimsby, Northampton, Ipswich, and Dudley, no serious attempt was made to enforce the Acts.[80]

Table 4.4 illustrates the variability of investment by local authorities in collecting samples for analyses. The LGB's initial target, published in 1880, was for one sample of food generally to be taken for each 1,000 people in an area.[81] This was eventually achieved in 1892–1893 but the report of the Russell Committee in 1896 decided that even this was not enough. It was not until 1930 that sampling achieved 3.5 items per 1,000 people.[82]

In 1894 the Foster Committee found that there were 99 public analysts looking after 237 districts in England and Wales. Of these, 49 were responsible for just one district each and 22 had four districts or more. Having a list of several districts did not guarantee a substantial business and only nine analysts in the whole country had a gross income for official work of more than £500 per annum. One hundred and twelve of them, including 29 in London, were paid a flat rate annual salary, varying from the miserly £5 laid out by Sunderland to £250 in Cardiff, and £200

---

79 Hildebrandt 2007.

80 Local Government Board Report: Twenty-first Report, 1891–92, P.P. 1892 (C.6745), xxxviii.142.

81 Liverseege 1932: 6.

82 Evans 1976: 131.

Table 4.4   **Urban population per milk sample, 1893–1897**

Town/city	Population per sample	Town/city	Population per sample
Salford	340	Sheffield	2,040
Cardiff	400	Newcastle	2,080
Manchester	480	Leeds	2,150
Liverpool	750	Leicester	1,210
Croydon	800	Birkenhead	2,970
London	1,040	Preston	3,140
Bristol	1,190	West Ham	3,170
Oldham	1,230	Blackburn	3,260
Bradford	1,270	Nottingham	3,870
Birmingham	1,310	Bolton	4,000
Portsmouth	1,410	Norwich	4,390
Hull	1,470	Sunderland	6,200
Brighton	1,610	England and Wales	1,730

Source: Liverseege 1899.

in Liverpool and Swansea. The average salary among the London local authorities was £92. Some localities preferred to pay fees for individual analyses, from which the analyst was meant to cover his raw material costs and laboratory overheads. These ranged widely from 2s 6d to three guineas a piece, with additional travelling expenses for attending court.[83] There is one interesting insight into the cost of overheads. The Foster Committee found that George Embrey of Gloucestershire had been provided by his County Council with a laboratory and the cost of chemicals. On top of that, his salary remunerated his own time, that of any assistants, and also his court expenses. Overall, he worked for a modest 2s 0d per sample.

In 1894 the busiest public analyst in England and Wales by far was Campbell Brown, whose laboratory in Liverpool processed 7,508 food samples for seven boroughs in Lancashire and also for the county council, at an average rate of 7s 0d per sample. Others who were fully-stretched included W. Morgan, the analyst for Swansea, Neath and six Welsh counties (2,762 samples, £1,532 income); Edward Bevan of Middlesex (1,315 samples, £979); and Alfred Hill acting for Birmingham and six other Midland towns and cities (3,005 samples, £628). The average cost per sample for the whole country was 8s 10¼d, but there were some local authorities that squeezed maximum value out of their analysts. Salford, for instance, was renowned for its active policy of food testing and it achieved this by

83   Foster Committee 1894: Appendix 9.

dealing with 1,010 samples at the very efficient rate of 2s 6d each. Eight hundred and fifty-four of these were milk samples. As a result of this intensive monitoring, Salford was said to have 'one of the best milk supplies of the kingdom'.[84]

Some analysts built up fiefdoms, taking on several local authorities in a region. John Pattinson was an example, with a portfolio comprising the urban districts of Durham City, Newcastle, Gateshead, South Shields, Tynemouth and the county of Northumberland.[85] For him this was not the path to riches, however, because no samples at all were collected in 1894 in Durham, and only six in South Shields and 13 in Tynemouth.[86] In fact, the North East was far from being a model of best practice in this regard.[87] Other analysts acted for authorities from around the country and relied upon the postal service for prompt delivery. Alexander Wynter Blyth, for instance, was the analyst for St Marylebone in London, but also for Barnstaple and the whole county of Devon, where he had to travel for time to time for court proceedings.[88] Another example was Charles Cassal, who looked after Battersea, Kensington and St George Hanover Square in London, and also Kesteven in Lincolnshire and Chipping Wycombe in Buckinghamshire. All of this was possible due to a clear division of labour between the analysts and the local authority officials who collected the samples, such as the Medical Officers of Health, Inspectors of Nuisances, Inspectors of Weights and Measures, or even the police.

At the quieter end of the market, the Foster Committee found that there were 39 analysts who dealt with 100 samples or less per annum, including five who looked at none at all.[89] Their services were retained for a modest fee by local authorities who had no intention of investing in the staff and systems necessary for a thorough food-testing regime. It was either not a priority for them or was actively opposed by local vested interests.

The initial reluctance, on the part of local authorities, to pay adequate salaries and fees was later replaced by the realization that investment in equipment and laboratory staff was unavoidable. The larger metropolitan authorities led the way but local political and cultural contingencies meant that there was a great deal of variation, even where financial resources were not a problem. For cities such as Liverpool and Manchester there was civic pride to be nurtured and political capital to be gained. Food quality, and its relation to sanitation and health, was seen to depend upon the skill of the analyst and the willingness of the authorities

---

84   Foster Committee 1894: Q.78, evidence of Herbert Preston-Thomas.

85   Tyne and Wear Archives: DT.PT/1–26, J.and H.S. Pattinson, Analytical Chemists.

86   Durham City was notorious for never having had a sample analysed. *Food and Sanitation* 17 February 1894: 56.

87   Sunderland insisted that its Medical Officer of Health double up as the borough analyst at a salary of £5. Foster Committee 1894: Q.288.

88   For more on Blyth, see Thornburn Burns 2007.

89   J.J. Beringer (Penzance), J. Bradburne (Hartlepool), W. Chattaway (Colchester), J. Haworth (Tiverton), and W.T.G. Woodforde (Newbury).

to prosecute those in the food industry responsible for carelessly dirty milk or milk reduced in quality by deliberate acts of adulteration.

Cattle Breeding: an Example of Distributed Expertise

Our discussion of analysts by no means exhausts the expertise that underpinned the milk supply system. It is labouring the obvious to stress that farmers deploy great skill in their struggle to capture and valorize the resources around them. My school geography text books spoke of specialist dairy farming regions such as Cheshire and the West Country as if the natural premium of climate and soil guaranteed successful pasture-based milk production. Yet such a high degree of specialism was really only true of the years 1850–1950, and historical geographers have revealed that, even then, it was under constant challenge from factors such as disease, soil exhaustion, and poor facilities.[90]

As more producers became involved in selling milk to urban wholesalers, they were entrained increasingly into a modern food system that required supplies of daily regularity and of a certain quality. This required sacrifices but most farmers seem to have found ways to bend the system to their own advantage. We have mentioned adulteration as one, but the most common, and perfectly legal, strategy was to translate the characteristics of nature. In this way their cows' metabolisms became entrained in the milk food system. It was clear for all to see that their efficiency as machines for producing milk differed according to breed. Some were renowned for the quantity of their milk and others for high butterfat content. The latter had been particularly valued for butter and cheese-making but the mid-nineteenth century was a time for reassessment. After that farmers were less sentimental about their local breeds and instead sought high-yielding cattle that produced a quality that would satisfy the market. There was an impressive degree of flexibility in the mindset of livestock farmers that enabled relatively rapid adjustment to circumstances.

Part of the calculation was the, as yet unquantified, relationship between feed and milk output. This was important because there were regional and seasonal variations of the availability of fodder, from pasture and arable, and each farmer had to calculate an informal cost-benefit ratio that might also include the purchase of concentrates or even of stimulants such as the spent grains from breweries or distilleries.

The point we are making here is that from the 1860s and 1870s onwards it would have been obvious to all but the most conservative of dairy farmers that it was in their interests to consider new, high-yielding genes.[91] For most farmers this meant the purchase or hire of a proven bull, but there was also a steady turnover in

---

90   Simpson 1957, Barnes 1958, Phillips 2004.
91   Brassley 2000.

some areas of bought-in cows and heifers, especially close to market where most would be fattened for the butcher but a few kept for breeding purposes.[92]

The shorthorn had risen to prominence in the nineteenth century as a result of its milk yield and the butchering potential of its substantial carcase. John Walton has mapped the spatial diffusion of pedigree shorthorns and shown how they spread to every part of the country where milk production was an important part of the agricultural economy.[93] The Ayrshire was also a well-liked specialist dairy breed and remained the animal of choice in Scotland. 'Dutch' or generic 'black and white' cows had been popular with London cowkeepers for a long period and their reputation grew gradually with the import of large numbers from the Netherlands. But it was not until the early twentieth century that a breed of stable and reliable quality emerged due to the efforts of the British Holstein Cattle Society, founded in 1909 and soon renamed the British Friesian Cattle Society.[94] By 1947 they composed 20 per cent of the herd, and 76 per cent in 1970, whereas shorthorns fell from 85 per cent in 1908 to 3 per cent in 1970.[95] This affected the compositional quality of market milk, as we will see later.

A more statistically-minded culture in the agricultural community was encouraged from the First World War onwards with the spread of milk recording. This was a grassroots movement at first, with farmers recording daily yields in order to help them eliminate cows that were unprofitable, but in 1943 it became an official, national scheme. In 1917 478 farmers were involved with 13,838 cows being recorded, expanding to 4,302 farmers and 161,077 cows in 1938. A *Register of Dairy Cows with Authenticated Milk Records* was published annually by HMSO from 1917 onwards.[96] The first edition listed 572 registered cows, which was a subset of those recorded, and included only those with annual yields of 8,000 lbs or over. Of these, 10.1 per cent were Holstein-Friesians and 69.2 per cent shorthorns. In 1929 15,065 animals were eligible for registration, of which 22.4 per cent were now Holstein-Friesians and 55.1 per cent shorthorns. No other breed exceeded 6 per cent. In theory these data have potential for an analysis of the spacings and timings of breed changes but there are issues of representativeness in the data set.[97] First, milk recording depended very much upon the enthusiasm

---

92　Pedigree herd books were established for the following dairy and dual purpose breeds: Shorthorns in 1822, Jerseys 1866, Red Polls 1874, Ayrshires 1877, Guernseys 1881, Dexters 1890, Kerrys 1890, South Devons 1891.

93　Walton 1984, 1986, 1999.

94　For more on the breed and the Society, see Hobson 1930, the British Friesian Cattle Society 1930, Burrows 1950, Stanford 1956 and Mingay 1982.

95　British Friesian Cattle Society 1930, Stanford 1956, Federation of United Kingdom Milk Marketing Boards 1972.

96　For more on the significance of herd books and milk recording, see Orland 2003, Nimmo 2008a, 2010.

97　An analysis of the data collected by the National Milk Records Scheme would allow an account of the spacings and timings of the shift to Friesian-Holsteins. Available at: http://www.nmr.co.uk/ [accessed: 29 April 2009].

of a relatively small band of farmers and, second, these elite herds were generally dedicated to breeds producing milk rich in fat.[98]

At the turn of the twentieth century scientific knowledge of dairy husbandry was limited. Witnesses to the Wenlock Committee spoke of pasture quality, feeding regime, breed of cow, her 'condition' (health, nourishment, stage of lactation), and interval between milking (Table 4.5), but there was little feel for the relationship between the variables until the work of Kellner in Germany and Armsby in America on animal nutrition, and on quantity and quality of milk by Crowther at Leeds and a group at Newcastle in the decade or so before the First World War.[99] What was not appreciated at this time was that each herd is made up of animals giving milk of different quantities and qualities each day, within surprisingly broad bands of uncertainty, according to many factors: sometimes cross-cutting, sometimes reinforcing. Looking at the supply of a city the size of London, it might therefore be made up of 130,000 different milks each day in 1900. There were never any precise repetitions, even from the same animal.

Table 4.5   Factors affecting milk composition

---

1. Breeding and selection
    *   breed and inheritance
    *   selection and culling
    *   age of cow
    *   stage of lactation
2. Feeding. Underfeeding can lower the protein content in particular.
    *   energy, fat, protein, minerals, fibre in diet
    *   quality of roughages
    *   supplementary feeding at grass
    *   seasonal variations in quality
    *   indigestion
3. Management and disease
    *   stockmanship
    *   milking technique and frequency
    *   animal's health
    *   milking machine efficiency
    *   whether the milk of several cows mixed

Source: Hall and Buckett 1969.

---

98   Mingay 1982: 63.
99   Wenlock Committee 1901: 380–86, Crowther 1939, Mackintosh 1939, Tyler 1956.

Conclusion

In one sense, expertise is about bringing the world to life, about making it knowable and known. As we have seen, this process can be fraught, with arguments about the necessary skills, about methods, and about ultimate authority over results. There is nothing unusual about such contestation but the level of vitriol and personal bitterness in our story is interesting because it indicates that food analysis in the second half of the nineteenth century was in the front line of debates about the formulation of professional opinion and where this stood in relation to the official knowledge of the state.

An important point to make here is that there was an historical geography of expertise as it gathered momentum from the 1870s. To a degree it was the large cities that were first in the field, providing their citizens with an analytical service that attempted to protect the public interest of an honest food supply. But the hint here of an urban hierarchy model of provision would not be sufficient to explain the local complexity of enthusiasm and inertia. This was more a matter of local political priorities that I do not have time here to discuss in detail.

Another point to make concerns the need for a distributed view of expertise. Our analysts were recognized by the law as experts, capable of revealing scientific truth in court, but the less formal skills of the farmer, the milk dealer, and even the consumer deserve recognition. In addition to their practical, day to day knowing of milk, they were all capable in their own way of influencing the direction of the system as a whole. This was possible through decisions about purchasing and also through the lobbying of others in order to achieve a translation of interests. An example is the decision of the New York City authorities in 1910 to lower their standards for total solids in milk from 12 per cent to 11.5 per cent as a direct result of pressure from Holstein breeders, who by that time had become sufficiently influential to be heard.[100]

Finally, if we are to take material seriously, we should also consider the innate qualities of cattle, milking machines, milk trains, and laboratory equipment in the process of making milk knowable. We normally vest in 'expertise' a sense of performed knowledgeability but the degree to which this is possible and successful depends upon the qualities of the objects that are in play, their acquiescence, resistance, or active participation. In as much as the cattle breeds that we discussed above have had a major influence upon our story of milk composition, and will also be central in Chapters 9–10 in our account of dirty and diseased milk, we may need to revise our human-centred narrative and see the cows themselves as experts.

---

100   Race 1918: 60.

Chapter 5
Standards

Quality in Rules, Standards, Grading, Certification and Labelling

> Standards and classifications, however dry and formal on the surface, are
> suffused with traces of political and social work.[1]

According to Busch and Tanaka, defining goods, standards and grades makes
nature and people 'more uniform, measurable, and controllable'.[2] Standardizing
the behaviour of both people and things makes markets possible; thus markets can
be seen as co-productions of nature and society.[3] Another way of seeing standards
and labels is as the dissemination of information, with the potential of building a
knowledge base to enable consumer choice and to establish or re-establish trust
through a degree of transparency.

But who are these standards for? One task they have is to classify and
therefore to differentiate. Their effect is one of mutual exclusivity. In hierarchical
classifications of food standards, the process of division is itself sufficient to
produce ideas about quality through status notions of desirability. French wine
is a good example of this, with legally enforced designations, from Vin de Table,
Vin de Pays, and Vin Délimité de Qualité Supérieure, to Appellation d'Origine
Contrôlée at the top end of the market. The AOC is then further divided into a
regional classification based on terroir, and, within this, Bordeaux wine has its
own hierarchy. The Napoleon III classification of 1855 lists chateaux in the Médoc
from first growths at the top, such as Chateaux Lafite Rothschild, Latour, Margaux,
and Mouton Rothschild, down to fifth growths. The St Émilion classification runs
from Premiers grands crus classés A and B, to Grands crus classés.

Simon Schaffer sees standards as potentially 'national treasures', and until the
last ten years or so the French wine AOCs had iconic value in this way and the
aura of commandments chiselled on tablets of stone.[4] One could even argue that
a part of French identity is bound up with the quality that these wines represent.
But classifications and certifications are often sites of conflict, negotiation and
power.[5] There have been many highly politicized disagreements in the past
about grades and labels for food and drink, for instance among those with vested

---

1   Bowker and Leigh Star 1999: 49.
2   Busch and Tanaka 1996.
3   Favereau et al. 2002.
4   Schaffer 1995: 136.
5   Mutersbaugh et al. 2005.

interests. Some have proved to be an opportunity for intervention by the state, and therefore a form of state-building.[6] Again the AOC system is an example and here our discussion overlaps with that about rules and standards.[7] The origins of the AOC lay in the fierce debate, following the devastation of phylloxera in the 1870s, among regional interests in France about each other's methods and the relative quality of their products. The shortages of the late nineteenth century turned into excess in the early twentieth century and this forced everyone to think of common cause. The first steps came with attempts to protect place of origin in 1919, and in 1935 the Ministry of Agriculture established an *Institut National des Appellations d'Origine*, which has been instrumental in developing definitions of quality in collaboration with the stakeholders.[8] In essence, the outcome has been an exclusionary system that builds quality and value out of scarcity, and scarcity from connexion with geographical locality.[9] In addition to the spatial logic of the AOC, there are clear, strict rules about grape varieties, yields per hectare, the size of lettering on labels, and much more. It is easy to see why both producers and consumers might wish to rebel against such bureaucratically-defined quality, especially since the rigidity of the system leaves space for lower cost producers from outside France to build their reputations and become competitive in other markets around the world that might previously have been loyal to French wine.

Standards are usually attached to sets of rules, which may be carefully articulated and based upon agreed and rational principles, or they may arise from tacit assumptions and customary practices. In the course of time rules may be taken for granted and therefore invisible without losing any of their power.[10]

In the 1920s in Britain, grading was officially sanctioned as a means of encouraging producers to take quality seriously. It created markets, first at the top end of the market, later spreading to make standard grades a basic requirement of customers. With some commodities, such as liquid milk, there had previously been no incentive to aim for quality or grade output. Now, with mark-ups, differentiation was possible and the 'better' producers and traders could distance themselves from the mass.

In order to facilitate grading, government established a number of marketing boards and charged them with organizing trade marks and monitoring the operation of grading. This was part of the politics of corporate agriculture that eventually led to the foundation of the most famous agricultural body of all: the Milk Marketing Board of England and Wales. This was dominated by the producers, especially those who belonged to the National Farmers' Union, and its monopoly of wholesale milk and close relationship with the Ministry of Agriculture had a profound impact upon the structure of dairy farming. Grading, then, stood for a

---

6   De Sousa and Busch 2006.
7   Barham 2003.
8   Moran 1993.
9   Guthman 2007.
10   Bowker and Leigh Star 1999: 319.

new form of governance that, in the case of milk, lasted from 1933 until a tidal wave of Thatcher-inspired neoliberalism overwhelmed it in 1994.

Over the last 150 years labelling has also come to represent a key dimension of quality construction. There is the element of corporate advertising, of course, that helps the consumer identify a product in general terms and then may provide some specific information, such as place of origin, ingredients and nutritional values. In addition, labelling is an opportunity to express values, such as the display of an endorsement by a campaigning organization or a celebrity, and there may be certification by, say, the Soil Association, to show that the contents meet certain rigorous standards. With certification there is a large slice of trust on the part of the consumer that the system of verification is sound, but there is a danger that this trust may be pushed beyond critical limits and get lost in a fog of fetishization.

In as much as labelling is often self-imposed by organizations and companies wishing to impress, it represents a form of governmentality, of self-government.[11] But who is this self? It may be corporate, and that is enough for the many consumers who trust iconic labels because of an association with a trusted manufacturer or retailer. It may be a national or international body, representing the globalization of standards, which again may satisfy in terms of label recognition.[12] But other consumers will want to know that such self-appointed guardians of food standards are not making a personal gain and that they espouse values that are worthy. In some countries there are several organic standards, a situation that not only leads to label confusion among consumers but may also offer opportunities for lowering standards, literally 'de-grading'.

Ilbery and Kneafsey's recent study of speciality foods in south west England suggests that relatively few producers have been quality certified or see the need for certification.[13] Instead they stressed, first, the specification of production methods, raw materials used and the close involvement of the owner, and, second, the attractiveness of their products in terms of design, texture, flavour, freshness, and appearance. For them quality was meaningful only in terms of market satisfaction and reputation.

Imposed standards have long played a role in international trade.[14] One reason is that they present an opportunity to filter imports and therefore impose strategic policies such as the protection of domestic production or the enforcement of phytosanitary sanctions to prevent the spread of disease.[15] Both of these motivations remain live even in the neoliberal era of the World Trade Organization but in the past they were at the cutting edge of international relations. Overcoming trade barriers has more than economic and political consequences. There may be

---

11   Mutersbaugh 2005a.
12   Mutersbaugh 2005b, Hatanaka et al. 2005.
13   Ilbery and Kneafsey 2000.
14   Bingen and Busch 2006.
15   Fagan 2005.

cultural pay-offs or penalties, as shown through the activism of figures such as José Bové.[16]

Standards have found a focus in agri-food studies in recent years as it has been realized that they are used strategically by food companies and supermarkets. These may be private in-house standards, including so-called 'quality assurance schemes', or there may be third-party certification.[17] Standards here may refer to the material qualities of the product, but possibly also to environmental sustainability or to the nature of employment (rates of pay, hours of work).[18] The growing number of standards can cause a clash of 'quality paradigms' between actors who define standards variously, whether in their own interests or reflecting broader values. Further complicating the world of standards are those negotiated by NGOs with and on behalf of small producers in developing countries. These have proved to be the backbone of the growing fair trade market and in the international trade in organic foods, but they pose a challenge for policy-makers attempting to accommodate a plethora of standards.[19]

Some of the actors in these networks or social circuits are more powerful than others. This has led Young and Morris to argue that the question of who is setting standards (i.e. constructing particular versions of what quality means) is an important element in the use of such notions for competitive ends.[20] Many quality assurance schemes, for example, are initiated by retailers, especially the large supermarkets, and relatively few are developed by producers. Furthermore, as Young and Morris emphasize, there is no national set of standards relating to quality assurance schemes in the United Kingdom, which means that 'quality assured' can have different meanings according to the directives adopted by each scheme.

Alessandro Stanziani has criticized the convention theorists for being vague about norms and paying insufficient attention to historical economic dynamics.[21] The role of the law and rules has been neglected and he proposes an economic-legal history that overcomes convention theory's self-imposed ignorance of the way in which norms are negotiated over long periods. Conventionalists focus on the agreements implicit in coordination and, while not being Panglossian, they do tend to emphasize consensus rather than conflict. Stanziani suggests a rebalancing of the work of those interested in quality standards to allow a hierarchical view of the legal rules that act in favour of some actors and against the interests of others. A problem is that historians are not in a position to criticize because they have not yet taken seriously the idea that the institutional and legal frameworks are

---

16   Bové and Dufour 2001.
17   Morris and Young 2000, 2004, Hatanaka et al. 2005, 2006.
18   Mutersbaugh 2005b, Hughes 2006.
19   Giovannucci and Ponte 2005, Muradian and Pelupessy 2005.
20   Morris and Young 2000.
21   Stanziani 2005. We will look at conventions in Chapter 6.

important for understanding food quality.[22] Their emphasis instead has been upon city regulations and enforcement.[23]

Stanziani's example of the French wine industry is the perfect illustration of his point.[24] The traditional conventions that applied under the *ancien régime* dissolved in the market expansion that followed urbanization and the growth of exports in the nineteenth century. By the mid century, trust was no longer a sufficient basis for the relationship between actors because the adulteration of wine had become widespread. This usually involved watering, plastering, or the addition of sugar. A new law in 1851 was inadequate in dealing with the uncertainty in the market and courts themselves were unsure how to settle disputes, given the contradictory scientific evidence adduced by the litigants. There was also phylloxera, which devastated production in the 1870s and caused confusion about both the quantity and quality of the surviving grapes. A close study of the archives shows that there were many different perceptions of the market and its products among the winemakers, traders, civic authorities, and consumers' groups. These many views were put forward at every opportunity and the period up to the law of 1905 was not one hamstrung by legal constraints, but rather 'a field of action' in which the actors tried to make the most of a chaotic situation. Eventually it was realized that adulteration was damaging the industry and creating a negative image of French wine across the world. Gradually a consensus developed that stable rules and new, legally-enforceable quality conventions would be in everyone's long-term interests. Stanziani comments that the new rules-based framework, which was later extended to other food products, was mainly about disciplining competition and not about consumers. As a result, the legal environment of the French food industry is not especially friendly to the concept of food safety and therefore not fully in tune with today's priorities.

Towards an Official Definition of Milk Composition

> If a dairyman cared to separate the yield of each quarter into 20 successive portions, he would find as many different qualities, and he is entitled to sell them all if he so chooses.[25]

Internationally, the definition of milk was contentious. The First International Food Congress in Geneva (1908), after much debate, saw it as 'the integral product of

---

22 But see Bourdieu et al. 2007.

23 For a collection that shows a good awareness of historical constructions of food quality, but which is still susceptible to Stanziani's criticism, see Atkins, Lummel and Oddy 2007.

24 Stanziani 2003b, 2003c, 2007b, 2009.

25 Composition of Milk Committee, NA: MAF/52/8, TD/1607, 10 May 1928, E.G. Haygarth-Brown.

the entire and uninterrupted milking of a female milk cow in good health and well nourished and not overworked'.[26] This was too vague for the British situation. Here the exercise of analytical expertise depended upon knowledge of the material characteristics of the food or drink in question, upon the technical means of analysis and measurement, and upon a normative approach to standards. In other words, someone had to decide upon the limits within which the composition of, say, milk was allowed to vary scientifically and legally. Given the palpable difficulties of understanding the physics and chemistry of milk, and capturing its 'natural' variations, a bold and confident step was needed to convert scientific consensus into legal certainty. In Chapter 8 we will discuss the ontological differences between science and the law. In the meantime, this section will explore moves towards an official definition of milk.

This part of the story begins in 1874 when Alfred Wanklyn suggested 9.3 per cent solids-not-fat and 3.2 per cent fat as the compositional minima of genuine milk. Then in a series of meetings in 1874 and 1875, the newly formed Society of Public Analysts decided that they would recommend to their members the enforcement of thresholds for liquid milk at 9 per cent by weight for solids non-fat and 2.5 per cent of butterfat. In addition, the minimum for genuine butter was to be 80 per cent of butterfat.[27] This decision and the subsequent revisions were attempts to provide a basis for the prosecution of adulteration. The felt need for standardization was a move towards what Andrew Barry has called a 'technological zone', or, more properly in this case, a 'zone of qualification'. Such zones of consensus are Deleuzian *agencements* that accelerate and intensify agency in particular directions.[28] The milk that met the analysts' criteria was not certified as having particular compositional characteristics except by default. Nevertheless, as consumers began to realize that monitoring by the local state was in their interests and, as it gradually drove up standards, so the zone of qualification was responsible for embedding information in milk as a commodity. However, this particular zone was to be one of contestation – a war zone over quality.

A problem faced by those in favour of standards was that the rapidly changing field of dairy chemistry produced a succession of tests, each with different levels of accuracy. Vieth, at the Aylesbury Dairy Co. in 1882 was using a benchmark of 8.5 per cent solids-not-fat and 2.75 per cent fat, but changed it the following year to 8.75 per cent and 2.75 per cent respectively. The Society of Public Analysts' adoption of Adams' coil in 1885 meant a reduction of their solids-not-fat standard to 8.5 per cent, with a corresponding increase to 3 per cent for fat.[29] The ingredients of milk treacherously revealed themselves differently to each

---

26   Bartlett and Kay 1950: 87. See also Douglas 1908.

27   By no means all analysts followed these standards, however. The adulteration rates recorded in different local authority areas are therefore not strictly comparable, but they are all we have.

28   Barry 2006: 241.

29   For more on public analysts, see Otter 2006.

observer and confusion reigned. As Table 5.1 shows, in 1900 the limits used by analysts varied from 2.5 per cent in Middlesbrough to 3.5 per cent in Birmingham. Internationally there was even greater variation. Interestingly, in Paris the official guidance was 4 per cent fat but the analysts actually used 3.25 per cent as a more practical threshold.[30]

Table 5.1 **The 'standard' of milk constituents current before the 1901 regulations**

	Fat (%)	Solids non-fat (%)	Total solids (%)
Middlesbrough	2.5	–	–
Bradford	2.6	–	–
Glasgow	2.75	–	–
Manchester	3.0	8.5	–
Edinburgh	3.0	–	–
Birmingham	3.5	–	12.0
Dublin	–	–	12.0†

†   This had been Charles Cameron's standard as far back as 1874. Foster Committee, 1874: QQ.4845, 4929.

Source: Wenlock Committee, 1901.

All of this is predicated on the assumption that standards serve a rational purpose. But not all analysts agreed that this was the case. Augustus Voelcker, for instance, the high-profile analyst of the Royal Agricultural Society, in evidence to Select Committee on the Adulteration of Food Act (1872) did not think standards to be a good idea because of the wide natural variations.[31] James Bell of Somerset House, wrote to the Local Government Board in 1876 along similar lines, although his point was that a knowledge base should be established before judgements could be made:

> It has long been known that genuine milks differ widely in composition ... and as chemists are not agreed as to the minimum limit in quality beyond which genuine milk has not been known, it was necessary in the interests of justice that this limit should be determined by ourselves. In furtherance of this object we have endeavoured to obtain what may be considered as representative samples of town and country milk both from dairies and individual cows, and also from churns of country milk as delivered in London.[32]

30   Foster Committee 1894: Q.2246, Long.
31   Read Committee 1874: Q.5514.
32   NA: DSIR 26/134.

Irrespective of these scruples, commercial dairy companies and local analysts began adopting their own standards in the 1880s and 1890s. There was a general assumption, it seems, that these were a fair, though provisional, reflection of some truth that would eventually be revealed. Some of the larger companies began writing minimum levels of fat and other solids into their supplier contracts, and these were enforced through elaborate laboratory-based checks. An example was given by S.W. Farmer, a large-scale Wiltshire producer, in evidence to the 1894 Select Committee on Food Products Adulteration. He had a contract with the Aylesbury Dairy Co. to supply 1,500 gallons of milk a day with a minimum of 8.75 per cent solids-not-fat, a standard he claimed to have little difficulty in meeting.[33] Similarly, the Great Western and Metropolitan Dairies, the second largest wholesale firm in the country, stipulated a 3.5 per cent butterfat standard in their contracts. This was apparently most difficult to meet in the month of May.[34] In summary, F.J. Lloyd, consulting chemist to both the British Dairy Farmers' Association and the Metropolitan Dairymen's Society, told the 1896 Select Committee on Food Products Adulteration that 'some of the large dairymen in London have adopted the standard [3 per cent fat and 9 per cent solids-not-fat] for the last ten years, and have had no difficulty in obtaining milk, during the whole period, which came up to that standard'.[35]

S.W. Farmer and other conscientious farmers benefited for the first time from the premia that quality attracted, and they therefore had a profitable alternative to watering. But others continued with the old ways and defended themselves against the accusation of adulteration with a variety of excuses. Take, for instance, the novel defence of a Wigton milkseller who was fined £4 for having added 10.6 per cent of water to his milk. He claimed that he had accidentally left the lid off the can overnight when it was standing in the rain. It was calculated, however, that twelve inches of rain, equivalent to one-third of the local annual rainfall, would have been required to make this plausible.[36]

The 1899 Sale of Food and Drugs Act gave the Board of Agriculture power to establish limits.[37] This power to make regulations governing the ingredients of milk was truly momentous. It ushered in a new era where government was charged with the definition of the 'natural' and the enforcement of that definition in a way that was to shape our view of milk for a hundred years. There was only token opposition to the need for such standards.

---

33 Foster Committee 1894: Q. 3433. The Aylesbury's fat standard was 3.25 per cent. Wenlock Committee 1901: QQ. 8388–91.

34 Wenlock Committee 1901: QQ. 5238, 7869–83.

35 Russell Committee 1896: Q358.

36 *Fortieth Annual Report of the Local Government Board, 1910–11, Part II*, P.P. 1911 (Cd.5978) xxxi.321.

37 Amongst other standards set and definitions made were spirits 1879, margarine (Margarine Act 1887), cheese (Sale of Food and Drugs Act 1899), butter (Sale of Butter Regulations 1902), and cream (Artificial Cream Act 1929).

On behalf of the Board, Lord Wenlock chaired a Departmental Committee to investigate the need and scope for regulations. Three of their number represented dairy farmers, two were analytical chemists, and one represented trade interests. Starting in March 1900, they heard evidence from 49 witnesses, including 20 milk producers; 17 analysts, medical officers and a sanitary inspector; and 12 from the milk trade. The committee's report stated that 'the overwhelming body of the evidence received ... tended to support the view that the setting up of some statutory limits for milk is necessary' and they concluded that 'it is imperative that a standard implying limits should exist'.[38]

Wenlock considered national standards and three alternatives. The latter were, first, to have limits varying by district, mainly due to accommodate regional variations of cattle breed. But this would have been impossible to enforce because cities drew their milk from many different districts. Second, his report looked at the grading of milk, which was already in practice where some creameries bought their milk by butterfat content. Third, limits differing by season were considered because butterfat content was naturally low from April to June.

The resulting recommendations (national, year-round), for minima of 3.25 per cent butterfat and 8.5 per cent solids-not-fat, cannot be said to have been truly representative of the views of the witnesses outside the large dairy companies and lacked authority.[39] George Barham produced a minority report objecting to a major feature of the 1899 Act, the notion of presumed guilt, which in his view was 'contrary to the traditions of English law'.[40] He was also sceptical of the idea of an 'appeal to the cow', which allowed farmers under suspicion to have their cows tested, to check whether they were giving natural but poor quality milk. The testing was never quick enough and in any case it would have been exceptionally difficult to trace the milk back to the herd, let alone to the individual cow.[41]

As it happened, Wenlock's recommendations were lightened in the following 1901 Sale of Milk and Cream Regulations to 3.0 per cent butterfat and 8.5 per cent solids-not-fat. The loss of the 0.25 per cent of butterfat was more significant than it sounds. Although the dairy companies were pleased with this decision, many analysts were contemptuous of what they considered to be political cowardice.

> Producers ... have been placed in a very unfortunate position by the new regulations, as the tendency of the trade will be to lower all milk to the official limits, with the result that those dealers who are still desirous of maintaining a

---

38 Wenlock Committee 1901: 377–8.

39 But it was close to the American Federal definition of milk as 'the fresh, clean lacteal secretion obtained by the complete milking of one or more healthy cows, excluding that obtained within 15 days before and 10 days after calving, and containing not less than 8.5 per cent of solids-not-fat, and not less than 3.25 per cent of milk fat'. Richmond 4e 1942: 1.

40 *Report by Mr George Barham*. P.P. 1901 (Cd.491) xxx.425.

41 Wenlock Committee 1901.

high standard of quality will have to compete in the matter of price with less conscientious traders, who, taking advantage of the protection afforded by the regulations, will be enabled to sell to the public genuine milk, from which all superfluous fat has been removed.[42]

The regulations were presumptive, implying that milk below these standard thresholds was to be considered adulterated until the contrary could be proved. They did, however, leave the door open to genuine milk falling below these levels and strictly speaking were therefore not normative legal standards, although this is how they were treated in practice. The wording of the accompanying Board of Agriculture Circular was somewhat conciliatory:

> Although the quality of genuine milk offered for sale will usually be well above the official limits of milk-fat and non-fatty solids, there may occasionally, and especially in certain seasons of the year, be cases in which a sample of genuine milk may fall below those limits. To meet cases of this kind it is suggested that in the absence of any special circumstances indicating that, the case is a fraudulent one. The local authority might, in the first instance, call the vendor's attention to the analyst's report, and ask him whether he desires to offer any explanation, and if the explanation is one they are able to accept, they might, in the exercise of their discretion, refrain from the institution of proceedings or withdraw any summons.[43]

The 1901 Regulations did not absolutely fix a standard of genuineness, but merely shifted the burden of proof from the prosecution to the defendant in cases to which they applied.

In Chapters 7 and 8 we will discuss the consequences of the 1901 regulations. Suffice to say here that 3 per cent fat soon proved to be a challenge for many farmers and was the most disputed aspect of food quality legislation in Britain for decades to come. Inevitably the interested parties argued for a review and, eventually, they persuaded the Inter-Departmental Committee on the Production and Distribution of Milk, which had its origins in wartime concerns about food shortages, to note their worries in its 1919 report.[44] This long-standing dissatisfaction among farmers and milk traders was the main reason for the appointment of the Mackenzie Committee soon after, in 1922.[45] Dairy interests in effect argued that the regulations were formulated with a composition of milk in mind that did not match practical reality and that, as a result, there had been many miscarriages of justice in which producers and retailers had been wrongly accused

---

42   Editorial, *British Food Journal* 3: 281.

43   Board of Agriculture, 28 December 1901, Circular: Sale of Milk Regulations.

44   Astor Committee 1919: 659.

45   Although this was a Scottish Inter-Departmental Committee, its findings were of general interest.

and convicted of adulteration. In consequence, part of Mackenzie's brief was to look at milk in the light of its chemical components.

The Committee called 52 witnesses, representing a broad range of interests and expertises (Table 5.2) but no consumers. They concluded that it was possible under certain circumstances for genuine milk to fall below the expected 3 per cent fat, but they realized that their knowledge of the full complexity of composition was incomplete. One of their main recommendations was therefore that a large-scale study of milk composition should be undertaken, along the lines of the detailed investigation under Tocher that they commissioned and was published soon after.[46] This was a remarkable statistical summary of the analyses of 676 Scottish milk samples, one of the world's first, large-scale, official studies specifically aimed at the public exploration of a food commodity's material qualities.[47]

Table 5.2   Witnesses called by the Mackenzie Committee, 1922

Milk producers	12	Representative of Agriculture Organization Society	1
Wholesale and retail dealers	7	Representatives of local authorities	6
Representatives of creameries	3	Medical Officers of Health	4
Representatives of Cooperative Societies	2	Sampling officials	5
Solicitors	1	Public Analysts	5
Representatives of agricultural colleges and societies	3	Other scientific witnesses	7

Source: Mackenzie Committee 1922: 843.

Another key recommendation, ignored by government, was to change the presumptive limits into legal minima, with the possibility of allowing local standards to be different from those at the national level, where support could be demonstrated from civil society. This would have been accompanied by a central register of milk producers, and anyone transgressing would have been struck off, thereby depriving them of that part of their livelihood, a heavy punishment indeed for a dairy farmer. The Committee also wanted an extensive system of sampling, which would have meant regular visits to producers, wholesalers, and retailers. This would have been backed up by uniform methods of analysis.

Mackenzie was thoughtful about quality. The report noted, for instance, that the Committee was evenly divided on whether the toning of milk down to such a standard should be allowed. They also mooted the possibility of payment to

---

46   Tocher 1925.
47   Tocher 1925.

farmers for the butterfat content of their milk.[48] In this they were in tune with the zeitgeist because, soon after, Edwin White, the Managing Director of the Midland Counties Dairy Ltd, despairing of his inability to persuade farmers to send rich milk to market, started bonuses for those who supplied above a predetermined threshold of fat.[49] White's scheme was then discussed as a model of best practice at the World Dairy Congress of 1928 and the idea gradually diffused around the United Kingdom in the years up to the outbreak of war.[50] At the same time the idea also spread of penalizing producers who failed on quality grounds, although this did not affect the Milk Marketing Board of England and Wales until 1957, when payments to farmers producing poor milk were reduced if they were unable/ unwilling to make the necessary improvements.[51]

The Mackenzie report was one input in a busy year for milk politics. As we will see in Chapter 10, the Milk (Special Designations) Order (1922) addressed three other dimensions of quality and thereby established a new basis for the retail sale of clean and disease-free milk. It introduced a number of grades which attracted a retail premium: Certified, Grade A, Grade A (Tuberculin Tested), Grade A (Pasteurized), and Pasteurized. These were defined in terms of their bacteriological content, whether the herd was free of tuberculosis, and whether the milk was heat-treated or raw. A retail premium for compositionally richer milk was not introduced until 1956, when the milk from Channel Island breeds (Jerseys, Guernseys, South Devon) was rated at four per cent fat.

Tocher's large-scale survey of Scottish milk and other research in the 1920s marked a sea change in thinking about milk. The old focus on butterfat and solids-not-fat was beginning to look rather narrow. Reporting to the Ministry of Health in 1931, Monier-Williams commented that there had been two major developments. First, the fat content of milk had been proven to be much more variable than had been appreciated in 1901 at the time when the Sale of Milk Regulations had been made. Second, much more now was known about the minor constituents. As a result, more attention was being paid by analysts to (a) the ratio of lactose, protein and ash; (b) the freezing point, the lactose-chlorine figure, and the refractive index; (c) nitrates, on the premise that milk adulterated on farms would contain well water with traces of fertilizer leachates.[52]

The 1920s and 1930s were a curious mixture of change and inertia. Further technological developments in the analysis of milk made no impact on the standard, which remained as it had been in 1901. There were further official enquiries into compositional variations, such as the 1928 Interdepartmental Committee on

---

48   Mackenzie Committee 1922: 911.

49   Jones [1924]: 8, National Institute for Research in Dairying 1931, Enock 1943: 45, Davis 1983: 5, Cook Committee 1960: 6.

50   For more on White, see Chapter 9.

51   Cook Committee 1960: 86.

52   NA: MH 56/91, 17 April 1931, G.W. Monier-Williams: Some notes on milk analysis.

Variations in the Composition of Milk, which covered the whole United Kingdom. There were also several calls for the presumptive standard to be upgraded to a legally binding commitment. Leighton pointed out that most countries had such definitions.[53] Although its remit did not include milk, the Willis Committee (1931–1934) recommended the extension of legally-binding composition regulations to other foods and this was taken up in the Act of 1938 and the Defence (Sale of Food) Regulations (1943).[54] Willis shows how thinking was going. The politics of food at this date was shifting its centre of gravity increasingly to the needs of the consumer and labels on food packets declaring the ingredients was a move in this direction. It made no statement on fresh, unprocessed foods such as milk, and therefore cannot be taken as a commentary on the natural.

After the war, concerns about adulteration receded. Milk quality took on a new edge that seems to have arisen from the wartime introduction of a sanitary monitoring system.[55] In 1950 Bartlett and Kay suggested that quality was more than chemical composition. Dimensions such as keeping quality, hygienic quality, flavour, and pasteurizability were more important at that date.[56] With regard to the last of these, there had been a long tradition of heat treating old milk to prevent it from souring, but in the post-war era pasteurization was becoming the main means of preventing the spread of bovine tuberculosis. As such, pasteurizability referred to milk that was not heavily contaminated and therefore of low initial quality.

In 1951 the Ministry of Agriculture returned to the politics of milk quality. There were widely expressed concerns that milk was compositionally poorer than it had been 30 years earlier. In a Presidential Address to the Royal Sanitary Institute a few years before, Herbert Kay had pointed this out in forceful terms. He was confident that watering was not the reason but immediately after the war there had been instances for the first time of whole herds producing milk of below 8.5 per cent solids-not-fat and, in a few instances, even whole creameries bulking the milk of many herds were also this low.[57] The reasons given included the spread of Friesians with their low fat milk, also poor feeding regimes during the Second World War which lacked concentrates, and also the neglect of pastures during the agricultural depression of the 1930s.

Interestingly, the fat values for most breeds remained constant in the first half of the twentieth century, except for Channel Island cattle where it fell by 0.5 per cent. However, the solids-not-fat values fell for all breeds, except for a marginal increase for Friesians. By 1950 the earlier divergence in their solids-not-fat between Friesians and Shorthorns was down to 0.1 per cent.[58]

---

53 Leighton 1929: 97–8.
54 Willis Committee 1933–4, Food and Drugs Act 1938 (1 and 2 Geo VI, c.56).
55 See Chapter 8.
56 Bartlett and Kay 1950: 87.
57 Kay 1947: 515.
58 Provan and Jenkins 1949, Davis 1952: 503.

A working party was appointed, chaired by Sir Reginald Franklin. Their report, published in 1953, found from a large and complex data set that there had indeed been some deterioration over the previous 30 years, with solids-not-fat declining more rapidly than butterfat. They urged the need for remedial measures that included quality payments and structural changes in the national herd by the encouragement of breeds with a better milk quality profile. Similarly, in 1952 Davis reported an even clearer decline in both fat and solids-not-fat, the hinge point being about 1930.[59] His conclusions were similar to those of Provan and Bartlett a few years earlier.[60]

One problem with this debate was that it was difficult to compare the British experience with other countries because very few had detailed historical records of changes in milk composition.[61] Among those that did, there had been no obvious change in Ireland and Sweden, an increase in fat content in New Zealand and Norway, and a fall in South Africa. A possible reason for the lack of comparative data was that milk was already standardized in countries such as Denmark (3.5 per cent fat) and the Netherlands (2.5 per cent).[62] Both had important butter industries and so the surplus butterfat could be put to profitable use. Since milk in the United Kingdom was largely sold to the liquid market, standardization was not an issue.

The Franklin Committee concluded that the Ministry should look into producers' prices with a view to incentivizing improvements in quality. They recognized the difficulty of a national scheme, however, because of the lack of adequate testing facilities to allow weekly or fortnightly testing.[63] This was not insuperable, however, as shown by schemes already operating in Australia, New Zealand and some other European countries. In 1956 a scheme was introduced, under the control of the Joint Milk Quality Control Committee. Suppliers of below standard or marginal milk were put on a year's probation with the warning that if there was no improvement they would be reported to the Milk Marketing Board, who then had the option of cancelling their contract.

Soon after Franklin, the Cook Committee was appointed in 1958, 'to consider the composition of milk sold off farms in the United Kingdom from the standpoint both of human nutrition and of animal husbandry and to recommend any legislative or other changes that may be desirable'. Reporting in 1960, it found that that the average fat content of milk had declined in the first half of the twentieth century from 3.7 to 3.6 per cent and solids-not-fat from 8.8 to 8.6 per cent (Table 5.3).[64] As much as 30 per cent of farm milk supplies were found to be below 8.5 per cent

59   Davis 1952.

60   Provan and Jenkins 1949, Bartlett and Kay 1950.

61   Davis 1952: 517.

62   Later farmers in the Netherlands opted for payment by protein content as well as fat. Cook Committee 1960: 15.

63   Franklin, 1953.

64   Cook Committee 1960.

solids-not-fat at certain times of year.[65] Cook agreed that the spread of specialist dairy breeds was a factor and his report also pointed out that the new Milk Marketing Scheme introduced in 1933 would have taken in poor quality milks because it was obliged to accept all supplies offered wholesale. The sale of high fat milk from the Jersey, Guernsey, and South Devon breeds was separate from the Scheme and in 1956 had become the first British milk to have a legal standard.

Table 5.3   **Weighted average composition of milk, 1901–1950**

	England		Scotland	
	Butterfat (%)	Solids-non-fat (%)	Butterfat (%)	Solids-non-fat (%)
1901–05	3.61	8.61	–	–
1906–10	3.53	8.70	3.59	8.77
1911–15	3.60	8.78	3.56	8.84
1916–20	3.66	8.83	3.55	8.83
1921–25	3.65	8.91	3.59	8.79
1926–30	3.70	8.83	3.54	8.77
1931–35	3.67	8.81	3.62	8.75
1936–40	3.65	8.79	3.65	8.76
1941–45	3.64	8.76	3.75	8.73
1946–50	3.62	8.74	3.76	8.76

Source: Cook Committee 1960: 164.

Cook advised against the standardization of milk composition, preferring instead the continuing sale of milk 'as it came from the cow'. However, he did argue for the abolition of the presumptive minimum for solids-not-fat, followed within ten years by the legal fixing of a minimum at 8.5 per cent. For fat, within five years a legal minimum standard of 3 per cent fat content was to be introduced and the abstraction of fat to be made an offence. The official adoption of the Hortvet freezing point test as a means of detecting added water was another of the committee's recommendations.

Since the Second World War there have been a number of significant pieces of legislation, such as the Food and Drugs (Amendment) Act (1955), the Food Act (1984), the Food Safety Act (1990) and the Food Standards Act (1999).[66] The foundation of the Food Standards Committee in 1947 was also important,

---

65   Cook Committee 1960: 20.
66   Giles 1976.

achieving at last something approximating to the independent body of experts that had been suggested by the public analysts over 70 years earlier.[67]

The concept of milk in the United Kingdom stayed unchanged until accession to the European Community. As we have noted, the standardization of fat was by then already practised in the other member countries, and it was formalized under Regulation 1411/71. The UK government wished to protect its own traditional view of natural milk 'as it came from the cow' and so negotiated a change in the rules so that, with effect from 1976, two definitions of whole milk would be permitted: EC milk standardized to a minimum 3.5 per cent fat and UK milk with a natural fat content of at least 3 per cent. This arrangement lasted until the implementation of the Single Market in 1993 when the UK at last became fully European, accepting the 3.5 per cent fat minimum and the legal possibility of standardization. In theory there remained a means of maintaining the tradition of natural milk because all labelling had to say whether the milk was standardized or not, but in practice most British dairies began to market standardized milk from that date onwards.

Conclusion

Joseph O'Connell argues that the metrological practices of technoscience are implicated in the construction of what he calls 'material collectives'.[68] These are 'communities of persons and institutions mutually exchanging the same representations and material representatives for abstract scientific entities'. He concludes that these are responsible for the creation of universality by the circulation of particulars, and, as the logic of the present book unfolds, we find that our discussion of food standards is on similar lines.

The debate about milk standards lasted for 30 years before the official one eventually emerged in 1901. The new material collective that was created stretched beyond the shores of Britain, to the many scientists and regulators around the world who admired British science and looked to it for a lead on the interpretation of quality. The new internationally distributed actor network of lactometers, laboratory experimental equipment, analytical techniques, and equations was connected through conferences, publications, and even the export of pedigree dairy cattle.

As we have seen, the 1901 Sale of Milk Regulations changed the world and we can trace the various adjustments that were made to fit its vision. First of all, farmers began to take the composition of their cow's milk seriously and monitoring regimes spread, and the ontology of milk became a matter of popular concern among producers and other milk marketeers. Second, nature itself was reflexively remodelled as the bodies of its bovine actants were modified to suit

---

67   Ward 1976.
68   O'Connell 1993.

the state-prescribed outcome. The national herd was thoroughly restructured in the following decades and the practice of dairy farming was never quite the same again.

What is astonishing is that the new world of 1901 had such a powerful and long-lasting impact upon milk and the milk industry in Britain for over nine decades, until the advent of the European Union's Single Market in 1993. This long stability was due to a path-dependent lock-in. Institutional and political investment in a consensus, despite the arbitrary nature of the compromise and the many challenges that came with further scientific developments and from court cases that indicated the need for greater differentiation and flexibility. The standard in 1901 enforced a kind of moral naturalism. Milk 'as it came from the cow' was assumed to be natural and therefore authentic, and the construction of a national consensus was so successful, and suited so many vested interests, that change seemed to be impossible. This might still be the case if it were not for the practical politics of Europe on the one hand, and the ideological iconoclasm of British right wing politics on the other. It was the Conservative government that smashed the long-standing monopoly of the Milk Marketing Board in 1994 and unleashed market forces that were to lead to a re-evaluation of every aspect of the milk industry.

Interestingly, the 1980s and 1990s also saw the 'moral technology' of milk composition moving on. The monitoring of fat content was no longer the major quality issue that it had been for so long, with protein and lactose becoming additional target variables for payment in 1984. Ironically, fat at this time came to seem like a bad ingredient, with widespread publicity about a link between cholesterol and heart disease. The consumption of liquid milk has been declining steadily since the mid-1960s in Britain as a whole, although other dairy products such as yoghurt have increased sales dramatically in recent years, encouraged by the market-forming power of the large supermarket chains. Since 1981 retail milk has been widely differentiated into whole, semi-skimmed, and skimmed, introducing complexities in a market that for two centuries had assumed that only one product was possible.

Milk standards are not just about chemical composition. Fat and solids-not-fat were dealt with on a presumptive basis from 1901 but bacteriological standards became legally binding maxima, for graded milk from 1923 and for ordinary milk supply from 1949. This was a different approach, as we will see in Chapter 10.

PART III
Disciplining Milk

Introduction to Part III

Taming a commodity to make it congenial to the needs of a branch of state administration took (and takes) a great deal of subtle manoeuvring.[1]

There is something deeply satisfying about the illustrative tools of realist history. Despite their known shortcomings and biases, maps and graphs and diagrammatic models continue to entertain us with their complex representations of the world and we are tempted to believe that their visual patterns may provide explanations, or, at the very least, inspiration, for further avenues of exploration. Figure III.1 is an example of this. We know, or think we know, from triangulation among several sources, that dairy products were heavily adulterated in Britain in the middle of the nineteenth century but that the worst excesses fell away in the period up to the Second World War. Various laws were passed against adulteration and methods of detection were refined during the period of improvement, so the graph is a whiggish story frozen in what appears to be the lower slopes of a mountain forest.

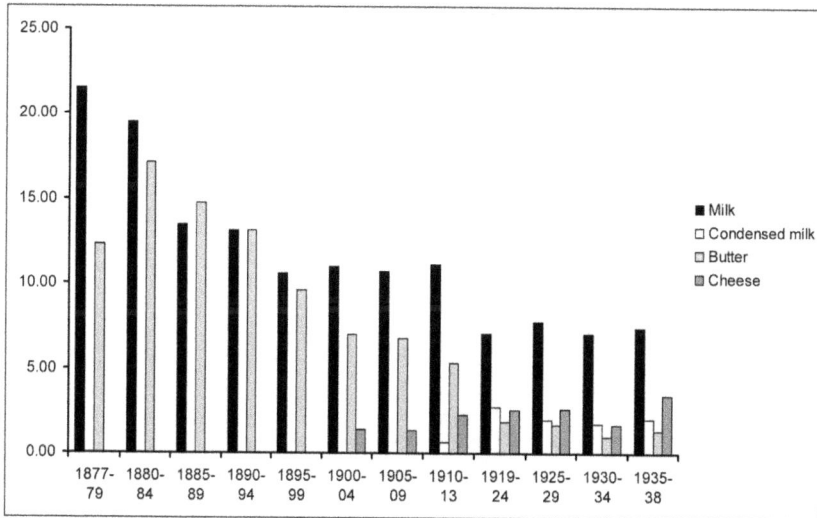

Figure III.1 Dairy products: percentage adulterated, 1877–1938

Source: Local Government Board Annual Reports.

---

1   Ashworth 2001: 42.

The full picture is so much more complex than this, however, that we will devote Chapters 6 to 8 to reviewing both the data and the interpretation. I will attempt this in four stages. First, I will introduce the notion of 'good' milk, reflecting the live concern in the late nineteenth and early twentieth centuries with the moral dimension of the food supply, namely honesty and the trust that is based upon it. Second, there will be a commentary on the governmentality of food adulteration and the interaction between society and the mode of identification and suppression of fraud. This has something quite profound to tell us about the organization of Victorian and Edwardian Britain. Third, a connexion will be made with the evolution of scientific and administrative expertise, making a link with the argument of Part II. Fourth, we will analyse the contention that the law courts held an increasingly key position in the establishment and reinforcement of everyday social norms in the nineteenth and early twentieth centuries. Dicey is a classic source of such assertions.[2] His position, in short, is that the common law was a filter for public opinion, and therefore, ultimately, a fount of democracy. We will not be making any such bold claims here. Our test, rather, will be the impact of judge-made law upon food quality and, through a review of various appeal cases in Chapter 8, we will have an opportunity to reflect upon courtroom visions of nature and the natural.

---

2   Dicey 1905.

Chapter 6
Moralizing Milk

Introduction

This chapter represents a change of gear. So far, the emphasis has been on the physical properties of milk and attempts to make them visible. Chapter 6 is also about material but its interpretation of quality is rather different. Here we will investigate what was meant by good milk and other value judgements, often with moral overtones, that were deployed at one time or another. The argument is that milk acquired some of its qualities by association with those who sold or consumed it, and those who manipulated it and adulterated it. This chapter is an important precursor to Chapters 7 and 8, where a full consideration will be given to bad milk and society's response through regulation and the law.

Material Quality as What is Good, Worthy, Moral, Fair, Honest, Trusted, Healthy and Lawful

The discussion here unfolds through three dimensions. First, for much of our period quality in the milk market quality was defined in terms of its 'goodness' or fitness for purposes such as infant feeding. The criteria deployed varied but, in the case of milk, they have included the presence of butterfat and the absence of dirt, germs and toxic substances.[1] Further organoleptic qualities are taste (sweet not sour) and smell. Also under this dimension of goodness, we will discuss in detail, here and in subsequent chapters, the morality of the milk trade's attitude to its product. Adulteration was common in the mid-nineteenth century and the period up to the 1930s was occupied with a struggle to eliminate cheating such as watering the milk and skimming off its cream, and to prevent the addition of chemicals such as preservatives and colorants.

In addition to honest milk, there was also 'decent' milk.[2] This was a phrase used by Ben Davies, a London dairyman and pioneer of quality milk in the 1920s, who argued that financial incentives should not be necessary for a farmer to produce clean, disease-free milk.

---

1   See Chapters 4 and 9.
2   See the introduction to Part III.

I have used the word 'decency' rather than hygiene, as my scientific friends and colleagues would do. Really it comes down to nothing more than that, if you are decent you cannot help producing clean milk.[3]

Second, a way to discuss good milk and adulteration is to consider 'orders of worth' in society generally and the dairy trade particularly.[4] The work of Boltanski and Thévenot is especially relevant here and it will be applied to the values that we can recover from the trade press and other sources. This approach also has clear relevance to food studies generally, where there is presently interest in fair trade commodities.

The third dimension is trust. This much neglected aspect of food systems is at last receiving attention. A recent study by Unni Kjaernes and colleagues has revealed the outlines of a geography of trust in Europe but we do not yet have an historical geography. The best that can be offered at this stage is a commentary on the evolution of trust through notions of social capital and embeddedness in food systems, in the work of Granovetter and his fellow travellers.

Good Milk

Despite the many food and drink scares that have hit Britain over the last 20 years, most consumers still take the quality of their diet for granted. There has not (yet) been a return to the uncertainty of the Victorian period when a vast portion of the nation's energy, wealth and engineering expertise was devoted to improving the quality of water and, to a lesser extent, that of food. But what was that struggle about and was there consensus about the quality criteria to aim for? What was good water? What was good milk?

Christopher Hamlin's research has shown that the interaction between Victorian science and public health was 'complicated and contingent' and entangled with the issue of the authority of expertise and with the politics of intervention.[5] The evolution of water standards was not just a matter of the scientific discovery of better methods of laboratory analysis but rather the result of dispute, even amongst the scientists themselves. Before about 1860 there was no consensus about analytical procedures. In the eighteenth century the use of chemical reagents was common, based upon anticipated reactions with known chemicals, but the results were uncertain and qualitative at best. Evaporation was an alternative, yielding a number of salts that, with care, could be quantified, giving at least the impression of accuracy. In the

---

3   NA: CAB 58/186, Economic Advisory Council, Report, Proceedings and Memoranda of Committee on Cattle Diseases (EAC (CD) Series, 88–109A), 1932–34, vol. 3, Memorandum no. 97: Stenographic notes of the evidence of Mr Ben Davies, 6 February, 1933.

4   See the introduction to Part III.

5   Hamlin 1990: 3. This paragraph is heavily dependent upon Hamlin's narrative.

early nineteenth century the notion of water quality became increasingly bound up with contamination, especially with organic and faecal matter, as revealed by the power of the compound microscope. The 1828 Royal Commission on the Metropolitan Water Supply heard evidence about the unsatisfactory situation of water companies drawing water from the heavily polluted River Thames but the chemists were unable to agree on the nature or degree of that pollution. Gradually a concept of bad water did emerge but, confusingly, it had many dimensions, including debates about softness/hardness; whether water was a vector of diseases such as cholera; the micro-ecology of water-borne organisms; and the relationship between dissolved nitrogen and sewage contamination. From the 1860s experts such as Edward Frankland, acquired authority as greater agreement became possible about appropriate modes of chemical analysis.[6] Practical administrative measures were also significant, such as the trend towards the end of the century of the filtering of drinking water. Coupled with the bacteriology that was coming forward at the same time, this last change helped to accelerate the improvements in quality that became widespread in the twentieth century. All told, this story was certainly not one of steady progress, but of a winding path that included many uncertainties and the mutual hostility of various vested interests. It could have had a different outcome, and there were local and regional variations to prove that point. There were also diverse water histories in other countries.[7]

Hamlin argues that ideas about water quality were pinned to the symbolic authority of individual scientists rather than to their laboratory processes.[8] Large numbers of water analyses were undertaken and this contributed confidence in the rational process of decision-making and the manufacture of certainty. Science was a 'symbolic technology' which assisted in the achievement of social goals that had been set often without any contribution from scientific knowledge, scientific expertise, or even scientific advice.[9]

Ideas about food quality in the nineteenth century were heavily influenced by the example of water in the context of three general issues. The first was cleanliness and sanitation, a theme that was an extension of the Victorian obsession with the environment of their cities. We will come back to this in detail in Chapter 9, when the point will be made that dirt was a discourse that was lost in the interwar years of the twentieth century, having gone from the most to the least important strand

---

6 Frankland was the government's retained water analyst and gave evidence to several official committees and commissions.

7 For instance Goubert 1986, 1988.

8 Gooday 2004 makes a similar argument for electrical expertise. He argues (p. 22) for the contingency and heterogeneity of trust and makes a study of 'which experts, which measurements, and which techniques were proposed as trustworthy and why judgements on such matters could be accepted or rejected'.

9 Hamlin 1990: 5, 303.

of quality within 100 years.[10] Instead of dirt, the mycobacteria of tuberculosis then became the popular dread, along with toxic substances. In addition, the germs of food poisoning and other diseases could be transmitted through milk. For milk, this was interpreted in terms of a moral economy that had a spatial dimension.[11] Rosenau, for instance, was convinced that 'infection flows from the country into the city' although 'where the blame lies' was less interesting to him than the implication that society had yet to come to terms with the radically new circumstances of modernity:

> There is a tendency on the part of the consumer to blame the producer. On the other hand, the producer blames the consumer; whilst the middleman points an accusing finger at both the consumer and the producer. The health officer blames all three ... Society must blame itself; that is, we are confronted with a situation that has been evolved naturally as a result of the existing state of civilization. The milk question is simply one of the difficulties of a complex age – one of the difficulties of an artificial civilization to which we have not yet adjusted ourselves.[12]

Rosenau was writing at a time when there were few remaining urban producers of milk in America. Some British cities (Table 6.1) also had very few and relied on railway milk or supplies from the immediate locality. Manchester and Bradford were examples of the latter, using small dairy farms on the flanks of the Pennines. Others were less favourably placed, such as Liverpool, and relied for longer upon city-based and suburban cowkeepers.

The second point is an extension of the first. While the intensification of food production took place on the urban fringe, and sometimes in the interstices of the cities themselves, after a time these resources proved inadequate for growing demand and eventually even the most perishable foodstuffs had to be brought over long distances. Table 6.2 illustrates the point. By 1900 milk for London consumers took about five and a half hours to come from an average of 140 miles, requiring sophisticated logistical solutions because this was a traffic that operated every day.[13] Special railway wagons and slots in the timetable for milk trains had to be provided and highly geared wholesale and retail trades gradually evolved to lubricate the various articulations that joined the distant farmer to the doorstep delivery.[14] In terms of supply management, there were some inefficiencies identified by various academic surveys and government enquiries, for instance

---

10   Only in the last 10 years or so has it returned, this time in the guise of debates about acquired immunity.

11   Morgan et al. 2006.

12   Rosenau 1912: 20–21.

13   This is based upon a calculation using the data in Table 6.2. The original survey asked for maximum, not average distances, so this estimate is on the high side.

14   Atkins 1978.

Table 6.1 **The proportion of milk sourced from cows kept within the urban boundaries**

Town/city	Date	Urban production (% of total supply)
Belfast	1929	20
Birkenhead	1919	20.6
Bootle	1918	41
Bradford	1920	61.5
Edinburgh	1921	52
Folkestone	1915	25
Liverpool	1927	29.5
London	1850	80.0
London	1880	28.3
London	1910	2.8
Newcastle	1925	15
Sheffield	1929	24.8
Weymouth	1914	33

Source: Mainly from Medical Officer of Health Annual Reports.

overlapping retail delivery rounds in the big cities.[15] But it was nevertheless a machine of awesome power for the timely provision of customers' needs. It also generated substantial profits for large-scale corporate dairy entities such as United Dairies – 'The Combine' – which grew steadily after the First World War.[16]

At times there was a mismatch between supply and demand. In hot summer weather, for instance, there were two problems. Pasture occasionally became scarce and milk yields fell, causing shortfalls in supply, which could be as much as 25 per cent lower in the late summer than in the spring.[17] Where there was an absence of cooling facilities, hot weather made it difficult to deliver such a perishable commodity in perfect condition. There was often a temptation on the part of farmers, middlemen and retailers, especially if they were subject to contractual targets, to stretch the milk available to them on any day when supplies were short, either by adding water or skimmed milk. Chemical preservatives were also common before the First World War as a means of prolonging the product's shelf-life.

Third of the food issues, the quality of some, not all, foodstuffs was compromised in this period by the widespread problem of adulteration. Not only was there

15   Forrester 1927.
16   Maggs 1924, Pocock 1933.
17   The years of highest July–September temperatures were 1857, 1859, 1865, 1868, 1884, 1895, 1898–99, 1906, 1911, 1914.

Table 6.2   The average distances that milk was brought to London by selected railway companies in about 1900-1904

Company	Maximum distance (miles)	Journey time (hours)	Amount carried (million gallons)
Great Eastern	130	4.00	7.00
Great Northern	150	5.42	7.50
Great Western	130	6.00	17.50
London, Brighton and South Coast	86	3.50	3.13
London and North Western	80	8.75	5.25
London and South Western	150	4.50	7.63
Midland	200	8.00	6.00

Not included: GCR, LT&SR, Metropolitan, SE&C&DR.

Sources: Maxwell Committee 1902: Appendix IX, Pratt 1906.

thought to be a danger to health from poisonous colourings that were added to food, but also adulteration was a common means of cheating the consumer by modifying the composition of foods in the pursuit of profit. For decades between the 1850s and 1880s it seemed that the food supply could not be trusted. It was tainted by the thought of institutionalized deception.

Free commerce in food had been limited by occasional checks on weight or volume, to guarantee fair measure – this had been the basis of the Assize of Bread until its abolition in the early nineteenth century – and to assess the revenue payable to the government on certain dutiable items.[18] From the 1870s onwards further restrictions were imposed. Rather than the onus being upon the purchaser of food items to judge quality and, in the act of payment, confirming her satisfaction, from this point onwards the responsibility shifted to the seller to guarantee that the quality was as had been requested by the customer. This was a seismic shift in the law and its view of food quality. There was one further layer when the identification and elimination of diseased food was appended to the duties of public health officials. It is important to distinguish this food safety function from that of adulteration, for, as Alessandro Stanziani has pointed out, adding water to wine had no health implications for the French public in the nineteenth century but it was taken so seriously by the authorities that it influenced the food law of that country and prevented what in another era might have been regarded as a legitimate form of commercial innovation.[19]

---

18   Davis 2004.
19   Stanziani 2005: 7.

At the beginning of the nineteenth century it was common for London milk retailers to buy from suburban cowkeepers, and for their own milkmaids to do the milking. The milk was set for about 12 hours and the cream taken off. The skim milk was then watered – 'bobbed' or 'washed' in the trade jargon – and sold in the streets by the same milkmaids. Middleton calculated the profit of this combined abstraction and adulteration to be about 22 per cent, or 4d on every gallon of milk sold.[20] These practices were the source of wry comment and satirical humour among contemporaries but little concern was shown by the authorities, and it seems that, by default, 'milk' had been redefined as a commodity. Excessive watering creates a bluish tinge but on the whole milk was very forgiving of manipulation and most consumers could not tell the difference. They were duped because the curious veiling effect of the colour white distracted them. Milk was the most adulterated of foods – because it could be.

The only way for customers to avoid this new, redefined, 'thinner' milk was to visit one of the cowsheds where warm milk was sold. The implication was that it had just come from the cow, although even this was open to deception. Until 1885 there was also a 'milk fair' in St James's Park in central London for the same purpose, and there were goats and cows that were brought to the doorstep for live milking if the customer wanted the ultimate reassurance.[21] The more sceptical among the bourgeoisie kept their own house cow in the back garden, usually one of the small breeds, such as the Kerry.

Orders of Worth

For many Victorians, the almost universal adulteration of milk in big cities was not only a question mark against trust. It also suggested a worrying acceptance of low-level cheating that undermined one of the moral assumptions upon which a healthy society is based. Thomas Carlyle summed this up in a tongue-in-cheek commentary on the mind-set of the English men and women of his day by comparison with the 'golden age' of the eighteenth century.

> Now, all England – shopkeepers, workmen, all manner of competing labourers – awaken with an unspoken but heartfelt prayer to Beelzebub – 'O help us, thou great lord of shoddy, adulteration, and malfeasance to do our work with a maximum of sluriness, swiftness, profit, and mendacity, for the Devil's sake. Amen'.[22]

---

20   Pre-metric pence. Middleton 1807.

21   *The Times* 2 September 1885: 9F, Fussell 1954. The author has seen doorstep milking still in operation on the streets of south Asian cities.

22   Carlyle in letter to Sir James Whitworth, an industrialist, quoted in *Local Government Board: Fifth Report, 1875–76*, P.P. 1876 (C.1585) xxxi.234.

There was an urgent mission in the mid-century to develop tools to detect cheats and exact punishment. Neither was easy. As we saw in Chapter 3, the available technologies were not especially effective. Microscopes could distinguish between coffee and chicory but were unable to tell natural, whole milk from watered milk. Nor were the techniques of bench-top physics or of gravimetric or volumetric chemistry much help in their early stages because they were well-known to traders, some of whom were sufficiently resourceful and skilful to find ways around all of the simpler tests of adulteration.

The first anti-adulteration legislation was passed in 1860, and subsequent employment of Local Authority analysts from the 1870s onwards was based on expertise that was fragile at first. Bruno Latour has reminded us that 'no human is as relentlessly moral as a machine', and eventually it was the technological morality of the laboratory and procedural probity of the court room that was responsible for a new phase of quality-making for milk in the last quarter of the nineteenth century.[23] The parable of Daniel Schrumpf in Chapter 2 was about technological uncertainty but it was also a moral fable, because, as Busch and Tanaka have observed:

> People care about tests of things because they are also tests of people. Tests are measures of nature at the same time as they are measures of culture … Tests, then, are best understood as measures of the value of our world. Indeed, the readings on the meters, dials, and gauges of machines designed to measure facts are known collectively as values. Put differently, we value facts because they show the facticity of values.[24]

The impression is sometimes given that the recent so-called 'moral turn' in consumers' attitudes to food is somehow novel. A moment's reflection makes this seem unlikely; indeed morality, *sensu lato*, was central to thinking about food in the nineteenth century and was responsible for mediating early definitions of quality.[25]

The Victorian and Edwardian worlds of food were under increasingly close scrutiny and there was a steady stream of articles and editorials in the press, and speeches in parliament, expressing value judgements that in many ways were the precursors of our own estimates of 'quality'. Food safety scares since the 1980s have been about high profile events such as the possibility of BSE being spread to humans through infected beef or of food poisoning originating in poorly managed abattoirs. But there is more to the story than that. Both in the media and in the discourse of the consumer, we seem to feel that someone is to blame.

---

23   Johnson 1988: 301.
24   Busch and Tanaka 1996.
25   As Andrew Sayer 2006: 88 shows, Adam Smith's *The Wealth of Nations* was written within a moral frame.

The Victorian model of morality is still with us: 'it presumes that agency is a vector of blame, shame, and guilt, and that causal explanation is a prerequisite for motivating responsible, other-regarding action.'[26] An example is the 1996 E. Coli O157 outbreak in Wishaw, near Glasgow, when 500 people were affected and 20 died after consuming infected meat products from a local butcher's shop.[27] John Barr, the butcher whose meat was involved, was found by the Pennington group to have neglected the risk of cross-contamination between cooked and uncooked meat: the same knives were used for both.[28] Yet he had support in the community that he served, even among the relatives of those who died. Despite this, the press coverage was relentlessly determined to apportion blame and to imply that his premises had been dirty and unfit. Although he had been Scottish Butcher of the Year, this is now only mentioned with bitter irony.

What, then, makes a 'good' farmer, butcher, dairyman or consumer? From 1750 to 1850 a good farmer 'improved' the land by draining it, reclaiming unproductive land, and adding soil conditioners and fertilizers; and in the mid-twentieth century farmers were paid to plough out hedges and to maximize production. But recent governments have made a hand-brake turn in policy and what is now good husbandry seems to be eerily different. Gradually good farmers are becoming ecologically-minded custodians of the landscape rather than agents of public food policy.

Nowadays we value animal welfare but this was certainly not a priority of most cowkeepers in the nineteenth century. Nor was cleanliness, although, as we will see in Chapter 9, attitudes on this changed gradually in the early twentieth century, to the extent that there was a significant improvement in the average bacterial loading of milk.

The obituaries of prominent milk wholesalers and retailers give some clues about contemporary perceptions of virtue (Table 6.3). While the trade press and obituaries in *The Times* tended to stress the positive – for obvious reasons – the selection of adjectives is often quite revealing. The benevolent and paternalistic qualities of Edmund Tisdall and George Barham in the nineteenth century gradually gave way to characters who were said to be 'direct', 'determined' or 'stubborn', and then on in the mid-twentieth century to those were 'severe' or even 'authoritarian'. Although this is too small a sample to be of significance, one does have the sense that perceptions of worthiness were open to change as the dairy industry became increasingly competitive and the stakes higher.

According to Steven Shapin, there was something similar when the evaluation of scholarship in the nineteenth century went through a transition from virtue to expertise. Scientists were seen as commendable scholars but also strange and even contemptible in some of their habits. Their unworldly asceticism, and their sacrifice of personal advantage for a life of discovery and intellectual rewards, were both

---

26 Barnett and Land 2007: 1070.
27 Atkins 2008.
28 Pennington 2003.

Table 6.3   The virtues of prominent dairymen, 1850–1950, as noted in their obituaries

Name	Personal	Business
Edmund Charles Tisdall (1824–1892), Tunks and Tisdall	Teetotal, respected, trusted, benevolent, able speaker, integrity	Eminent, esteemed, good administrator
Sir George Barham (1836–1913), Express Dairies	Striking personality, charming, fluent, humorous, energetic, self-confident	Anti-adulteration, knowledge of science, innovator, opportunist, outstanding, distinguished
Sire Reginald Butler (1866–1933), United Dairies	Direct manner, generous	Able, talented organizer
Sir William Price (1864–1938), United Dairies	Confident, assured, determined, unflinching, approachable	'The General', not a man for detail
Titus Barham (1860–1938), Express Dairies	Sympathetic, calm, genial, stubborn	Keen businessman, organizer, expert, hard bargainer, straight, experienced
Arthur Barham (1869–1952), Dairy Supply Co.	Brave, just	Severe
Edwin White (1873–1964), Midland Counties Dairy	Impulsive	Authoritarian, tough, exciting
Joseph Maggs (1875–1964), United Dairies	Retiring disposition, strict standards, honest	Knowledgeable
Leonard Maggs (1890–1959), United Dairies	Warm, generous, humane, patriarchal, countryman, punctual	Successful

Sources: Trade press and Atkins 1984a, 1984b, 1984c, 1984d, 1985a, 1985b, 1985c, 1986a, 1986b, 2004b, 2004c.

admirable and their reputation for rejecting luxury and showing little interest in delights of the flesh, such as food and drink, gave them a spiritual aura. As the century wore on, however, more scientists found employment in industry and scientific societies increasingly recognized the need for links with the mercantile and manufacturing classes, whom Shapin calls 'the new men'.[29] By the end of the century the transition was complete, with science by then institutionalized in universities and expertise certified as a fuel of activities as diverse as court proceedings and government committees. The hair shirt of the ascetic had been replaced by the white laboratory coat of the insider. Science was now constitutive of modernity, not an optional extra.

---

29   Shapin 1991: 312–15.

Lest we make the classic error of assuming top-down values to be the only ones that are truly worthy, it is interesting to note the 'underground' approbation that seems to have been attached to 'disreputable' behaviour in the milk trade. George Barham, founder of Express Dairies, 'thought at one time that the farmers were all honest men, but when I went into business [in 1858] I was astonished to find that some of the farmers *took the greatest delight in the world in deceiving the Londoners'.*[30] From the producers' point of view, this is resonant of those 'little victories of daily life' described by Michel de Certeau.[31] Cheating their customers was also a coping mechanism for some small, financially-challenged corner shopkeepers.[32] It was one of the many forms of petty crime that, in societies where the power of the regulating authorities is weak, may be seen as 'inevitable' by traders and consumers, either due to the meagre resources allocated for inspection, or because of technical problems with detection. Such falsification became so universal and so routine in the nineteenth and early twentieth centuries that, for the farmer, trader and retailer at least, it was an everyday means of making a marginally legal living. By way of example, a dairy manager in London with twenty years experience reported in 1914 that:

> It is almost impossible to secure a carrier who will give proper measure ... Giving short measure is considered by the carriers to be almost their right and a carrier is not considered a Milky by [his] fellows unless he can make 3d per barn gallon ... by giving each customer a shortage.[33]

In 1905 Hassard adopted the role of an investigative reporter in order to discover why the milk trade had such a bad reputation.[34] He took employment in a retail dairy and his hair-raising story is one of unscrupulous tradesmen and their dishonest employees; an organized system of kick backs to servants to guarantee continuity of supply; and corrupt inspectors whose price was a guinea to overlook irregularities.

In recent years Luc Boltanski and Laurent Thévenot have sought a conceptual framework for understanding values and qualities, whether elitist or vulgar. They have attempted generalizations about the forms of evaluation that agents use in various types of coordination in economic and social contexts.[35] In a fascinating essay on French society, *On Justification,* a pragmatist manifesto arguing for a sociology of action, they found a means to classify the ways in which reputations are won and lost. Through their account of 'worthiness' and 'orders of worth',

---

30 Read Committee 1874: Q.2434. Emphasis added.
31 De Certeau 1984.
32 Anon 1894.
33 Letter from C.F. Green to Waldorf Astor, 7 April 1914. University of Reading Archives, Astor Papers, MS1066/1/1018.
34 Hassard 1905.
35 Thévenot 2006: 111. See also Thévenot 2002.

we come to understand better the nature of trust and the social construction of goodness. They address the tensions between the modes of coordination rather than tensions between social groups, one of social theory's traditional concerns. Its subject matter is the reasons people give for their behaviour, not in institutional settings but more often in disputes.[36] Boltanski and Thévenot have identified six common 'worlds' or dimensions in which there are widely known and accepted bases for actors to judge each other's behaviour. These are: the inspired world, the domestic world, the world of fame, the civic world, the market world, and the industrial world. In broad terms, estimates of worth in each are as follows:

- the inspired world – expressed in feelings and passions that cannot be measured;
- the domestic world – a function of one's position in chains of personal dependence that can only be grasped in a relational sense. Understood as respect and responsibility. Expressed as respectability, authority, honour;
- the world of fame – repute, success, dignity;
- the civic world – unity, collective, authorized, representative, legal;
- the market world – value, quality, luxury;
- the industrial world – functionality, reliability, efficiency, expertise.

The precise details of each context of worth are, of course, debatable, but Boltanski and Thévenot have given us pause for thought in our quest to understand quality. No doubt we could increase the number of *cités*, as these worlds are called, and there would surely be variations according to national context, but it is possible to see ways in which all six dimensions of worth are applicable to food in nineteenth- and twentieth-century Britain. Table 6.4 illustrates this for liquid milk in 1930, where I have selected three Boltanski orders and added another: the world of science. The table would be different if drawn up for periods 50 years earlier or later.

As we saw in Chapter 3, the orders of worth in the science and civic columns were dependent to a substantial degree upon the resources available at the local level for chemical analysis, which in turn were dependent upon political priorities.

In the domestic and civic orders, good retailers and good consumers have had varied identities through time. First of all, although the good consumer is assumed to have become more knowledgeable about the foods she eats, particularly in recent times with regard to nutrition and health, Jukka Gronow reminds us of the state of uncertainty about the canons of taste in food and therefore of the difficulty in establishing a scale of goodness.[37] Antoine Hennion argues that the answer is to think of taste as the reflexive activity of amateurs, and not the élitist activity we tend to associate with experts. In this understanding, quality in the object consumed is a matter of individual judgement and may therefore be associated with lowbrow

---

36   Boltanski and Thévenot 2006. The French language original came out in 1991.
37   Gronow 1997.

Table 6.4   Orders of Worth in the British milk industry, *circa* 1930

	Market	Domestic	Science	Civic
Value	Price	Reputation	Meets standard, healthy	Collective interest
Format of relevant information	Monetary	Word of mouth	Statistics	Official reports
Qualified object	Goods, services	Assets	Sample	Court summons
Elementary relation	Exchange	Trust	Expertise	Rules, standards
Human qualification	Disposable income	Authority	Professionalism	Citizenship
Time	Daily repetition	Customary path	Timely	Perennial
Space	Commodity chain	Locality	Traceability	Detached

appreciation and passion as much as with the supposed intellectual judgement of the connoisseur. It also sees taste as performative and closely dependent 'on its situations and material devices: time and space frame, tools, circumstances, rules, ways of doing things. It involves a meticulous temporal organization, collective arrangements, objects, and instruments of all kinds, and a wide range of techniques to manage all that'.[38]

The good shopper has been gendered in most western literature since the mid-nineteenth century. She has been a member of the local Co-operative Society, or a frugal, price-comparing shopper, or, more recently, a pleasure-seeking flaneuse who stalks the retail malls in search of retail therapy.[39] The very latest version of the good consumer is no longer a commodity fetishist but is said to be increasingly well-informed about healthy food and perhaps about minimizing her carbon footprint and her food miles in the name of environmental sustainability or to show support for alternative food networks.[40] She is also a consumer-citizen activist in these and other causes, although Danny Miller has warned against making this consumer 'a rhetorical trope in the critique of capitalism'.[41] In addition, ethics have begun to intrude into debates about food, for instance with regard to animal welfare and fair trade.[42] The latter is especially interesting because it provides development opportunities in the global South and establishes previously unimaginable communities of common interest across vast distances

---

38 Hennion 2005: 138. See also Teil and Hennion 2004.
39 Sassatelli 2006, 2007.
40 Weatherell et al. 2003.
41 Quoted by Sassatelli 2004: 179.
42 Barrientos and Dolan 2006.

for commodities such as coffee and bananas. It is in these channels and these relationships that the 'quality' of fair trade resides.[43]

Frank Trentmann excavates early notions of consumer-citizenship, starting with the widespread early nineteenth-century boycott of goods produced by slave labour.[44] He finds late Victorian/Edwardian activists both for and against free trade.[45] And most interesting from our point of view, he describes what he calls 'the new consumer politics of milk'. From the First World War onwards, he claims, this was centred in women's movements campaigning for pure milk. He cites the example of the Women's Co-operative Guild, which, with its broad social base, was representative of thinking in many walks of life.[46] It is wrong, it seems to me, to place these groups at centre stage when their particular concerns were post-war price control and the concentration of retail capital, as against the much wider agenda of grass roots milk politics. Nevertheless Trentmann does us a service in showing that consumers were not passive dupes and that a vigorous debate was underway by the turn of the century.[47] Although in his own work he glides elegantly between time periods, Trentmann is an advocate of specifically historical consumer studies.[48] He particularly objects to the blindness of present-day consumption literature to the 'troubled genealogy of consumption and power' and its 'indifference to the moral imaginaries of the past'.[49]

Up to now most research on European consumer-citizens has been on the period following the Second World War, but there is an emerging a body of empirical work that is revealing the depth of organized consumer activism in the nineteenth and early twentieth centuries.[50] There were consumers' leagues, for instance, in Paris in 1902, Switzerland 1903 and Germany 1907, followed by Italy, Spain and Belgium.[51] Kroen aligns these developments with the maturing of democracy and particularly the empowerment of women.[52] Even further back, in the 1870s and 1880s in Germany, a movement of citizen self-help associations was mobilized against the adulteration of food.[53] The first was founded in Leipzig in 1877. It soon

---

43   Renard 1999, 2003, 2005, Jones et al. 2003, Goodman 2004, Bacon 2005, Lyon 2006, Alexander and Nicholls 2006, Lockie and Goodman 2006.

44   Sussman 2000.

45   Vincent 2006.

46   Trentmann 2001, 2006b.

47   Trentmann 2007.

48   Trentmann 2006a.

49   Trentmann 2007: 1097, 1080.

50   There is a large literature on consumption and consumer culture, but here I am referring to consumerism, consumer activism and consumer politics. See Furlough and Strikwerda 1999, Daunton and Hilton 2001, Cohen 2003, Hilton 2003, Chatriot et al. 2004, Hilton 2006.

51   Furlough 1991, Furlough and Strikwerda 1999, Chessel 2006. See also Athey 1978, Sklar 1999.

52   Kroen 2004.

53   Hierholzer 2007.

had 500 members and within a year or two the idea spread to 20 or so cities. In Britain also, food quality was a central concern, leading to the formation of groups in civil society such as the Anti-Adulteration Association (1871) and the Food Reform Society (1877, merged with Vegetarian Society 1885).[54] More research is required before we can form a comparative perspective of such voluntarism in different countries but it seems likely that the history of consumer citizenship is more complex than we have hitherto grasped.

Attention has been paid recently by social scientists to the part played by consumption in shaping quality conventions, which have played a central role in the governance of supply chains.[55] Geographers, in particular, have also drawn attention to the explicitly *political* ways in which some groups of consumers have influenced the governance of transnational trading relationships via forms of direct protest and resistance. It is suggested in the literature that this has been achieved through alliances with civil society organizations, social movements and trade unions, as well as being influenced by critical journalism on the subject of trade injustice. Meanwhile, Freidberg and Crewe have focused on the role of the media and NGOs in pressuring UK supermarket chains and clothing retailers respectively to trade more ethically.[56]

Sally Eden found that her present-day focus groups used a vernacular 'sorting out' of bad and good that amounts to a moral ordering.[57] Processed, pre-packed and fried food, takeaways, and food beyond its sell-by-date or containing preservatives, were all said to be bad, whereas fresh, organic, home-made food was good. But, since we all lack detailed information about the sources of our food and the conditions in which it is produced and processed, a series of knowledge fixes is required to provide surrogates of quality. The first is space: farm shops and farmers' markets were trusted places to buy but not large supermarkets. A second is labelling and other forms of quality assurance, such as certification or endorsement.

Not surprisingly, some retailers have realized that it is in their long-term interests to follow the new ethical agenda, adjusting their sourcing policies and subjecting themselves to auditing.[58] Time will tell whether this amounts to more than window dressing. In the past, the 'good retailer' was more problematic, with fewer cosy stereotypes. They may have shared profits with their customer members (Co-operative) or have had very high standards in sourcing their milk (Aylesbury Dairy Co.). But the retail sector of the milk trade, certainly up to the First World War, and into the 1930s in the case of small corner grocers, were notorious for bad practice: poor sanitation, short measure, adulteration. The stocking of branded and certified goods was one means of reassuring customers of

---

54   Oddy 2007.
55   Gibbon and Ponte 2005, Ponte and Gibbon 2005.
56   Freidberg 2004a, 2004b, Crewe 2004.
57   Eden 2008a, 2008b.
58   Hughes 2005.

a continuity of quality standards but this was not always sufficient. In the twentieth century, the sale of bottled milk became another material sign of quality, although the 'pasteurized' milk they contained, for instance, in the 1920s and 1930s caused one of the greatest storms in the history of food safety. At the time it represented a fundamental disagreement between opposing parties about the technological component of modern food quality.

Coming back to the good retailer, and the present day, Susanne Freidberg identifies the paradox of British supermarkets that offer their customers a range of cheap exotic foods that satisfies the modern need for 'gustatory tourism'. At the same time, these retailers are *apparently* becoming more transparent, with new packaging models that emphasize the ethical nature of their sourcing policies and make declarations about environmental impact. She argues that 'supermarkets are now racing each other up to the moral high ground'.[59] It seems that there is money to be made from ethical trading and also important brand-enhancing prestige to be gained from activities in the area of corporate social responsibility.[60] Unlike 'fair' trade, ethical retailing is more difficult to measure, although in recent years the public has been responding more to revelations about the conditions of employment of the producers in poor countries, whether it be the Kenyan farm labourers responsible for the baby vegetables on our supermarket shelves, or the sewing of footballs by children in India. Freidberg finds a neo-colonial 'civilizing mission' among some supermarket wholesale buyers who regard their imposition of new technologies, hygiene, workplace discipline, and quality control, as 'good' for development and ethically sound. They provide employment at the minimum wage or just above but even that may be taken away now that imports from Africa are being questioned as having too high a carbon footprint.

Three last points on worthiness. The first is that we have implied an association with individuals or companies, but we shall argue later that quality can also be a property derived from networked association, in other words a distributed attribute. Second, apart from the practical considerations stressed here, quality may have ideological foundations in ethics. Animal welfare, fair trade and environmentally sustainable foodstuffs are examples. Third, trust replaces active with passive goodness. There is no need necessarily to buy into the motives of a retailer, nor indeed of the food system as a whole, but trust indicates an anticipated satisfaction, from one's own point of view, with a relationship and the quality of goods/services that it delivers. In the quality classification discussed by Ponte and Gibbon, for instance, most aspects of food are still based upon credence attributes.[61]

59   Freidberg 2003.
60   Tallontire 2007.
61   Ponte and Gibbon 2005.

Trust

Trust reflects contingency. Hope ignores contingency.[62]

It is common to argue that the unfettered deployment of both trust and scepticism is essential for democratic governance, the functioning of liberal markets, and the very cooperation upon which our social and cultural institutions depend. Francis Fukuyama, for instance, argues that a high level of trust is a necessary condition of economic development and can be used to explain some of the curiosities of economic history at the national scale.[63] But arguing that trust is the bedrock of a modern society and the fuel of social and economic interaction has attendant dangers of reductionism. Niklas Luhmann was firm in his rejection of trust or distrust as means of characterizing modern society. He preferred instead two other interdependent structural changes: 'firstly, the increasing diversification and particularization of familiarities and unfamiliarities; and secondly, the increasing replacement of danger by risk, that is by the possibility of future damages which we will have to consider a consequence of our own action or omission.'[64]

The real world of trust has a history and a geography – in other words, trust cannot be taken for granted; nor can it be manufactured easily. Recent research has found that public trust in food, food producers and manufacturers, food retailers, and the institutions supposed to guarantee food standards, is quite variable. Some consumers are more trusting than others, and there are variations within and between different elements of the food system, shifts through time, and remarkable divergences among national and cultural contexts.

It seems that there are high trust and low trust societies. In Europe the citizens of northern countries such as Finland are both trusting and resilient in the face of the many tests of trust that are a feature of the modern world. By comparison, consumers of former communist states are less trusting. A recent Eurobarometer survey showed, for instance, that 64 per cent of Finns trusted their public authorities to inform them about a food risk, but in Estonia, Lithuania, Poland and Slovakia the figure was only 5 per cent.[65] Having said that, each food and each type of risk attracts different responses.

What are the origins of this geography of trust? There are cultural explanations, where social capital looms large and trust is seen as a function of the nature of the interpersonal relationships that an individual experiences in relation to their socialized expectations. Alternatively, institutional explanations stress the relative performance of the various institutions that an individual has come into contact with over the recent past. Their attitude may be coloured, for instance, by the

---

62   Luhmann 1979: 24.
63   Fukuyama 1995.
64   Luhmann 1988: 105.
65   European Commission 2006.

quality of the risk communication and other information flows that go to make up their risk perception.[66]

Unni Kjaernes has found that institutional trust is differentiated in Europe. Germans trust experts and consumer organizations, whereas Norwegians trust their government agencies. Generally speaking, Nordic respondents are more trusting of the various actors in the food system than those from southern Europe, such as Italy and Portugal, and also Germany, where there is a great deal of public concern on a number of food safety issues.

We may draw out a number of dimensions of trust from this and similar works that are relevant for our purposes.

- The texture and depth of knowledge in the various food filières, including tacit knowledge and 'common knowledge'. These are visible to the consumer in the form of customary practices, skills, service and maybe even expertise.
- Well-organized, predictable, reliable and repeatable transactions, to which can be attached valuations such as reputation. Over time, such relations become grooved and possibly institutionalized.
- The monitoring and audit of the rhetoric of transactions with regard to quality, price, to measure truth-telling and trustworthiness.
- Networks or 'worlds' of trust that go beyond individual transactions and are the building blocks of successful markets. These may be based upon institutions such as supermarket or restaurant chains or on branded goods from a trusted manufacturer.
- Confidence in the strategic dynamic of a food system, its constituent actors and contributing technologies, including the efficiency and stability of institutions.
- The openness of actors and institutions to critical challenge and the transparency of process. Rigidity in the face of consumer disquiet, for instance that coming from the unresolved debate about genetically modified foods, may lead to distrust, especially if a narrow market without real competition locks consumers into a relationship with a particular supplier. In an open market they have the options set out in Hirschman's *Exit, Voice and Loyalty*.[67]
- Ultimately, trust is a means of coping with complexity.[68] We need more research on its scalability in modern societies. An example of why is identified by Luhmann, in discussing what he calls 'system trust', when he notes the paradox of our learning to trust – in an almost abstract sense – long chains of connexions, but at the same time the control of trust at

66 Poppe and Kjaernes 2003, Kjaernes et al. 2006, Kjaernes 2006.
67 Hirschmann 1970.
68 Luhmann 1979.

higher scales has become more difficult.[69] There is a link here with Ulrich Beck's *Risk Society*, as we will see later.

Given the lengths gone to by some consumers in France and Italy to establish and maintain chains of trust with personally-known suppliers, it would be extraordinary if they were to express anything other than satisfaction with their sources and the qualities of the individual items of food and drink. Isabelle Téchoueyres writes about this for Bordeaux, where her interviewees not only return to trusted stallholders in the city's various markets, but also drive out into the distant countryside to visit 'reliable' farmers for fat ducks, fruit, eggs, meat and wine.[70] Although dwellers of a large, modern city, they remain connected to the origins of their 'quality' food in a way that gives satisfaction and constantly renews and reproduces valued, small-scale sources. Research now needs to be done on whether this represents a long-running survival of such links from the days of France's peasant heritage or whether these links have recently been rediscovered and are part of the Europe-wide revival of alternative circuits of food that include organic production and farmers' markets.

The meat industry in Umbria demonstrated similar characteristics, at least until recent changes in EU abattoir regulations threatened its very structure.[71] In the 1990s the regional beef market in this part of Italy was dominated by small butchers' shops and by farmers slaughtering animals for their own consumption and to pass the meat on to friends and neighbours. Local consumers demanded local meat and made an association between place specificity and genuine-ness.

Similarly, Henk Renting recalls a rather telling Dutch proverb: 'what the farmer does not know, he will not eat.'[72] This not only indicates the importance of information flows but also harks back to an age in the rural Netherlands when face to face contact was important in establishing trust. In other words, there is an equation of economic transactions and social relations that may operate in the local grain or livestock market, but it also has relevance to many other situations in which food is passed along a chain.

This idea has been popularized in economic sociology by Granovetter, and it is now commonly accepted that 'social embeddedness' is a key feature of any economic behaviour that depends upon respect and reputation.[73] Under such circumstances, trust is embedded into a chain of mutual reliance, in which it is in everyone's interests to maintain standards. Cheating a partner would mean undermining the whole network and compromising one's own interests, so it is less likely to happen than in an anonymous, unregulated system. Despite its

69   Luhmann 1979: 56–7.
70   Téchoueyres 2007.
71   Ventura and Milone 2000.
72   Renting et al. 2003.
73   Granovetter 1973, 1985. See also Polanyi 1944, Dasgupta 1988, Hardin 2002, Hess 2004, Wilkinson 2006.

appeal to 'common sense', Granovetter's idea may have been taken too far by his enthusiastic followers. This is Andrew Sayer's point when he remarks that 'the focus on embeddedness can inadvertently produce an overly benign view of economic relations and processes'.[74] In reality, local food economies that are fuelled by interpersonal ties, trust, and reciprocity may also have undercurrents determined by relations of power, inequality, conflict, and personal gain.[75]

The verb 'to trust' can imply physical activity – the 'doing' of trust. After all, we trust food with and through our bodies on a daily basis. We put ourselves in the front-line of environmental risk, or so the media would have us believe, because of a long list of hazards that are associated with food: BSE, salmonella, E. Coli O157, pesticides, saturated fats, salt, and so on. An example is the trust that individuals place in others to send them signals about quality. Researchers at the California Institute of Technology have shown that wine drinkers say they like a wine more if it is said to be expensive. We often take it on trust that 'you get what you pay for', so cost is a surrogate of quality. In addition, the CALTEC experiment showed that neural activity in the medial orbitofrontal cortex of volunteers' brains, the area which registers pleasure, was activated, so our brains can easily be tricked into the false appreciation of quality.[76]

Kjaernes, Harvey and Warde reject such an approach based on a cognitive and individual analytical framework. Instead they define trust in social and relational terms.[77] The setting for the performance of these relations may either be the socially embedded, practical routines of supply and consumption, or the institutional setting of political and administrative organizations. This is an important insight for understanding the growth and decline in trust in food during our period and it provides a platform for discussing trust and quality as network concepts.

Quality and the 'Impure Dynamics' of the Law[78]

In the context of a market awash with adulterated products, trust needed an institutional setting. From 1860 onwards this was provided within the framework of the law, weakly at first but the gaze of the state gradually intensified. An important dimension of quality, then, became what judges said it was. We will explore this in depth in Chapters 7 and 8.

But who gets to say what food quality is before laws are passed and malefactors are sent to court? We can best illustrate this by referring to the situation in France. As we saw in Chapter 3, Paris was an early centre of organic chemistry and various methods of analysis were brought to bear on that central symbol of French food

---

74   Sayer 2001: 698.
75   Goodman, 2003, Hinrichs, 2000.
76   Plassmann et al. 2008.
77   Kjaernes et al. 2007: 37.
78   The phrase 'impure dynamics' is from Pickering 1995: 54.

culture, wine. The French preoccupation with wine was an important incentive to investigate the full range of its frauds, for instance plastering and attempts to imitate regionally-specific wines using imports and blending. This explains why French discourse employed words such as falsification, counterfeiting and forgery, with subtle differences of meaning, and was less concerned with adulteration, a term more appropriate to milk.[79] It is also the reason why the early analytical methods were not so much targeted at milk. In Britain the use of toxic food colours and the watering of milk were the priorities, and for the latter at least the chemists, legislators and lawyers were lagging far behind the fraudsters by the 1870s.

The logic of the famous French food law of 1905 was again driven by wine. It was widely regarded as a model in other European countries and eventually became the basis of the single European market regulations on standardized food quality in 2003.

In England and Wales, with their very different legal system and without the intellectual template of wine, it seems that almost everyone from the local authority analyst right up to the President of the Queen's Bench was frustrated by the fog of uncertainty that prevented a full understanding of milk and therefore the dispensing of justice to those who manipulated it illegally. The quality police knew that better equipment and more efficient techniques would help, along with experts who could make definitive statements. But in the nineteenth and early twentieth centuries advances in dairy science were always closely followed by adjustments in the mode of adulteration and on each occasion any temporary clarity faded once more.

Magistrates and judges were well aware of the problems that we have laid out in previous chapters. They knew that designing a fair system of sampling and analysis was a challenge and on a number of occasions the Court of Appeal felt obliged to throw out prosecutions on technicalities arising from a mismatch between the law and a particular characteristic of milk. The judge-made law that we will discuss in Chapter 8 starts here. The courts were also mindful of the practical difficulties, mainly issues of cost, facing local authority analysts. Not every sample of milk could be subjected to the rigorous, time-consuming, and expensive laboratory tests that were the state of the art at any one time, and therefore almost every case tried was open to methodological and interpretative debate. Finally, the clash of wills between provincial analysts and their government-sponsored colleagues in London was well-known. Magistrates were, generally speaking, in awe of the 'official' opinion of the latter and deferred to it in their judgements. This caused much friction and recalls the questions about the nature of professional expertise that we touched on in Chapter 4.

What I am saying here is that legal disputes and judgements about food were always embedded in the less-than-ideal, and often messy, circumstances of public

---

79 Since the early eighteenth century there had been a distinction between 'falsification', the substitution of one ingredient for another, and 'fraud', which is the result giving short measure or using false weights. Stanziani 2005: 41.

science. Although based in common law precedent, this type of justice had many features in common with Andrew Pickering's 'mangle of practice' in laboratory science.[80] It emerged from a dialectic of resistance and accommodation and so was often unpredictable in its direction. Above all it was performative, in the moment, without much opportunity for deep reflection. Court time was, of course, a scarce resource and the deliberations and judgements were therefore also constrained in their quality.

Overall, then, mangle or sausage machine, the individual courts were processing many cases each year. A limited number will be discussed in specific detail in Chapters 7 and 8, mainly appeal cases. The distinct impression one has is of a struggle between vested interests that was mediated through scientific uncertainty and arcane legal procedure. Whether nature was tamed and better known as a result is questionable.

In Pickering's terms, the courts' ability to capture material agency was limited in the nineteenth century. In fact, the courts were involved *because* of the mismatch between milk and equipment. If there had been a better fit for the adulterators, their fraud would have gone undetected. If the public laboratories had been more efficient and authoritative there would have been no basis for legal argument. The courts existed because of the impurities in the dynamic relationships in human/ material hybrids but, more than that, they were responsible for consequent reconfigurations of human agency, the emergence of which is therefore the result of an engagement with the material.[81]

Conclusion

This chapter has provided us with some non-material analytical tools for thinking about material qualities. Chapters 7 and 8 will now proceed with the technical detail of adulteration, and with the legal and regulatory means of holding it in check. Given the depths of criminal degradation in Victorian and Edwardian cities, no-one thought that tampering with food was the priority issue of the age, and yet it is the everyday hazards, and the will of a populace to get to grips with them, that can tell us more about a society than penny dreadful headlines about Jack the Ripper.

---

80 Pickering 1995.
81 Pickering 1995: 53–4.

Chapter 7
Policing the Natural

The Priorities of Modernity

> Modernity ... is that which seeks to expunge all semblance of ambiguity by invoking or, more accurately, imposing its own particular 'will to truth' as the universal means by which to 'improve' society ... Modernity is the age in which all things contingent and particular are (re)presented as necessary and universal ....[1]

Knowing the world through classification and measurement has been important for moderns. Zygmunt Bauman called this the 'task of order'.[2] Foucault went further, showing that measurement is itself a form of government and, in reflexive terms, also one of the many strands of self-government, or 'governmentality'.[3] Considered in this way, even the most mundane analysis of a food sample is part of a web of power that stretches from Westminster to Wester Ross. Our account in Chapter 4 of emerging standards for milk in nineteenth- and twentieth-century Britain therefore has implications beyond the food economy. It shows how laboratories, analysts, inspectors, and metrologists were key agents in imposing their resulting view of the natural. This was achieved through state intervention, which began with legislation in the 1860s and 1870s and culminated in the regulation of compositional standards in 1901. The latter was a momentous threshold and an early example of government intrusion into material limits, representative of a new way of thinking that had already led to the formation of the National Physical Laboratory in 1900 and the British Standards Institution in 1901.

As we saw in Chapter 3, food science and the study of food quality played its part in the ever-widening scientific project of investigating the natural world and its interactions with society. In France particularly, the essentials of everyday life formed a core interest for natural scientists, especially the new breed of organic chemists. Wine, food, even blood and other bodily fluids, were all enthusiastically analysed as being, in a sense, 'new' to science. As we have seen, the first major publication in this vein, on milk, was by Deyeux and Parmentier in 1790, and there was a steady stream of further work on dairy products in Paris in the first half of the nineteenth century, making that city a hearth of dairy science. Research tended to be laboratory-based and small-scale; it did not reach a wider scale until the 1870s and 1880s when milk became worthy of the gaze of both state and industry.

---

1   Halsey 2006: 14.
2   Bauman 1991: 4.
3   Dean 1999.

Although the statistical movement that grew in parallel with, and facilitated, modernity was a fluid and complex phenomenon, it was late in illuminating the stubborn and silent rhythms of everyday practices.

Food production and consumption presented epistemological challenges for those who would know. In Britain there was no nationwide annual agricultural census until 1866 and no in-depth official data on diet until Edward Smith's one-off survey in 1862/3 of 'the food of the poorer labouring classes in England'.[4] The biopolitics of consumer health were influencing one wing of the sanitary movement; so why was the interest in food so late in comparison with, say, the population census, or the statistics associated with the new Poor Law or the Registrar General's series on mortality, which started in 1837? The answer lies in epistemological history, which reveals a genealogy of thought.

The various elements of the food system were either difficult to distinguish from the background noise of everyday life and therefore not worthy of critical analysis, or faced seemingly insurmountable problems of cost-effectiveness in data collection. The late appearance of adulteration data – first published nationally in 1877 – was typical of these uncertain conditions and of course worked in favour of the fraudsters. It is probably no coincidence that the 1870s are thought to be the all-time peak of adulteration in Britain. After that, the publicity of court cases in the newspapers, and the talking point of definitive data, gradually galvanized public opinion and forced both retailers and inspectors to be on their guard. From this perspective, adulteration data eventually became part of the 'technologies of government'.[5]

Agriculture is a good example of the epistemological history of public knowledge about food. General observers such as Arthur Young, and the regional authors of the Board of Agriculture's *General Views* (1793–1815), were interested in descriptive accounts of land use and agronomy but deploying sufficient resources for a full statistical account at the national level was problematic.[6] The very idea of agricultural statistics was treated with caution by land owners and farmers, for whom quantities of any sort were associated with value and therefore with the possibility of taxation.[7] This was a realm of reluctant facticity. There were several early experiments in collecting data, for instance 1796–1804 (livestock), 1801 (arable), and 1854 (crops and livestock), but it took crisis conditions to overcome the collective doubts.[8] The danger to livestock farming of the cattle plague (rinderpest), which swept through the

---

4   Smith 1864.

5   Rose and Miller 1992: 183.

6   Scotland was better served, with the 'Old' Statistical Account (1791–9) and the 'New' Statistical Account (1834–45).

7   See, for instance, *Copy of the circular sent by the Secretary of State to the Chairmen of Quarter Sessions in England and Wales with reference to the collection of agricultural statistics, of the resolutions adopted thereon, and of correspondence relating thereto*, P.P. 1862 (205) lx.1–31.

8   Turner 1981, 1998, Turner et al. 2001.

country in 1865–1866, finally tipped the balance in favour of inventories of animal numbers and crop acreages, both starting in 1866.[9]

If there was conservatism about agricultural statistics, then there was even less interest in the collection of information about food – supply or demand – once it had passed through the farm gate. Surveys of food consumption were confined to academic and activist studies of household budgets, particularly those linked to studies of livelihoods and poverty in the style of Frédéric Le Play and Seebohm Rowntree. These faced difficulties in devising methods of representative sampling in space and eliciting a per capita dietary profile based upon food availability or purchases. The latter was slippery and remains so today for those nutritionists and dieticians who work in the difficult conditions of developing economies.[10]

The state's curiosity in food consumption was somewhat limited but a couple of developments held its attention. First, strategic planning necessitated discussions about the United Kingdom's ability to keep its shipping lanes open in time of war because of its high level of dependence upon food imports. Second, there were worries expressed in parliamentary enquiries in the early twentieth century about an underfed nation that could not produce enough healthy recruits for its armed services. Yet successive governments were wary of academic and polemical works about undernourishment. Until the 1930s the politics of food were about individual rather than collective responsibility for nutrition. The best example of this was the 'hungry England' debate of the 1930s.[11] The annual National Food Survey only emerged as a priority in the wartime conditions of 1940.[12]

This is not to say that the biopolitics of the period excluded food. Aaron Bobrow-Strain argues for a reading of Foucault that sees healthy bodies in terms of diet, influenced by the self-discipline of choice, but also by the societal control of food standards regulation.[13] For white bread in the America of the early twentieth century, standards meant the ratio of ingredients used by bakers, the quality of inputs, and hygienic considerations. There was controversy especially about the increasing whiteness of bread and the nutriment extracted in the roller milling process. The elimination of adulteration was also a priority but there was resistance from bakers, just as they fought the wrapping of bread and other 'improvements' to their trade inspired by the state or by consumer activists.

If the state was not particularly interested in what each person ate, why then did it become involved with collecting and publishing statistics on food adulteration? Let us consider five reasons. The first is that this era was part of the prehistory of what Engin Isin calls 'neurotic citizenship'.[14] As he notes, the legitimacy of governments increasingly came to depend upon the management of risks on

---

9   Matthews 2000.
10   Pacey and Payne 1985.
11   Mayhew 1988, Smith 1997, 2000.
12   Baines 1991.
13   Bobrow-Strain 2008.
14   Isin 2004.

behalf of their citizens. Hassall's work in the 1850s had reminded consumers and politicians of the danger of poisoning from the chemicals added to food. Select Committees and several Acts of Parliament followed. In a sense, a small industry relating to adulteration was created out of the state's deployment of resources to collect and analyse data. Arguably the courts over-litigated with regard to milk because lawyers earned a living from a heavy case load or by occasionally finding loopholes in the legislation, and so there were few incentives from the point of view of the legal profession to argue for a streamlining of the process. There were also inspectors, analysts, laboratory assistants, suppliers of laboratory equipment and chemicals, lawyers, and legal journalists, all depending to some extent for their livelihoods upon a phenomenon that had not existed in this way before 1877. This was part of a general trend towards governance in the hands of absent experts. Earlier, the hazard had existed but the risks had been unknown, and perceptions were even more uncertain.

There were other worries too. For instance there was a growing realization of the potential for the spread through the human food supply of animal diseases: the zoonoses. Knowledge of these had been growing in Britain since the 1840s with the discovery of diseases such as bovine tuberculosis in London's cattle and meat.[15] The panic associated with the 1848 cholera epidemic in London was further evidence of consumer concern. There was a sharp decrease in demand for fruit, vegetables and fish, in what amounted to a food scare, although understandings of the disease were, of course, pre-Pasteurian.[16]

There were other animal-related risks in the nineteenth century that in retrospect have the features of social construction. Rabies was one disease with a high profile from the 1870s because of the certainty of a painful death.[17] The incidence was low but the public outcry was such that governments over the following 100 years could not contemplate easing the quarantining of imported pets.[18] The bacterial horse disease, glanders, was a source of similar, 'irrational' public fear.[19]

Second, from about the 1830s, an environmental view of hygiene and health was emerging, most strongly advocated by Edwin Chadwick and his followers, which argued that a reduction in mortality and morbidity could only follow a concerted attack upon dirt and pollution.[20] Significant sections of the statute book in the second half of the century were devoted to sanitary measures, with housing, water supply and refuse disposal as the main thrusts. Central governments were

---

15   Perren 1978: 60. One reason that the hazard has grown is the increased consumption of livestock products.

16   That is, until the work of John Snow became well known. Hardy 1999: 296.

17   Walton 1979. Compare with Kete 1988.

18   So deep did this concern become embedded that the quarantine requirement was only lifted in the year 2000.

19   The symptoms of glanders are lesions in the nostrils, submaxillary glands and lungs and on the surface of limbs or body for farcy.

20   Wohl 1983.

becoming aware of the gradually increasing political power of the food issue, but they calculated that there were more votes in the supply of plentiful, cheap provisions, than in expensive bureaucracy that would be required to guarantee dietary minima.

Public health represented one aspect of the running repairs that modernity required. Official reports, amateur surveys, health statistics, and the evidence of the nose, all seemed to point ineluctably to the need for intervention. This was resented by those who were called upon to pay, but hygienism gradually acquired supporters, theories and plans of action with a collective moral and practical force that resembled the climate change movement today. And the challenge of cleaning Victorian cities seemed to be just as great as global warming. The state of the food system was just one strand within this broader movement, and its priority was certainly less than those of housing, drains and sewers. As a starting point for debate, the dimensions of adulteration were thought to be as follows:

- the addition of some foreign substance or inferior value but possessing plausible resemblance to the food itself;[21]
- the removal from the food in part of some valuable natural constituent;
- the addition, without the knowledge of the purchaser, of some foreign matter producing or preventing changes in the natural food.

The first four Sale of Food and Drugs Acts (1860, 1872, 1875, 1879) were passed at a time of civil and political ferment about sanitation.[22] Without this favourable intellectual context on public health matters, their passage through parliament would have been more difficult. The Public Health Act of 1848 and the Nuisances Removal Acts of 1855, 1860 and 1863 had already established the principle of intervention for the public good, although their permissive character led to a spatially complex mosaic of implementation or neglect according to local political circumstances. The Sanitary Act (1866) introduced the notion of centrally-guided compulsion and this was soon followed by the formation of the Local Government Board in 1871, absorbing the medical department of the Privy Council, and then by the passing of the Public Health Acts of 1872 and 1875. The Sale of Food and Drugs Acts were a minor part of this great shift in state medicine and public opinion and we should not overestimate their contemporary significance with the benefit of hindsight.

Was the movement to improve the quality of the nation's food supply autonomous of this sanitary surge? The evidence indicates a strong link, because the 1860 Act was motivated largely by the thought of the poisonous ingredients identified by Hassall in the 1850s. Attention to the element of adulteration concerning cheating was less intense.

---

21 Spottiswoode Cameron 1904: 2. See also Chirnside and Hamence 1974: 12, for the SPA definition of 1874.

22 For more detail on the legislation, see Chapter 8.

Third, consumer citizenship, defined as active participation in institution-
and market-shaping, was increasingly evident in the numbers of societies and
associations that in the late nineteenth and early twentieth centuries campaigned
for unadulterated food, wholemeal bread, vegetarianism, unpasteurized milk, or a
minimum dietary standard for children.[23] Some were inspired by mystic or political
ideologies, some by the new science of vitamins, and others by a romantic vision
of pre-industrial, wholesome food.[24] Trentmann argues that consumer politics
gradually shifted from that icon of free trade, the cheap white loaf, to a vision
of a state-regulated food supply, with the public protected from exploitation and
disease; but he oversimplifies a complex picture, particularly for the active and
contentious inter-war period in the twentieth century.[25] The situation by then was
fragmented and difficult to characterize because consumers did not necessarily
share common interests, modes of consumption, or health outcomes. Moreover,
within certain limits, the state was open to lobbying by pressure groups.

Consumer citizenship was nourished by statistics, some of which had
democratic potential. Those consumers aware of data on adulteration and informed
about the individual perpetrators were in a position to change their diet or withhold
their custom, both powerful critiques of a dysfunctional market. Patrick Joyce
sees statistics, through their objectivity, as having the potential to remake trust
on a new basis, 'so that it became one expression of the liberal balancing act of
knowing and not knowing, its deliberately limited and targeted style of knowledge
reflecting the sense of the limitations of what needed to be known'.[26] This was
achieved because 'judgements of quality were taken away from those who did the
measuring, as well as extracted from the instruments to measure themselves, and
what was measured'.[27]

Fourth, a point rarely made in the literature on adulteration is that the common
law, as practised in the courts, was evolving in the nineteenth century, especially
the law on contract with regard to the sale of goods.[28] Pat O'Malley suggests that
eventually it acquired incompatibilities with the theories of liberalism that had
been so influential in economic development. As a result, the two or three decades
before the First World War saw the beginnings of judge-made law in the area
of consumer protection.[29] In addition, we can point out that important new ideas
about naturalness and quality were being sedimented in the case law of the Queen's
Bench. This was an incremental process that was contingent upon a number of
factors and cannot helpfully be reduced to general principles. We will explore this
proposition further in Chapter 8, when it will be suggested that certain aspects

---

23   Kassim 2001, French and Phillips 2003.
24   Atkins 2000b.
25   Trentmann 2001, also Gurney 1996.
26   Joyce 2003: 25.
27   Joyce 2003.
28   Simpson 1975, Hamburger 1989.
29   O'Malley 2000.

of English common law changed in response to the structural make-over of the food chain as, increasingly, daily supplies of perishables were brought hundreds of miles to city consumers. This new system involved an 'always on', networked notion of trust that replaced the previous reliance upon occasional face to face contact and physical inspection of goods. The key change, after about 1850, was from obligation on the basis of promise, to the idea of contract as a formal consensus between parties to a transaction.[30]

Contracts are examples of what Latour calls the immutable mobile: 'an inscription device that moves within a network and its nodal points of passage but remains the same in different contexts.'[31] They illustrate 'how time, places and agents are reassembled through their association. It is neither humans nor the social which are central here, but the socio-technical, and this is the best explanation of how such worlds are constructed'.[32] In addition to their 'hard', legally-enforceable nature, contracts also act as cognitive resources in the sense of building the conventions of expectation and behaviour that we encountered in Chapter 6. As Simon Deakin points out, their role as reference points is their most important function. Even when the terms are breached, the parties only rarely resort to the law and tend to prefer other means of resolving their differences.[33]

Fifth, the monitoring and control of adulteration increasingly came to be seen as fundamental to the construction of stable markets. Food quality increasingly became a commercial priority that required administrative oversight. Confirmation of this may be found in the independent laboratory research and sampling undertaken by several large companies at the top of the London market for retail milk.

These, then, are some of the priorities of the modern state with regard to food quality. Let us now move on to discuss adulteration as a provocation for intervention.

Adulteration: The Early History

> It is not from the benevolence of the butcher, the brewer, or the baker, that we
> expect our dinner, but from their regard to their own interest.[34]

In England and Wales, juries of the Court-leet and, later, the justices in petty sessions were responsible for food quality under the pre-industrial legal system, but they were largely ineffective in preventing food frauds.[35] There were many

---

30   Hamburger 1989.
31   Latour 1990.
32   Austrin and Farnsworth 2005: 151–2.
33   Deakin and Wilkinson 2005, Deakin 2006a, 2006b.
34   Smith 1776.
35   Colquhoun 1829: 339.

Acts of Parliament, especially from the mid-eighteenth century onwards, for instance concerning beer, tea, coffee and cocoa, pepper, and bread, but, again, enforcement was weak in the absence of an institutional framework and reliable means of testing.[36]

The problem of poor food quality seems to have grown steadily with urbanization in the anonymity of the *stadtluft*, where retailers no longer necessarily knew all of their customers. But it would be reductive to claim that the process of commodification was alone responsible for this. Some late Victorians thought that the various actors in the agro-food sector were complicit through the lengthening of supply chains and the pace of industrialization, which together resulted in pollution and adulteration. In one sense, they were right, because the acceleration in demand as a result of rising real wages, along with the switch by some women away from breast-feeding to bottle-feeding, over-reached the farmers' ability to provide milk. As a result, large numbers of mixed farmers who happened to keep a cow or two were now encouraged to consign liquid milk to the urban markets, along with the former specialists in cheese and butter manufacture. Not all had appropriate facilities, such as a clean water supply and buildings that would have enabled clean milk production.

Eighteenth century pamphleteering on the subject of food quality was of a passionate but exaggerated nature, exemplifying a growing public awareness.[37] Even fictional characters such as Humphry Clinker encountered degraded milk quality on the streets of Georgian London and there were occasional comments in the *Annual Register*.[38] By the 1790s and the first decade of the new century, the authors of the four editions of the *General View of Agriculture of Middlesex* were reporting watering and skimming of milk in such a matter of fact way that they seemed to be assuming an established knowledge of this on the part of their readers. They reserved their complaints instead for the inability of science and the authorities to intervene:

> When it is considered how greatly it [milk] is reduced by water, and impregnated with worse ingredients, it is much to be lamented that no method has yet been devised, to put a stop to the many scandalous frauds and impositions in general practice with regard to this very necessary article of human sustenance.[39]

In 1820 Friedrich (usually anglicized to Fredrick) Accum, a well-known German consulting chemist and lecturer working in London, published a book that caught the public imagination.[40] He had a detailed scientific knowledge of the purity of food and drugs, which he used to analyse and itemize outrageous

---

36 Filby 1934 lists 36 relevant acts before 1860.
37 Filby 1934.
38 Smollett 1771/1823, vol. 1: 143, Hodson 1765: 129.
39 Middleton 1798: 335–6. See also Baird 1793, Foot 1794, and Middleton 1807.
40 Brown 1925, Burnett 1958, 1966: ch. 5.

adulterations, such as those involving the use of poisonous substances.[41] His book, *A Treatise on Adulteration of Food and Culinary Poisons*, received a great deal of attention. Filby called it a 'turning point'.[42] Unfortunately for Accum, however, his credibility and reputation were ruined in 1821 by a scandal over alleged thefts from a library and, as a result, the issue went off the boil.

> This is, indeed, a sad world; but if Mr Accum's reagents and retorts, backed by a dozen acts of parliament, could be rendered a substitute for honesty, it would go on well enough.[43]

Further literature on adulteration ensued but before the 1840s much of it was repetitive.[44] Among the more original insights were comments on the place of good quality milk in the diet of working families and the growth of a specialized branch of the law known as medical jurisprudence.[45] There were also a number of investigations into the possible use of lactometers to compare the specific gravity of a sample with that of 'pure' milk.[46] In addition, there were occasional writings about the state of the milk trade in cities such as Glasgow and Dublin.[47] Above all, by the mid-nineteenth century milk fraud had become so universal that it was the butt of satire.

> A suggestion has recently been made for the supply of the Metropolis with pure country milk, in lieu of that wishy-washy triumph of art over nature, which flows morning and afternoon into our jugs and mugs, from a thousand milk-cans … London, in fact, knows nothing of real milk, which differs as thoroughly as chalk is unlike cheese, from the spurious stuff we are at present contented with. Commercial milk is a compound which any conscientious cow would indignantly repudiate. The Londoner … knows literally nothing of milk; for of the stuff he has been taught to accept as milk, he knows it would be idle to attempt to skim the surface. We understand that the Chalk Market has already begun to show symptoms of weakness at the bare rumour of real milk being introduced into the Metropolis.[48]

In France the presence of milk fraud was also commented on from time to time in the eighteenth and early nineteenth centuries, for instance by Mercier

41   *Monthly Gazette of Health* 5, 1821: 445.
42   Filby 1934: 19.
43   Anon. 1821: 346.
44   Anon. 1799, Anon. 1800, Nicholson 1813: 69, [Williams] 1830, Chemist 1831.
45   Curwen 1809, Nicholson 1813: 69, Tuckett 1816, vol 1: 356, Colquhoun 1829, Ryan 1831: 236.
46   See Chapter 3.
47   Cleland 1816, Harley 1829.
48   Anon., Milk for the million, *Punch* 17 November 1849.

in the *Tableau de Paris* in 1783 and by Parmentier in 1803.[49] Desmarest and Barruel followed in the 1820s, with explanations related to structural features of the milk trade and lists of methods of adulteration. Then the 1840s and 1850s saw an acceleration of revelations once the lactometers designed by Dinocourt and Quevenne were readily available to detect changes in the specific gravity of milk, and technical improvements in microscopes enabled a closer study of the additives used. Parisian milk was now said to be 'très souvent falsifié' and there was a mawkish public interest in lurid accounts of the use of thickeners such as flour, starch and even liquidized brains from sheep or calves.[50] There were similar stories in Britain and Germany but Fleischmann later suggested that, in truth, such adulterations were rare.

> In former times, before much experience had been obtained in the supervision of the milk trade, it was customary to draw up a formal list of adulterants said to be found in milk … albumin, white of egg, caramel, artificial emulsions, meal, gum, dextrin, glue, bird-lime, soapy water, calcium and magnesium carbonates, the pulverized brains of calves, sheep, and horses … The large experience which has been gained in the course of the last twenty years has shown that in Germany, at least, hardly one of the above-mentioned and highly improbable adulterants has been used.[51]

Alphonse Chevallier's *Dictionnaire des Altérations et Falsifications* published in 1850 was particularly influential in the literature on the adulteration of food and drink in the second half of the nineteenth century. It remained in print for nearly 50 years, the seventh and last edition appearing in 1896/7.[52] The particular, French concern with the plastering and watering of wine gave momentum to both a scientific and a commercial investment in research in that country.[53]

In Britain the work of Barruel, Chevallier and other French writers was noted and sometimes excerpted and translated in journals from the 1830s onwards.[54] French-inspired books were published in mid-century by Mitchell and Normandy

49   Mercier 1783, vol. 6: 125, Parmentier 1817 (first published in 1803): 223.

50   In modern terms, the use of brains in milk was an 'urban myth'. There is a large French literature on it: Barruel 1829, Anon. 1830, Quevenne 1842, Gaultier de Claubry 1842, Chevallier 1844, Monfalcon and Polinière 1846: 292–7, Chevallier 1850, 1856, Chevallier and Reveil 1856, Husson 1856. In Britain, Professor J. T. Queckett did find one sample containing calves' brains. Select Committee 1856: Q. 396. A few other exotic adulterations (almonds and gum arabic, gum tragacanth, emulsion of hempseed, starch) were occasionally found: Scholefield Committee, P.P. 1854/5 (432), VIII, Q.5.

51   Fleischmann 1896: 66.

52   Chevallier 1850.

53   Stanziani 2005.

54   *Quarterly Journal of Agriculture*, 2, 1829: 304–6, *London Medical Gazette*, 5, 1830: 571–2, *Edinburgh Medical and Surgical Journal*, 33, 1830: 226–8, Dodd 1856: 17, *British and Foreign Medico-Chirurgical Review*, 20, 1857: 49–65.

that showed that there had been no improvement in the honesty of the food industry since the days of Accum.[55] Soon after, in the early 1850s, there was a public outcry when an 'analytical and sanitary commission' sponsored by Thomas Wakley MP specifically identified some traders in London as guilty of adulteration. The results were published in the *Lancet*, a medical journal founded and edited by Wakley. He had hired the doctor and chemist, Arthur Hill Hassall, and his microscope, to conduct a wide-ranging enquiry. This lasted four years (1851–4) and, by its outspoken yet scientific approach, was a milestone in the history of investigative journalism.[56] The daily press took up the story and revelled in Hassall's publication of the names and addresses of the culprits, along with the details of each analysis. The popular suspicion seemed to be confirmed that a good deal of the capital's food was adulterated. Hassall later extended his researches to provincial cities and found much the same problems. Table 7.1 presents the miscellaneous evidence of milk adulteration collected in London from 1850 to 1871. It seems that almost 90 per cent of supplies were affected. The retail product contained on average 25 per cent of water added after one-third of the cream had been skimmed.

Official reaction was slow. This was because of the growing free-trade mentality of the legislature. By mid-century the spring tide of liberalism meant a deep-seated reluctance to interfere in the food, or any other, industry. Yes, there were market bye-laws widely in force to seize and destroy rotten meat and other unfit foods, but not all food passed through markets and, anyway, this system was incapable of identifying processed foods with toxic ingredients. There were a number of laws on the statute books to control the adulteration of specific foodstuffs, but most of these were intended to protect the government's excise revenue and were of little benefit to either the consumer's pocket or her health.[57] Another restraining factor was the respectable social status and political lobbying power of middle-class shopkeepers. Successive governments preferred to give them the benefit of the doubt. Herbert Spencer later caught this particular mood when he complained that anti-adulteration legislation meant that one-half of the community would be occupied in seeing that the other half did its duty.[58]

In Paris a free market for food was slower to emerge. In the mid-nineteenth century there were still restrictions on butchers and bakers as providers of the basic foodstuffs, particularly with regard to controlling prices and therefore defusing potential civil unrest. The city had a long history of popular uprisings and the authorities were nervous that food riots about high prices might have wider political consequences. According to Husson, there were only 600 authorized

---

55   Mitchell 1848, Normandy 1850.
56   Hassall 1855, 1857, 1876, 1893, Clayton 1908, Smith 2001.
57   The Board of Excise was responsible for preventing the adulteration of tea, coffee, tobacco, hops and sugar. There were also acts against the adulteration of bread in London (1822) and the rest of England and Wales (1836). Paulus 1974: 16.
58   Spencer 1872: 367–8.

Table 7.1   **Evidence of the adulteration of London milk before 1872**

Item	Samples	Adulterated (%)	Average dilution (%)	Average fat abstraction (%)
1	24	52	12	–
2	–	–	25	33
3	–	–	20	25
4	20	–	20–50	–
5	–	–	25	–
6	100	74	–	–
7a	5	80	33	36
7b	82	73	–	–
7c	20	40	–	–
8	10	90	19	23
9	16	75	30	30
10	15	80	21	30
11	617	93	–	–
12	97	90	26	46
13	33	94	–	–
Total	1,040	–	–	–
Weighted mean	–	87	25	35

Sources: 1. *Lancet* 1850–4; 2. Rugg 1850: 34; 3. Wynter 1854: 293; 4. T. Hillier, *Annual Report of the Medical Officer of Health on the Sanitary Condition of the Parish of St Pancras* 1856; 5. Anon. 1856a; 6. Scott 1861; 7. Morton 1865, quoting estimates by (a) Voelcker, (b) Druitt, (c) Whitmore; 8. Voelcker 1867; 9. Divers 1868; 10. Hassall 1871: vol. i; 11. *The Milk Journal*, 1871–2; 12. *Report of the Superintendent of Contracts*, PP 1872 (275) li; 13. Islington Medical Officer of Health, *Annual Report 1871*.

bakers and their prices were fixed each fortnight, according to the price of flour.[59] Eventually, in 1863, licensing and price control were replaced by a tax on flour and the bakery trade quickly adapted to a market reality much more like that of London. Similar restrictions on butchers were abolished in 1858.

In Britain, the utilitarian philosophy of the mid-century, which had validated notions of state intervention in public health in the name of the general social good, especially as expressed in work efficiency, was equally relevant to the food supply. However, such rationalities were still subordinated in the 1850s and 1860s to the same vested interests which had earlier smeared Accum and forced him into exile. Adulteration was a direct route to windfall profits – the greater the undetected

59   Husson 1856.

fraud, the greater the return – and food traders were at first unwilling to relinquish what they regarded as a perquisite. In the late 1850s there does, however, seem to have been a change of commercial tactics, with a few astute traders at the top of the market setting out to satisfy the new consumer demand for pure food.

Sampling and Compositional Data

The series represented by Figure III.1 was compiled by the Local Government Board from the annual reports sent in by local authorities as required under section 19 of the Sale of Food and Drugs (1875) Act.

At first, in the 1870s, local authority inspectors collected samples of a wide variety of foods and drinks, but they soon realized that certain items were rarely adulterated and so specialized in those most at risk, in order to maximize the cost-effectiveness of their work. Milk was one of the items chosen and this made the food supply as a whole appear to be more subject to fraud than was really the case.[60]

In 1877 milk samples constituted 27 per cent of all food and drug analysis in England and Wales, or 33 per cent if one includes other dairy products (cream, condensed milk, butter, cheese) (Table 7.2). By 1907/8 these proportions had risen to 48 and 71 per cent, amounting to a domination of the system. Bread, beer and tea had been virtually free of adulteration for some time, but milk was still causing concern.[61] There was therefore an even greater concentration in the samples that were deemed to be adulterated, the peak coming in 1921, when 70 per cent of these were milk and 79 per cent samples of dairy products more broadly.

The methods used to gather these samples were lacking in subtlety. In the early years, officials became well-known locally and were of regular habit, sometimes even announcing their business when entering a shop. A few counties employed the local police as inspectors, mostly constables in uniform. As a result, the samples of loose milk collected would not have been typical.[62]

One of the greatest difficulties that the public have to contend against is that their food shall be pure, although the Acts exist, because every one must know who has had any experience of the matter that under the present system it is very largely a failure – namely, that the inspectors are not able to obtain adulterated samples when those adulterated samples exist.[63]

---

60 Redwood, speaking to a conference on food adulteration, as reported in the *Analyst* 9, 1884: 160.

61 Collins 1993.

62 *Local Government Board: Twelfth Report, 1882–83*, PP 1883 (C.3778) xxviii.111.

63 F.J. Lloyd, speaking at a conference on adulteration. *Analyst*, 9, 1884: 161.

Table 7.2 Samples collected by county and local authorities and the percentage that were judged to be adulterated

Commodity	1877–79		1880–84		1885–89		1890–94		1895–99		1900–04	
	Samples	%	Samples	%	Samples	%	Samples	%	Samples	%	Samples	%
Milk	15,012	21.6	39,146	19.5	52,471	13.5	69,599	13.2	98,277	10.6	148,321	11.0
Butter	3,126	12.3	6,889	17.1	12,912	14.8	23,247	13.1	43,459	9.6	64,589	7.0
Cheese	–	–	–	–	–	–	–	–	–	–	5,778	1.4
Margarine	–	–	–	–	–	–	–	–	–	–	3,209	6.2
Bread	3,206	7.3	5,595	4.4	4,672	2.2	3,643	0.7	3,144	0.9	2,553	0.4
Cocoa	–	–	–	–	–	–	–	–	–	–	1,304	12.3
Coffee	3,027	18.5	6,468	19.1	6,781	14.5	8,700	13.9	9,527	9.2	11,280	6.8
Confectionery	1,023	3.4	1,525	3.2	1,517	3.0	1,759	1.5	2,484	4.2	8,646	3.5
Flour	1,799	3.4	2,385	0.6	2,210	0.2	1,786	0.2	2,963	1.1	–	0.7
Lard	–	–	–	–	3,794	10.3	6,733	4.6	7,909	0.8	9,879	0.6
Mustard	2,335	19.2	4,272	14.9	4,080	11.6	3,514	7.8	3,844	4.6	3,839	5.2
Sugar	795	1.9	1,406	0.1	958	0.0	1,318	6.4	2,605	4.8	4,086	8.8
Tea	–	–	–	–	1,838	0.2	2,100	0.0	2,316	0.8	2,338	0.0
Wine	191	14.1	296	8.1	200	3.5	255	4.3	330	3.6	739	17.1
Beer	2,099	6.1	2,157	3.5	2,543	2.4	1,726	7.6	1,362	1.2	11,503	6.6
Spirits	4,420	43.1	10,335	24.4	11,897	19.2	17,597	18.7	23,173	14.4	29,996	12.3
Drugs	1,607	25.3	1,936	15.1	2,210	11.0	4,268	14.3	8,899	13.0	13,176	11.5
Other items	9,306	7.6	15,124	4.5	16,571	8.2	19,444	6.6	28,692	5.0	42,250	5.0
TOTAL	47,946	18.9	97,534	14.8	124,654	12.0	165,689	11.8	238,984	9.2	363,486	8.6

Commodity	1905–09		1910–13		1919–24		1925–29		1930–34		1935–38	
	Samples	%	Samples	%	Samples	%	Samples	%	Samples	%	Samples	%
Milk	217,811	10.7	203,549	11.1	363,595	7.1	323,568	7.7	363,927	7.1	321,138	7.4
Condensed milk	–	–	152	0.7	3,712	2.7	5,982	2.0	5,964	1.7	4,916	2.0
Butter	93,317	6.8	85,282	5.3	54,350	1.8	54,405	1.7	48,527	1.0	30,323	1.4
Cheese	9,756	1.3	6,978	2.2	6,588	2.6	6,769	2.6	7,368	1.7	6,226	3.4
Cream	–	–	83	57.8	9,202	13.7	12,045	8.0	10,930	2.1	7,309	1.7
Margarine	6,673	4.2	7,194	2.7	23,228	0.7	17,911	1.3	16,694	0.9	10,294	1.7
Bread	2,110	0.6	1,359	0.1	1,064	0.2	1,374	0.2	1,345	0.4	1,093	0.7
Cocoa	2,441	6.2	3,136	6.9	10,458	0.7	6,338	0.2	5,648	0.2	3,969	0.2
Coffee	12,408	5.4	8,907	4.8	14,273	0.8	8,682	0.7	7,839	0.6	5,329	0.7
Confectionery	6,163	3.9	4,768	5.0	8,030	2.4	11,112	3.6	14,269	4.9	12,111	5.8
Flour	–	0.4	–	2.9	7,119	1.2	6,453	0.9	6,655	0.4	5,295	0.7
Lard	11,134	0.6	14,040	0.9	3,169	0.2	9,140	0.2	14,103	0.3	10,792	1.1
Mustard	3,719	3.8	2,763	2.6	4,497	3.3	2,901	3.0	2,406	2.5	883	2.6
Sugar	3,528	5.0	2,955	5.8	4,576	0.5	4,176	0.7	5,112	0.3	4,926	2.2
Tea	2,264	0.0	1,470	0.5	3,359	0.2	4,224	0.5	9,004	0.4	6,438	0.5
Wine	765	9.2	456	9.6	356	5.3	655	2.0	698	1.6	931	2.4
Beer	3,270	2.5	2,409	4.1	3,083	3.6	2,402	1.8	2,326	1.8	2,070	6.4
Spirits	27,157	10.4	19,923	9.9	8,436	12.5	11,025	11.7	10,034	8.0	6,580	5.8
Drugs	16,618	7.8	13,432	8.3	31,914	5.6	25,690	5.1	27,774	4.5	25,287	5.0
Other items	43,035	6.8	37,533	8.1	111,752	4.4	111,578	5.0	128,796	4.0	124,791	4.0
TOTAL	462,169	8.4	416,389	8.4	672,761	5.5	626,430	5.8	689,419	5.1	590,701	5.5

Source: Local Government Board and Ministry of Health *Annual Reports*

To overcome the absence of surprise, in 1905 the Board of Agriculture recommended the practice of two-stage sampling and this was taken up in some areas. Children were sent in first to purchase some loose milk. This was tested and, if adulteration was identified, a formal sample was later taken for the purpose of legal proceedings.[64]

Later, sampling was systematized, not just for the detection of adulteration, but also for bacteriological analysis. For adulteration this was important in fairness to the supplier of the sample and, for disease, a scientifically reliable indication of the degree of contamination was essential.[65] Standardization of sampling meant the use of plungers, sterile sampling bottles, and stoppers or discs, all of an officially acceptable design. An agreed quantity was to be drawn, and the bottle or container had to be transported and stored as specified. Under the 1875 Act samples in retail outlets had to be divided into three portions, one of which was offered to the shopkeeper for independent analysis and use, if necessary, in evidence for defence against an action alleging adulteration.

Samples were also taken by the commercial dairy companies. In 1881 Paul Vieth, chief chemist of the Aylesbury Dairy Co., started his company's famous, long-running series of analyses of country milk received from their suppliers (Table 7.3).[66] This was in the company's own interests, since their business was heavily dependent upon a reputation for quality. It helped them to identify those farmers who were unable or unwilling to produce milk of a certain standard, but, rather than treating them as confidential, the data were published and soon they became the dairy world's most definitive indication of what could be expected statistically. It was realized that earlier research about standards had not been firmly grounded in scientific fact. Arguments had often been based on the output of small and possibly atypical herds. Now Vieth (1881–1892) and his successor, Henry Droop Richmond (1892–1920), were able to satisfy the demands of statistical representativeness. From 1881 onwards they took 10,000–20,000 samples a year and, due to the forcefulness of its sheer scale, their data became the definitive series.[67]

Richmond went further and established a reputation as a global leader in laboratory research. Few textbooks on milk were available at the turn of the century and he led the field with his *The Laboratory Book of Dairy Analysis* (three editions: 1905–25) and his *Dairy Chemistry* (five editions: 1899–1953). The latter was described as '*the* reference book' for all analysts.[68]

---

64   Liverseege 1932: 12.

65   See British Standard No. 809 (1938) and Memo. 36/Foods issued by the Ministry of Health in 1927, 1929 and 1939.

66   Vieth was a German dairy scientist.

67   Another famous series was that of C. A. Cameron who analysed 100,000 specimens of milk over a 40-year period, and found that 13–15 per cent total solids was the average for the cows supplying Dublin. Wenlock Committee 1901: Q.2414.

68   Hughes 1960.

Table 7.3   Average monthly composition of milk supplied to the Aylesbury Dairy Co., 1897–1916

	Specific gravity	Total solids (%)	Fat (%)	Solids not-fat (%)
January	1.0322	12.75	3.79	8.96
February	1.0322	12.67	3.72	8.95
March	1.0322	12.62	3.67	8.95
April	1.0321	12.54	3.65	8.89
May	1.0323	12.50	3.56	8.94
June	1.0322	12.42	3.52	8.90
July	1.0317	12.39	3.63	8.76
August	1.0315	12.51	3.76	8.75
September	1.0318	12.70	3.85	8.85
October	1.0320	12.84	3.91	8.93
November	1.0321	12.94	3.98	8.96
December	1.0321	12.87	3.91	8.96
Average	1.0320	12.64	3.75	8.89

Source: Richmond 1920: 306.

Both Vieth and Richmond were sensitive to criticisms of their series. They regularly reviewed their methods, and were fully self-reflective.[69] One issue was that the Aylesbury Dairy Company bought only milk produced under good conditions, and did not see much milk from poorly fed animals. So, extrapolating the findings of their laboratory to the milk trade as a whole was questionable. Another potential difficulty was that the different dairy cattle breeds gave milk of varied composition. The Channel Island breeds, in particular, are renowned for their rich milk, whereas Friesians, which became very popular in the twentieth century, have a high yield of milk low in butterfat. It is not known whether the Aylesbury drew their supplies from farms with a bias to one or other of these breeds, or indeed the Dairy Shorthorn or Ayrshire, which were also common at the time.

Not to be outdone by their rivals, in 1884 the Express Dairies imported scarce expertise in the person of Harald Faber, formerly of the Royal Danish Laboratory in Copenhagen, to be their analytical director.[70] The following year Dr Shirley Murphy was also added to the team as sanitary director.[71] A new laboratory was

---

69   For instance, see Richmond 1904.

70   These appointments, of Vieth and Harber, a German and a Dane, to such senior analytical positions, suggests a shortage of home-grown talent in the area of dairy science.

71   Anon. 1914. Murphy was a prominent Medical Officer of Health and later was knighted for his services to the public health.

expensive and required a culture of vigilance, not an easy task when the work was not too different from routinized, semi-skilled, factory-based, mass production.

In the mid-nineteenth century there was a debate across Europe among statisticians about the significance of variation in data sets such as those compiled by the dairy companies, and their relation to nature, human freedom and state control.[72] From the 1850s, German statisticians searched for systematic covariation rather than mere description of regularities.[73] In Britain, the administrative locus of statistical data compilation and analysis was more devolved than elsewhere, particularly France, but state biopolitics nevertheless drew strength and legitimacy from high quality work on the population census and public health statistics.[74] By 1900 'exact measurement was advertised as a vital accompaniment of commercial, military, and thus imperial triumph' and the resulting statistics were 'cognitive commitments' to thinking of phenomena in the way decided by the collector.[75] Solid and trustworthy facts were crucial in the emergence of norms and thus the analysis of milk sample data put commercial dairy companies on the moral high ground with their self-proclaimed attempt to use the modern sciences to solve the social ills of a dishonest and polluted food supply.

Commercial research revealed a number of points vital for dairy science. First, it was realized that genuine milk was a great deal more variable than had initially been thought, on scales from the daily, to the seasonal, to the annual. Second, the early, rather simplistic, focus on fat had distorted the industry's understanding of genuine milk and encouraged farmers to engineer a regression to an annual mean for that ingredient, to the neglect of other factors. For instance, Figure 7.1 shows a noticeable improvement through time in the fat profile of samples taken in Birmingham but in Figure 7.2 the problem of seasonality remains. For fat, up to 10 per cent, and for solids-not-fat, up to 20 per cent, of samples in the 1920s remained below the 'normal' compositional threshold. There is no reason to think that Birmingham was unusual in this regard: indeed Milk Marketing Board data as late as 1957/8 showed something very similar for the whole of England and Wales in the proportion of herds producing milk at a monthly average of below 8.5 per cent solids-not-fat. The worst month was usually April, with the North and South Wales regions over 30 per cent, and only the Far West and South East being much below 20 per cent.[76] In short, well into the twentieth century, a substantial minority of farmers was unwittingly sending presumptively illegal milk to market for a portion of every year.[77]

---

72   Hacking 1990.
73   Porter 1986: 179.
74   Desrosières 1998: 170.
75   Schaffer 1995: 135, Starr 1987: 53.
76   Cook Committee 1960.
77   Cook Committee 1960.

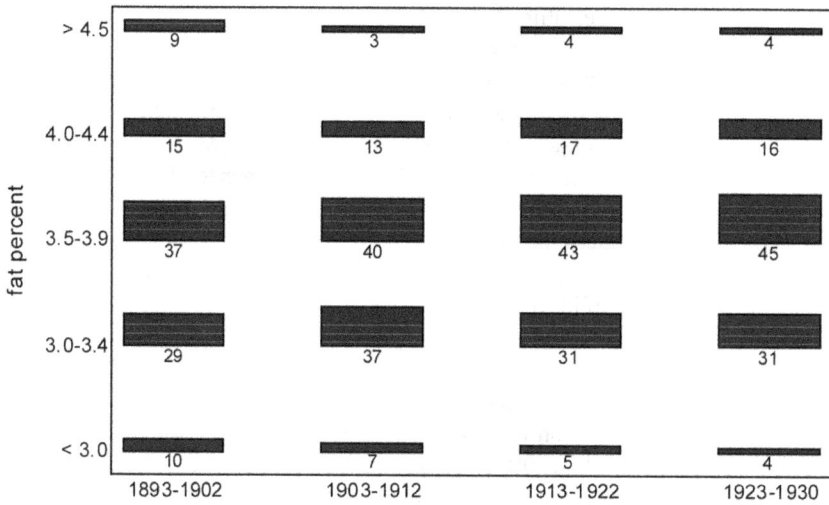

Figure 7.1   The fat content of Birmingham milk samples

Source: Liverseege 1932: 195.

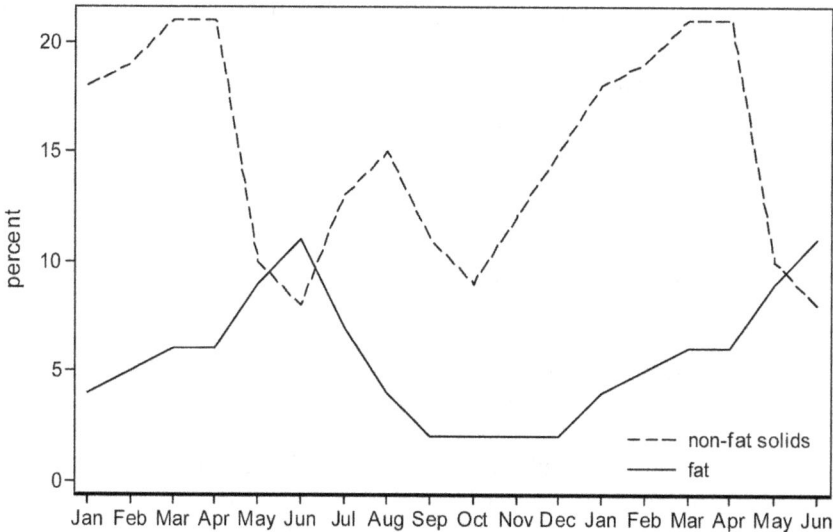

Figure 7.2   Birmingham milk samples below standard, 1899–1930

Source: Liverseege 1932: 196.

As mentioned above, official data reporting on the adulteration of food and drink began in 1877. It was a compilation by officials in the Local Government Board from reports sent in by local authorities. The dataset was incomplete at first and heavily biased by the large numbers of samples collected in a few pro-active cities, such as London, Manchester and Liverpool. Moreover, we should remember the universal resistance of shopkeepers and other food vendors towards this activity. The majority seem to have deployed every stratagem within their power to deflect or undermine the efforts of the inspectors.

But there are samples and there are samples. At first, in the 1870s, the uncertainty about sampling methods was compounded by a lack of consensus about appropriate modes of laboratory analysis. The Victorian art historian and social commentator, John Ruskin, apparently did not approve of the use of microscopes because they challenged the 'mystery of everyday life'.[78] But such doubts about the penetration of laboratory science into the spiritual and organic realms were swept aside in the empirical rush to understand natural variations through the techniques of physics, chemistry and statistics, and then to modulate them within a normative framework.

Once underway, the collection and analysis of commercial milk analytical data seem to have been taken seriously by the scientific community, although they could easily have been dismissed as biased given the overwhelming dominance of corporate sector investment in this area and pivotal role of their laboratory staff in the evolution of dairy science.

The adulteration of milk constituted what Callon calls an 'overflow'.[79] Although in the mid-nineteenth century consumers had a general unease about the state of their food supply, there was no way for them to calculate either the level of watering or the possible effect upon their bodies if there was any contamination. Adulteration was outside the frame of normal economic calculation. Later, once official monitoring was in place and there was at least some chance of cheats being caught and fined, the calculability of the situation changed.

> Through measurement, overflows become calculable. The costs of such overflows become factored into specific economic transactions, in ways the immediate participants may not always be aware … Calculation increases reflexivity about the organization of the market, but it also effects a reduction in the potential space of political conflict.[80]

This industrial-scale creation of data calls to mind Theodore Porter's comment that modern society demands what he called 'trust in numbers'.[81] In a Foucaultian sense, the advent of mass-produced scientific data put power into the hands of

---

78   Davis 2002: 79.
79   Callon 1998: 250–55.
80   Barry 2002: 273.
81   Porter 1995.

company analysts. They were now in a pivotal position in the reduction of risk for the public and for the company balance sheet because, under modernity, risk is a means of eliminating indeterminacy or uncertainty, through calculability and, along with their Local Authority colleagues, they had a virtual monopoly of the truth that was generated. There should be no surprise at all that the first official standards, when adopted in 1901, were those already implemented for years by the large dairy enterprises.[82] They in turn had looked to America for inspiration. New Jersey (1882) and New York (1884) were the first states there to enforce legal compositional standards: 3 per cent fat and 9 per cent solids-not-fat.[83]

A new world and a new nature was visible in this metrology. As Barry and Slater remark, 'metrological practices (such as those associated with, for example, quality control, audit or environmental monitoring, etc.) do not just reflect reality as it is. They create new realities (calculable objects) that can, in turn, be the object of economic calculation'.[84]

The Technicity of Adulteration

For Simondon 'technicity' refers to the achievement of a material presence from an ensemble that includes poietic, cognitive, emotional and political dimensions.[85] This 'bringing forth' can refer to objects and also anti-objects or material perversions such as adulterations, and to ways of thinking about them and representing them. This last point includes etymology.

According to the *Oxford English Dictionary*, the earliest evidence of the word 'adulteration', in the sense of 'the action of adulterating; corruption or debasement by spurious admixture' stretches back to 1506. 'Falsify', the alternate word nowadays more popular in France and much of Europe, meaning 'to alter fraudulently; to introduce false matter into', was even earlier at 1502. Even before the emergence of this vocabulary, the concept of fairness in retail food purchases had been fundamental to public order, as shown in the Assize of Bread in London, which commenced in 1266 to adjudicate honesty in weight and price.[86]

But what means were used to adulterate milk? We can identify six (Table 7.4) and this sub-section will be devoted to understanding their application and the anti-adulteration measures adopted by the authorities.

---

82   Russell Committee 1896, Q.358, F.J. Lloyd.
83   Aikman 1899. The Belgian state laboratories used guideline standards from 1895 onwards. Wenlock Committee 1901: 386.
84   Barry and Slater 2002: 181.
85   Mackenzie 2006, 208.
86   It was abolished in 1815.

Table 7.4   The types of milk adulteration in the nineteenth and early twentieth centuries

- Watering
- Extraction of cream
- 'Toning' with separated or skim milk
- Addition of thickening or frothing agents
- Colorants, usually yellow
- Chemicals added as preservatives

Neither the 1860 nor 1872 Sale of Food and Drugs Acts were really concerned with the types of adulteration shown in Table 7.4. Their main preoccupation was with the use of poisonous chemicals in food, although none were ever much found in milk. It was the 1875 Act that introduced the notion that the sale of goods might be to the prejudice of the purchaser without actually injuring her health. The wording used was 'not of the nature, quality, and substance demanded' and this remained at the conceptual core of discussions about adulteration in Britain for the next 100 years.

Small traders would 'set' the milk for 12 to 24 hours in shallow pans with a view to skimming the rising cream, but this was an unpredictable process leaving up to half of the fat still in the milk. The introduction of mechanical separators in the 1870s and 1880s greatly increased both the speed and efficiency of abstraction and therefore put a weapon in the hands of the adulterator. The year 1875 was a threshold because the skimming of cream became an offence for first time. Previous to this only the addition of a substance, usually water, was actionable.

Based on their trade's experience with the law, dealers knew by the 1870s that blatant misdemeanours would be punished but that they stood a good chance of escaping with a conditional discharge or nominal fine if they scientifically 'toned' whole milk with skim milk or condensed milk.[87] This was hard to detect[88] and few magistrates were willing to convict when there was reasonable doubt.[89] One Medical Officer of Health commented in 1908 that

> The amount of detectable adulteration of milk is now less than at any preceding period, but it is very doubtful if adulteration is actually less. At one time it was the custom to add water in considerable quantities, but there is reason to think

---

[87] Toning was already common in the 1850s. Anon. 1856b: 53. For the later period, see Foster Committee 1894: Q.601, Bannister, Wenlock Committee 1901: Q.586.

[88] Methods of detection were proposed but, due to expense, could only be used selectively. Faber 1889, Richmond and Boseley 1893b, 1897.

[89] Many magistrates were unwilling to convict their status equals, and others simply refused to accept the spirit of the anti-adulteration legislation by adopting their own informal standards for milk composition. The lenience and inconsistency of magistrates was a major factor in the persistence of food fraud. Paulus 1974: 108–10, 124.

that the practice has been superseded by adulteration with separated milk which amounts to the same thing as depriving the milk of its fat or cream.[90]

To hide the 'thin' or 'bluish' appearance of watered milk, some dairymen restored its 'natural' look by the addition of various substances. In America lead chromate was used to restore a rich yellow colour to watered milk.[91] In Britain the vegetable dye annatto was the colorant of choice and by 1871 advertisements for it were appearing in the dairy trade press; later various aniline and sulphonated azo-dyes were also used.[92] The synthetic chemical dyes were particularly strong and could therefore be used in cheap dilutions of 1:200,000. By the turn of the century about half of the milk sold in towns seems to have been coloured, a proportion which was higher in London if the data from Lewisham in Table 7.5 are representative.[93] A survey by the Central Chamber of Commerce in 1900 showed that, of those traders who used colouring, 65 per cent preferred annatto, the rest using brands such as 'Silver Churn' and 'Cowslip Colouring'.[94] London consumers were said to be especially keen on a rich, yellowish look to their milk.[95] J.A. Hattersley, Managing Director of the Aylesbury Dairy Co., reported in 1901 that he had been trying for years to stop the practice but had received complaints from his customers.[96] The artificial colouring of milk was forbidden in 1912 under the Public Health (Milk and Cream) Regulations,[97] a prohibition that was reinforced by Section 4 of the Milk and Dairies (Amendment) Act of 1922.

The other dairy foods were also routinely coloured: cheese and butter, and their surrogates margarine and margarine cheese. Imported cheese set the trend and English producers were able to earn £10 more per ton if they could match its uniformity of flavour and distinctive colouration. The yellowness of butter was enhanced to give the impression of 'richness', although there seem to have been regional variations in demand for this characteristic.[98]

Detecting colouring in foods does not seem to have been a priority at any time in our period. It rarely appears in analysts' reports and in the ontological sense we

---

90 Medical Officer of Health to the Borough of Islington, *Annual Report* 1908.

91 McKone 1991: 30.

92 Scholefield Committee 1854–55: Q.5, Hassall, Q.4215, Postgate, Foster Committee 1894: Q.1351, Russell Committee 1896: Q.260. Annatto was added at the rate of one teaspoon to two gallons of milk. Anon. 1894: 180.

93 Foster Committee 1894: Q.1383. The *Annual Reports* of the Medical Officer of Health for the Metropolitan Borough of Lewisham show an incidence of colouring matter in milk which peaked at 77.4 per cent of samples in 1906, falling away to 21.9 per cent in 1913.

94 Maxwell Committee 1902: Q.4458.

95 Foster Committee 1894: QQ.2542–6.

96 Maxwell Committee 1902: Q.5861.

97 Made by the Local Government Board under the Public Health Act (1896) and the Public Health (Regulations as to Food) Act (1907).

98 Foster Committee 1894: Q.3653.

Table 7.5   Colouring matter in Lewisham milk samples

Year	%	Year	%
1903	60.0	1909	68.1
1904	48.6	1910	63.0
1905	67.3	1911	65.0
1906	77.4	1912	30.9
1907	66.4	1913	21.9
1908	51.5	1914	34.3

Source: Medical Officer of Health to Metropolitan Borough of Lewisham *Annual Reports.*

can say that it was not much of a fraud. One reason was the cost and complexity of the laboratory tests.[99] Although the intent may have been one of deception, our performative definition of adulteration means that a value judgement is not appropriate. It is, after all, still the case in the twenty-first century that British consumers will not buy tinned, processed peas unless a bright green dye is added to make them look palatable.[100] And, of course, other food additives continue to be used extensively, with our complicit assent.

The keeping quality of milk was another major issue for the trade in the century before the Second World War. As we will see in Chapter 8, this was, firstly, because of the poor sanitary conditions of production. Milk contaminated with dirt and dust was unlikely to last long in a fresh state because of its bacteriological load. Second, even clean milk was at risk of souring because of transport arrangements that meant supplies from the furthest stations might take a full day, or more, to reach their destination. In hot weather this meant that consumers received their milk in a deteriorated condition.

A priority for most cowkeepers was to cool their milk as quickly as possible. Recommended temperatures varied: 12°C seemed realistic to Fleischmann in 1893 but lower figures were commonly recommended.[101] Some farmers placed their milk cans in a pond or stream to achieve this but technologies gradually became available, the first of which was the Lawrence capillary refrigerator, which was introduced in a small way in 1872.[102] It was an adaptation from the brewing industry where a similar device was used as a means of cooling wort. This 'British cooler', as it was sometimes called, was such a simple yet functional design that it remained the industry standard until the 1960s when temperature-controlled bulk tanks took over.[103] It consisted of

99   Richmond 1920: 212–14, Blyth and Cox 1927: 257–9.
100   The dyes Tartrazine (E102) and Green S (E142) are still used for this purpose.
101   Fleischmann 1896: 60.
102   Roberts 1872: 477, Knight 1884: 603, Fussell 1966: 178–80.
103   Hall 1951: 10.

a pair of corrugated metal sheets (usually tinned copper) between which the cooling medium circulates. The pair of sheets is surmounted by a feed trough with a series of perforated outlets. The warm milk is either tipped from a pale or piped into the feed trough and trickles through the perforations over the exposed services of the cooled plates. The cooled milk flows by gravity into a lower trough having a single outlet, which may be closed by a plug.[104]

One problem was that the early water-tube surface coolers used well or mains water, which was rarely cold enough, and at times in summer was over 15°C.[105] Another was the serious risk of infection if they were not kept clean. The alternatives were expensive. Ice was in use to a limited extent, imported from Norway, but more efficient from 1877 onwards was the Bell-Coleman compressed air machine, which was gradually adopted in butter factory cold stores. By 1950 the majority of dairy farmers were cooling in one way or another (Table 7.6). The alternatives at that date were: surface coolers; brine coolers (from the 1890s);[106] direct expansion coolers; plate heat exchangers (factories only); churn immersion coolers; spray coolers; and coil immersion coolers.

Table 7.6   Results of a survey on milk cooling by the Milk Marketing Boards and the Ministry of Agriculture, 1950 (percentage of farmers)

Type of cooler	England and Wales	Scotland
Refrigerated	14.8	48.9
Non-refrigerated	53.1	43.5
No cooler	32.1	7.6

Note: Refrigerated coolers used brine, ammonia or some other coolant.

Source: Hall 1951: 10.

The Milk and Dairies Order (1926) required the cooling of milk to within 2.7°C of any water available on the farm, and from the mid-1930s Milk Marketing Board contracts required cooling to 15°C.[107] But most farms lacked a mains water supply and had to rely on well water, rainwater, or surface supplies in ponds or streams. Several official reports pointed this out in the 1930s and the problem was

---

104   Brissenden and Shepheard 1965/66: 60.
105   Cameron Brown 1936: 5.
106   Douglas 1904.
107   The Milk and Dairies Order was simply ignored in those parts of the country where farm water supplies were problematic. Cooling was treated as an aspiration, not an instruction.

analysed by the wartime National Farm Survey.[108] Sixty per cent of farms were said to have been hampered by this problem, and 'shortage of water has been the chief single cause of the unsatisfactory quality of milk in the cases investigated'.[109]

Apart from the cosmetic aspect of milk, chemicals were also added to keep it 'sweet' as long as possible. The earliest record of this is from Paris in the 1820s, where bicarbonate of soda was used to neutralize the lactic acid which powered the souring process.[110] In mid-nineteenth century Britain such 'preservatives' came to be regarded as essential in a market increasingly dominated from the 1860s by distance and speed because they delayed the process of decomposition. The supply of milk to a rapidly urbanizing society meant the drawing of supplies from further and further afield. The urban cowsheds which had once been dominant were unable to meet the surge in demand and it was necessary to push the 'milkshed' outwards, at first to dozens, and later to hundreds of kilometres. This acceleration was facilitated, first, by the railways, and then by the addition to the milk of chemical 'preservatives'.

The extent of chemical use is not easy to quantify. In 1886 the *Metropolitan Dairyman's Directory* listed the names of five manufacturers in London, but we have no idea of their individual output. Fulwood's patent 'Antiseptic' was the market leader in the 1870s, but dairymen seem to have been experimenting widely, with compounds based on borax and/or boric acid gaining many adherents.[111] A British patent for a boric milk preservative had been taken out in 1869, but it took several years to make an impact. In 1876 'E.K.', or 'Kimberley's Food Preservative', was advertised in *The Dairyman* on free trial, and by mid-1877 it had 'found its way into some of our largest dairies, where [it was] in constant use, more or less throughout the year'.[112] The year 1876 also saw the launch of another borax product, 'Glacialine', which, along with its rivals 'Preservitas' and 'Arcticanus', held a large portion of the dairy preservatives market for the next thirty years.[113]

Other chemicals were also used. In 1871 Alfred Wanklyn proposed a solution of two parts glycerine to one hundred of milk and in 1878 Burgoyne, Burbidge & Co. marketed salicylic acid.[114] The latter was especially effective but proved

---

108   NA: JV 6/67, Water supply on farms 1933–1950, Ministry of Agriculture and Fisheries, Economics and Statistics Division (September 1945), Reports on the Economic Position of Agriculture (Farm Survey) Number 23: Farm Water and Electricity supplies.

109   NA: MAF 38/475, 'Farm Survey. Supplementary report on water supplies and electricity, Part I'.

110   Barruel 1829.

111   Wigner 1879.

112   *The Dairyman*, 2(4) 1871: 56. *The Dairyman, 4*, 1876. Kimberley's Food Preservative in 1876 cost three guineas for a 12 pound jar, reduced in 1877 to 33 shillings for seven pounds or two shillings for a gallon of fluid.

113   Glacialine was marketed by the Antitropic Co. of Stamford Street, London S.E..

114   Patent no. 1,861, 1871.

to be uncompetitively expensive by comparison with borax and it also had an unpleasant taste in doses over 10 grains per quart (75 centigrams per litre).[115]

Over 40 per cent of the samples examined by E.G. Clayton from 1893 to 1898 contained some boric acid, as did 27 per cent of the milk analysed in Islington in 1899.[116] In evidence to the Maxwell Committee investigating the use of preservatives in food, Leonard Bosely, estimated that about half of London dairymen used one or more compounds,[117] and the same Committee heard evidence from Alfred de Hailes that 'the milk trade as it is at present ... cannot be carried on without the use of preservatives'.[118]

Advertisements in the trade press of the day show the wide range of preparations available. In 1898 Wallace counted 21 brands of milk preservatives in four dairy papers and in 1901 the Central Chamber of Agriculture prepared a list of 18 then used in milk.[119] Ever eager to experiment, dairymen seem to have been willing to try not only the chief patent compounds, generally based on boric acid or formalin, but also the new chemicals developed to confound the tests of public analysts. Thus G.A. Stokes in 1912 tested a brand named 'Mystin' and found it to consist of a mixture of sodium nitrate and formaldehyde, the former interfering with Hehner's test for formaldehyde.[120] Dextrin, as a dilute solution in milk was first detected by A.W. Stokes in 1897, while Richmond noted the use of a mixture of formic acid and glucose, with a little ester to disguise the formic smell.[121]

Manufacturers' instructions were given on all of these products for the optimum amount to use but there were many dairymen in the marketing chain and it is possible that preservatives may have been added more than once. These chemicals, although they inhibited the visible souring of milk, killed microbes only selectively and therefore may have lulled the unsuspecting consumer into a false sense of security.[122] The physiological effects of treated milk were not clear at first. The first medical article on the subject was published in the *Lancet* in January 1877. Later there was evidence from France of the injurious effect of salicylic acid,[123] and several witnesses called by the Maxwell Committee testified that boric acid exacerbated diarrhoea in infants,[124] especially when the milk was

---

115 Grotenfelt 1894: 145–6.

116 E.G. Clayton, *Report of the Medical Officer of Health to the Islington Vestry, 1899.*

117 Boseley was analyst to Keiller and Sons. Maxwell Committee 1902: Q.956.

118 De Hailes was analyst for the Dairy Trade Protection Society. Maxwell Committee 1902: Q.3928.

119 Wallace 1898: 79, Maxwell Committee 1902: Q.4458.

120 *The Analyst*, 37, 1912: 178.

121 *The Analyst* 22, 1897: 321, 33, 1908: 116.

122 Hill 1899: 527–38.

123 Swinthinbank and Newman 1903: 532.

124 Maxwell Committee 1902: QQ.1436, 2374, 5492–4, 6787.

heavily impregnated with the chemical – as much as 80 grains of boric acid per pint in one extreme example.[125] The Committee's report was very critical:

> the use of preservatives in the milk traffic ... may be relied on to protect those engaged therein against the immediate results of neglect of scrupulous cleanliness. Under the influence of these preservatives milk may be exposed without sensible injury to conditions which otherwise would render it unsaleable. It may remain sweet to taste and smell yet have incorporated disease-germs of various kinds, whereof the activity may be suspended for a time by the action of the preservative, but may be resumed before the milk is digested.[126]

Several large dairy firms in the London trade made an early stand against chemical additives. The Aylesbury Dairy Co. stipulated in its contracts that no preservatives were to be used, but Welford's, another prestigious London company, implied their own guilt by refusing to give evidence to the 1901 enquiry.[127] The smaller wholesale and retail dairymen were not fully in control of the condition of the milk sent to them and were in a difficult situation. Their consigning producer might have been on a branch line with an inconvenient train timetable, obliged to mix the cows' morning and evening 'meals', and therefore tempted down the preservative route.[128] Inappropriate rolling stock did not help.[129] Milk might take a day or more to reach the city and in hot weather this inevitably led to deterioration without some form of intervention. Neither cooling nor heat treatment were yet common, and the use of chemicals was a cheap and essential part of the system. In Professor Blyth's view:

> In the height of summer I should imagine that quite a third of the milk supply would be spoilt before it reached the metropolis, unless some preservative was used.[130]

Although tests for preservatives were available, and Somerset House was using them from 1894 onwards, the addition of chemicals to milk did not become a political health issue.[131] The chief concern seems to have been that it constituted

---

125   Maxwell Committee 1902: QQ.3439–40. The presumptive limits in the Local Government Board's 1906 circular were 1:40,000 of formalin, 1:100,000 of formaldehyde, and 1:1,750 of boric acid (40 grains/gallon). Trade practice had varied between 1: 625 and 1: 7300 for boric compounds and 1:32,000 and 1:82,000 for formaldehyde. Foulerton 1899: 1427, Rideal and Foulerton 1899: 9, Savage 1912: 319–20, 386–7.

126   Maxwell Committee 1902, Report: 605.

127   Maxwell Committee 1902, Report: 605.

128   Maxwell Committee 1902, 1902: QQ.3481, 3948.

129   Atkins 1978: 211.

130   Maxwell Committee 1902: Q. 3447.

131   Foster Committee 1894: Q.1684, Bannister. For methods of detection, see Richmond 3e 1920: 205–12, Blyth and Cox 1927: 251–7.

a form of cheating, rather than its toxicity.[132] The Russell Committee in 1896 and the Wenlock Committee in 1901 both recommended that preservatives should be banned, and the Local Government Board issued a circular to Local Authorities in 1906 suggesting action under section 6 of the Food and Drugs Act (1875), which forbade the sale 'to the prejudice of the purchaser of any article ... which is not of the nature, substance, and quality demanded'.[133] But formal action was delayed until the Public Health (Milk and Cream) Regulations of 1912 and for cream even later, 1928.[134]

Time, Space and Adulteration

Who was responsible for adulteration? In the three decades at the end of the nineteenth century it seems to have been common where the retailer was making casual sales and where the purchaser was unlikely to have been connected in any way with official sampling. This bore out the mid- century observation by the journalist Henry Mayhew that adulteration was especially common amongst informal sector dealers, such as those who worked in London's markets or in open spaces like Hampstead Heath on Sundays.[135] The latter was doubly illegal because of restrictions on Sunday trading. Watered, skimmed milk was sold by short measure as 'whole' milk. According to contemporary police accounts, many of these casual street retailers in the East End of London came from the various immigrant communities, for whom regular employment was difficult to find.[136] It was a similar story in other large cities. In 1905, 9 per cent of the Birmingham weekday samples were adulterated, but 24 per cent on Sundays.[137]

The suppliers of Poor Law institutions were also notorious for their cheating. The Guardians usually accepted the lowest tenders, for reasons of economy, and this was incompatible with the supply of genuine, whole milk. In 1856 Dr J. Challice found that the Bermondsey Guardians had taken a tender of five and a half pence per barn gallon when no other offer had been below one shilling. He concluded, not surprisingly, that 'all the milk was adulterated; it turned sour, and was altogether bad milk'.[138] In 1871 the *Milk Journal* investigated the Shoreditch and Holborn Unions, both of which were supplied by the reputable Express Country Milk Co. Analysis showed the milk to contain 47 and 29 per cent of added

---

132   Anon. 1897.

133   Local Government Board Circular, July 1906.

134   Under the Food and Drugs (Adulteration) Act 1928.

135   Mayhew 1861: 191–3.

136   Sweeney 1904: 313. Care is needed in reading this autobiography of 'Sweeney of the Yard'. His bombastic and self-serving account is at times racist in tone.

137   Liverseege 1932: 13.

138   Scholefield Committee 1856: QQ.1485–6.

water respectively.[139] Such was the public interest in this scandal that questions were asked in parliament. In the same year an official enquiry into London's workhouses arranged for samples to be sent to analysts Alfred Wanklyn and Henry Letheby. Of 57 institutions visited, very few received even 'best skimmed milk'.[140] In 1872 the *British Medical Journal* found that the milk supplied to the eight major London hospitals had, on average, 63 per cent of its fat extracted and 23 per cent of water added.

Amongst fixed retail outlets the highest incidence of chronic adulteration was in the small general provisions shops in slum areas. These 'corner shops' provided 'cheap' goods to people who were unable to exercise much choice. Many of the customers were in debt to the shopkeeper and therefore in a weak position to complain. Only small quantities of milk were purchased, for a farthing or a halfpenny at a time. Karl Marx was not the only contemporary to suggest that the poor were the worst sufferers from these frauds.[141]

A counterpoint to the 'poor suffer more' argument was the commonly-made observation that some subcategories of milk were a boon to the poor. As we have noted, butter milk was popular in some regions, and, because of its cheapness and ease of storage, condensed, separated milk became a common mode of consumption in the late century slums of the large cities. Skim milk was also popular in London where it was available at a fraction of the price of the 'whole' milk. There were even a number of well-established and respectable suppliers who argued that skim milk was a healthy, nutritious product that should be made as widely available as possible. Lord Vernon was the best known example, bringing good quality skim milk from Derbyshire to special depots in London.[142]

The third cluster of comments concerns the changing spatial pattern of adulteration through time. We must proceed with caution here because, at least before the official regulations on milk composition in 1901, there was a lot of debate as to what constituted zero adulteration. Analysts differed on this crucial point and at first the officers of the Local Government Board themselves expressed doubts about comparability of their own data between districts.[143] Quite apart from their opinion on the thresholds of genuine milk, the local authority analysts varied in their laboratory resources, to the extent that we can say that civic expertise was rationed.

---

139   *The Milk Journal*, 1 March 1871: 57.

140   *Report of the Superintendent of Contracts, Admiralty, Relative to the System of Supply of Provisions and Stores for the Workhouses of the Metropolis*, P.P. 1872 (275) li.619, *The Milk Journal*, 1 August 1872, Johnston 1985: 122.

141   Marx 1867/2007: 274, Foster Committee 1895: Q.578.

142   Rew 1892: 280.

143   *Eleventh Annual Report of the Local Government Board, 1881–82*, P.P. 1882 (C.3337) xxx, Pt.1.105.

The adulteration of milk seems to have been a more serious problem in London than elsewhere in England and Wales.[144] Table 7.7 shows separate series for the capital and the provinces and, although there were simultaneous fluctuations, their absolute ranges are varied and the periods of accelerated change are lagged. The most rapid decline in the provincial adulteration rate took place between 1883/4 and 1885/6, a fall of over 5 per cent in just two years, while the marked fall in the metropolitan data, of over 10 per cent in four years, was between 1893/4 and 1897/8. The period of greatest divergence was in the late 1880s and early 1890s, and the two curves did not meet until 1907/8, by which time the national rate was just over 11 per cent.

Apart from London, certain other cities were infamous for adulteration, for instance Dublin. Charles Cameron, who was Medical Officer of Health there, reminisced that before 1860 it 'had attained scandalous proportions ... 150 per cent of water being frequently added ... adulteration was universal; there was no such thing as getting good milk in Dublin at all'.[145]

Unfortunately we cannot confirm the implication that the provincial authorities were either more efficient or their dairymen more righteous, because there is an information gap. In 1891, for instance, it was reported that, despite the adoption of regulations by most councils, the law was virtually inoperative in 22 counties, 19 of the largest towns and two metropolitan districts.[146] The Local Government Board sent out a circular to stimulate action, but little happened in many jurisdictions until the 1899 Act insisted on sampling. This injunction was subject to the scrutiny of the Local Government Board and the Board of Agriculture who, in the instance of a default, were empowered to take samples at the expense of the local authority. The upward trend in provincial adulteration from 1894 to the First World War (Table 7.7) is probably the result of this improved monitoring system picking up previously undetected fraud.[147]

There was surprisingly little difference between the provincial urban (borough) and rural (county) experiences (Table 7.7). Neither managed to reduce their adulteration rates by much from 1885–1891 to 1909–1911, whereas London progressed significantly. Some towns were more successful than others, but there seems to have been no clear correlation with the type or size of city (Table 7.8).[148] The independence of action so jealously guarded by Victorian local politicians and administrators resulted in an almost anarchic complexity of legislative implementation for social betterment. In effect, local whim and political context were at least as important as the 'diffusion' of pangs of social conscience through space or down the urban hierarchy.

---

144   Hassard 1905: 74.
145   Read Committee 1874: Q.4593.
146   Twenty-Sixth Annual Report of the Local Government *Board,* 1896–7, PP 1897 (C.8583) xxxvi.138.
147   *British Food Journal,* 8, 1906: 41.
148   Foster Committee 1894: Q.78.

Table 7.7   Adulteration of milk, 1877–1913, by type of authority

	Population per sample	Adulteration		Population per sample	Adulteration
England and Wales			*London*		
1877–80	4191	21.62	1877–80	2202	25.59
1881–86	2099	17.12	1881–86	1785	22.77
1887–90	2325	13.18	1887–90	1329	21.59
1891–95	1911	12.72	1891–95	1180	21.89
1896–1900	1406	10.63	1896–1900	815	14.89
1901–05	999	11.02	1901–05	481	13.26
1906–10	522	11.21	1906–10	153	11.48
1911–13	694	11.11	1911–13	344	9.80
Admin counties			*13 cities*		
1877–80	3174	20.13	1887–90	893	8.89
1881–86	2201	15.28	1891–95	948	8.36
1887–90	4970	12.49	1896–1900	926	9.42
1891–95	3207	11.18	1901–05	843	10.30
1896–1900	2103	8.83	1906–10	607	11.13
1901–05	1585	9.35	1911–13	650	10.87
1906–10	1086	10.26	*Other boroughs*		
1911–13	972	11.88	1887–90	2524	9.45
			1891–95	2007	9.81
			1896–1900	1396	9.64
			1901–05	927	11.08
			1906–10	687	11.94
			1911–13	645	11.45

Source: Calculated from Local Government Board, *Annual Reports*.

Table 7.8   A comparison of milk adulteration in 14 cities

	1879–81		1889–91		1899–1901		1909–11	
	Samples/10,000 population	Adulteration (%)	Samples/10,000 population	Adulteration (%)	Samples/10,000 population	Adulteration (%)	Samples/10,000 population	Adulteration (%)
Birmingham	1.7	53.2	4.9	19.1	8.0	20.7	18.2	10.8
Bradford	2.3	13.4	6.1	7.8	6.6	2.7	16.9	4.5
Bristol	9.3	22.4	8.2	8.4	9.5	7.6	18.4	13.6
Hull	–	–	10.0	5.5	6.7	7.7	20.2	10.4
Leeds	1.1	11.0	4.8	8.8	9.6	10.6	10.9	28.8
Leicester	–	–	5.3	11.3	7.2	3.7	6.8	5.9
Liverpool	3.8	17.3	12.6	12.6	13.3	12.1	12.9	15.1
London	4.5	28.8	7.2	22.2	13.7	14.9	29.1	9.8
Manchester	3.0	27.6	24.2	6.6	16.7	2.7	16.7	4.4
Newcastle	–	–	4.0	9.3	2.6	11.5	17.1	10.7
Nottingham	–	–	2.1	15.8	4.9	23.8	11.2	11.9
Portsmouth	–	–	7.0	14.1	5.5	23.0	22.4	7.7
Salford	21.4	23.7	31.9	3.2	14.6	4.8	17.1	4.4
Sheffield	1.3	12.5	1.1	9.5	5.0	16.1	10.5	9.6

Source: Local Government Board, *Annual Reports.*

A final point is that the publicity surrounding the Sale of Food and Drugs Acts of 1872, 1875 and 1879 was probably responsible for a declining incidence of fraud in the 1870s, and in the 1880s a more general and effective legal enforcement took over. It is unfortunate that the Local Government Board data series does not begin until 1877, because by then some of the amelioration had already taken place.[149] The somewhat over-zealous action of a few analysts, coupled with a general public reaction against the abuses uncovered by both private and government enquiries, had created a new psychological context in which most dairymen and retailers felt less inclined to take risks. State intervention helped to puncture the resignation with which consumers fatalistically accepted watered milk. From the 1870s the 'image' of milk began to improve and this must have contributed to the documented increase in consumption. Some dealers had to continue adulterating milk in order simply to stay in business, but others found they could profit from the premium on fresh, 'healthy', whole milk.

Conclusion

Pure milk appears to be the exception rather than the rule in London.[150]

Given the Victorians' wariness about the sanitary environment in their cities, it is hardly surprising that they came to romanticize rural landscapes as repositories of healthiness and true values. Our contention is that the stinking drains and cramped housing of large industrial cities were not the only source of this feeling. The poor quality of some foods was also contributory.

The substantial noise that was made about the adulteration of food in the nineteenth and early twentieth centuries was largely about milk and dairy products generally, with a few sideshows such as the strength of spirits. Milk was significant because of the difficulty of comparing its composition to a 'natural norm' and might even be said to represent Victorian and Edwardian anxiety about the uncertainty of nature's intelligibility set against the contemporary urge to believe in the capability of science and technology to solve this problem. So powerful was the belief in universal cheating by traders at all stages from farmers to delivery men, that analysts and the judicial system identified and punished adulteration that, in truth, was probably within the bounds of the experimental error of the laboratory methods used. Our views have since changed so much that we have now legalized many of the practices that were regarded then, not only as fraudulent, but also as morally repugnant.

---

149   For example, the statement by George Barham of Express Dairies to the Read Committee (Q.2508) that the Sale of Food and Drugs Acts (1860 and 1872) had 'very much improved the supply of milk to London'.

150   Wigner 1883: 243.

By 1960 it was generally accepted that informal changes in milk composition due to deliberate adulteration had become extremely rare.[151] There is no doubt that dairy capitalists had played an important part in improvements in a range of quality criteria for the basic milk supply.[152] Since the structure of the trade made it difficult for them to compete on retail price, improving quality to the consumer was the main basis on which they could increase their turnover. Those dairy companies that were large enough to have negotiating muscle, backed up with laboratory facilities, were in a strong position to impose contractual obligations on their suppliers as to minimum standards of milk composition and cleanliness of production. In the last two decades of the nineteenth century they came to an informal decision that 3 per cent butterfat and 9 per cent solids-not-fat were appropriate threshold standards and they seem to have been successful in squeezing out of the system much of the adulteration that had previously been associated with farmers and wholesalers, in addition to putting compositional quality on the agenda. A Simondian 'collective' emerged here from the novel power of capitalist agro-industry to use the constituents and qualities of milk to its advantage in ways that previously had been beyond the reach of smaller-scale operatives.

---

151   Cook Committee 1960: 6.
152   Atkins 1984a.

Chapter 8
Legal Ontologies and the Performative Realm of the Law

Introduction

For historians of science, the courts beckon.[1]

A key feature of the ontogenesis of commodities is the contestation of the knowledge that is deployed in their identification, objectification, measurement and classification. With commodities this is crucial because of the commercial imperative of marketing a consensus vision that can be bought into by potential consumers. With milk in the nineteenth and early twentieth centuries, there was a hybrid forum of debate in which views from all of the various interest groups were heard, but it lacked what Rip has called a 'forceful focus'.[2] One reason was that some in the industry preferred an indefinite commodity that maximized their opportunity for profits at the margins of what was acceptable. Another reason was that, although voices in civil society were raised in favour change, the slow crystallization of their intent meant that lobbying of government was ineffective until after the First World War. Then, as we saw in Chapter 3, the lack of a suitable means of compositional measurement, and of a scientific conceptual context, undermined the desire, on the part of hygienists, to find an enforceable set of natural limits. Chapter 4 then illustrated the contestation implicit in expertise and opened up another dimension of complication for those who wished to specify legally-binding compositional thresholds. Chapter 7 coloured in the detail of the physical act of adulteration, as understood and measured by contemporaries. But our discussion so far leaves a substantial problem hanging in the air. How was the law to enforce its vision of a whole and natural commodity?

The purpose of the present chapter is to investigate the potential of the legal sphere to provide the necessary forceful focus. A reasonable expectation might be that the law played a constitutive role in our commodity ontogenesis.[3] Not only did it implement the statutes but it also moulded their influence through judicial interpretations that at times stretched the spirit of the parliamentary measures to the limit. This story runs from the magistrates' courts right up to the Court of Appeal. But there were problems. First, remember that the Schrumpf case had to be settled

---

1   Burnett 2007: 314.
2   Callon et al. 2001, Rip 2003.
3   Sorensen 1982.

by an instrument because no-one had observed the accused adding water to his milk. Second, analysts did not have a chemical signature by which to distinguish the natural water in milk from added water.[4] Third, given the general adherence in Britain to the idea of natural milk, right up to the 1990s, the only solution was to compare samples of suspicious milks with a model of what was thought to be natural. Since finding consensus on the latter was exceptionally difficult, a fourth alternative, as adopted in some countries, was to allow the manipulation of milk to a predetermined standard.

The present chapter will carry forward the third alternative, the British consensus, into the realm of the law. One possible narrative is in terms of the formulation of the legislation, its passage through parliament, and the inevitable lobbying back and forth as the various vested interests jostled for influence. I will reserve a discussion along these lines to Chapter 10 when we look at the material politics of clean milk, and the remaining milk politics will form the core of my next book. The style of the present chapter will instead be to describe the legal context within which the will of parliament was actioned, an element of the implementation of food quality that has been neglected in previous work. First, I will give a brief summary of the Sale of Food and Drugs Acts. This will be followed, second, by an investigation of what I call legal ontologies, in other words the conceptual approach of the law to this new misdemeanour. This includes speculation about difficulties in distinguishing between adulteration and free market product innovation. Third, the law of contract will be explored in as much as the concept of warrantied quality was the outcome of network connexions. Fourth, the presentation of a number of appeal cases will serve the purpose of subverting our assumptions about legal definitions of natural foodstuffs. Fifth, there will be a discussion of the self-image of the milk trade vis-à-vis the law and, finally, I will conclude with some comments about a way to relate the jurisprudence of Oliver Wendell Holmes Jr to the technological philosophy of Gilbert Simondon.

Legislation and Regulation

> Law ... has no solidity of its own, it merely adds 'legitimacy' to the hidden strength of power.[5]

Filby, Burnett, Paulus, French and Phillips, and others have provided narrative histories of the Sale of Food and Drugs Acts of the nineteenth century.[6] These

---

    4   Having said that, milk is relatively free of nitrates whereas water nearly always contains traces. Nitrates show up in the Gerber test, the chemistry of which produces a different colour from normal milk.

    5   Latour 2003: 29.

    6   Filby 1934, Burnett 1958, 1966, Paulus 1974, French and Phillips 2000.

statutes were enacted in 1860, 1872, 1875, 1879, and 1899.[7] Paulus in particular is interesting for her account of the back story of the clash of interests between the anti-adulteration lobby of middle class consumers and medical campaigners on the one hand and powerful commercial interests on the other, represented by journals such as *The Chemist and Druggist* and *The Grocer*.[8] To this she adds speculation about legislative cycles that were, she claims, to a degree dependent upon the political strength and motivation of successive administrations.[9] She then investigates the proceedings of the parliamentary Select Committees on adulteration, chaired successively by William Scholefield (1854–1856), Clare Sewell Read (1874), George Sclater-Booth (1879), Sir Walter Foster (1894–1895), and Thomas Russell (1896). These were published in the parliamentary blue books, including full transcripts of the evidence collected. The somewhat quirky selection of witnesses in each case ensured that these were not balanced or disinterested enquiries but they do provide the historian with some wonderfully detailed raw material.

The politics of Scholefield was about parliament's felt need to react to the scandal generated by the *Lancet*'s revelations about food adulteration in London. Hassall's laboratory observations had achieved the spectacular double effect of media exposure followed by the official enquiry. But, despite its airing of the issue, mostly in the sense of additives to food and drink that were thought to be somehow 'injurious to health', the Scholefield Committee had no immediate effect. Maybe this is not too surprising when one reads the tone of uncertainty in Scholefield's report:

> It is impossible to frame any enactment on this subject which shall rely on strict definitions. The object of the law is to strike at fraud, and wherever a fraudulent intention can be proved, there to inflict a penalty. What constitutes fraud must be left to the interpretation of the administrators of the law. Thus mixtures of an innocuous character, made known to the seller, or used for the preservation of the article, cannot be forbidden without danger to the needful freedom of commerce, and ought not to be interpreted as coming within the provisions of a penal law.[10]

Five years later, the 1860 Act was passed, making it an offence to sell any article of food or drink which injurious to health, adulterated or not pure. But this Act was little more than a 'temporary truce' between the opposing lobby groups and its vagueness and lack of teeth meant that it was soon seen on all sides as a

---

7   23 and 24 Vict., c.84, 35 and 36 Vict., c.74, 38 and 39 Vict., c.63, 42 and 43 Vict., c.30, 62 and 63 Vict., c.51.

8   Paulus 1974: 28.

9   Liberals (Palmerstone 1855–58, 1859–65, Russell 1865–66) and Conservatives (Derby 1858–59, 1866–68, Disraeli 1868, 1874–80).

10   Scholefield Committee 1856: 7.

dead letter. Only seven public analysts were appointed in the whole of the United Kingdom and Ireland in the next 12 years, and few of them were active.[11] Only Dr Charles Cameron, who became joint Medical Officer of Health and Public Analyst in Dublin in 1862, managed to secure many convictions.[12] The number of milk samples taken by his department in the next decade surpassed fivefold the total for the rest of the United Kingdom put together.

When at last it came, the 1872 Act was more positively framed, allowing for sampling by local authorities and testing by public analysts, but the definition of adulteration remained problematic. The key offence was now said to be mixing any article of food or drink injurious or adulterating material, which, as far as milk was concerned, omitted the skimming of cream. Any admixtures had to be declared, and some of the bolder hawkers did advertise 'milk and water' in small type on the side of their perambulators.[13] Contemporaries saw this as a loophole in the Act, and a number of other technical faults in the drafting were also soon exposed in the courts.

The 1875 Act that followed was to a certain extent a corrective measure, repealing the previous legislation, and starting again. As a government-sponsored measure rather than arising out of a private member's bill, as had been the case with the previous two Acts, it was intended as a more forthright normative statement about quality, in that manipulators of food and drink were warned that crude adulterations would be punished. The two key points were, first, the creation of an offence of selling 'to the prejudice of the purchaser any article of food or drink which was not of the nature, substance and quality of the article demanded'. This sentence proved to be pivotal in the history of British food quality and became the mantra of state's anti-adulteration effort for the next century. Second, as discussed in Chapter 4, the Laboratory of Inland Revenue at Somerset House was made a 'court of chemical appeal' to which magistrates could despatch disputed samples. There were also a number of minor adjustments, such as making it an offence to refuse to sell a sample to an inspector, and the establishment of the warranty defence. The latter became a major issue in trying to locate responsibility for adulteration and we will return to it later.

The 1879 Amendment Act that followed soon after was necessary due to a Scottish appeal case, Davidson v. McLeod (1877),[14] which seemed to suggest that an inspector could not be prejudiced in purchasing a sample, because s/he was buying it on behalf of the public.[15] The Act smoothed out this wrinkle and, at the same time, a few new powers were added, including the sampling of milk in transit

---

11   Most were summary actions in the magistrates' court.

12   Read Committee 1874: Q.4590.

13   *Twentieth Annual Report of the Local Government Board, 1890–1*, PP 1890–1 (C.6460) xxxiii.156.

14   42 JP 43, 5 Rettie (JC) 1, 3 Coup 511.

15   Foster Committee 1894: Q.15, Preston-Thomas.

at railway stations because of the frequently quoted worry that the servants of the railways or of the wholesalers were adding water on the platform.

Did all of this legislation attain its objectives? By the end of the 1880s bread, beer and tea were virtually free of adulteration, but milk continued to cause concern.[16] The journal *Food and Sanitation* complained in 1894 that

> pure milk in London is at the present-day practically unobtainable, save from the Aylesbury Dairy Company ... the article vended generally as pure milk consists of some seven or eight gallons of separated milk practically deprived of its fat and a gallon of water added to each 20 gallons of genuine milk.[17]

Such a product would have contained an average of 8.5 per cent solids not fat and 2.5 per cent fat and, as we will see later, this would have been approved by the government chemists. However, it compared unfavourably with the Aylesbury Dairy Company's average analysis of its milk at 8.77 per cent and 3.91 respectively.[18] Part of the problem seems to have been a lack of political will at the centre. According to Henry,

> When we began this journal nearly two years ago, it seemed hopeless to arouse Parliament, local authorities, or even the penny dreadful press to the realization of the enormous importance of the subject. No member of the House of Commons knew or cared a rap about the question ....[19]

The Sale of Food and Drugs Act (1899) was productive for milk, adding a number of new features. It became an offence, for instance, to import adulterated or impoverished margarine, butter, milk, cream, condensed, separated or skimmed milk. Second, street vendors of milk now had to have their address displayed on vehicles and receptacles, to reduce the anonymity of this casual trade. Tins of condensed, separated and skimmed milk had to be labelled with a statement of the correct contents in lettering of a sufficient size to be legible. Finally, the Board of Agriculture was given power to make regulations about the composition of milk, cream, butter and cheese. This last provision was quickly followed by the Wenlock Committee and the Sale of Milk Regulations (1901).[20] As we saw in Chapter 5, presumptive standards were made − 3 per cent butterfat content and 8.5 per cent solids not fat − to provide a basis for the prosecution for adulteration. Skimmed and separated milk had to contain a minimum of 9 per cent milk solids, although this was reduced to 8.7 per cent by the Sale of Milk Regulations (1912).

---

16   Collins 1993.
17   *Food and Sanitation*, 12 May, 1894.
18   *Food and Sanitation*, 12 May, 1894.
19   *Food and Sanitation*, 26 May, 1894: 161.
20   See Chapter 4.

The Milk and Dairy (Consolidation) Act (1915) and its Scottish sister Act were both delayed in their implementation by the war and only came into effect in 1925; but the Milk and Dairy (Amendment) Act (1922) made interim arrangements for certified milk and the banning of the colouring of milk and its toning with water or separated milk. After that, the Food and Drugs (Adulteration) Act (1928), the Food and Drugs Act (1938), the Food and Drugs (Milk and Dairies) Act (1944), the Food and Drugs Act (1955), and the Food Act (1984) all made contributions to the cause of milk quality, especially in response to technical and legal developments, but none of them made fundamental changes to the principles that had been established in the last few decades on the nineteenth century.

As we saw above, there were frequent struggles leading up to the passage or rejection of laws on food quality. But receiving the royal assent was, again, just the beginning of contention. There were debates among the various interest groups, as evidenced by their respective trade and professional journals, about what constituted illegal behaviour and how it should be treated. There were also legal representations and judgements in the courts about the application of the law. From time to time these created confusion with novel interpretations of the wording of the acts and raised questions about the very definition of milk and the standards of quality imposed upon it by parliament and by local prosecuting authorities.

Despite the different legal context and legislative timetable, the situation in France with regard to milk was similar. From 1811 onwards the courts had applied the Civil Code in commercial matters.[21] This was the early source of jurisprudence with regard to adulteration and there was technical support from the Conseil de Salubrité, an expert panel comprised of doctors, chemists and engineers, one of whose duties was to analyse samples. Although the Conseil's activity was not great, the attached publicity made fraudsters think twice.[22] Further to this, falsification of food became a specific offence under the law of March 27, 1851 and Article 423 of the French Penal Code, with a conviction meaning a fine and three months to one year in prison.[23] There was a further law in 1855 and a general law against adulteration in 1884. In Paris the medical and chemical press began reporting court cases much earlier than in Britain and there were many more science-based convictions.[24] However, it was the law of 1 August 1905 that was the most important in the history of French action against adulteration, and which proved to be of international significance.[25] This was a general law and there was no specific measure aimed at milk in France until 1935, when at last we can say that

---

21   Stanziani 2006: 66.

22   Scholefield Committee 1854–55: Q.786, Normandy.

23   Chevallier 1854, Anon. 1855.

24   Anon. 1845, Anon. 1846, Anon. 1853, Chevallier 1854, Anon. 1860.

25   Direction Générale de la Concurrence, de la Consommation et de la Répression des Fraudes 2007.

a concerted crack-down was underway against fraud for this commodity.[26] Among various provisions was a ban on the sale of skim milk, although the reasoning was less a matter of scruples about the composition of milk than an attempt to reduce a market oversupply.

Legal Ontologies: Is Adulteration Really a Crime?

> What is adulteration, and what does it mean? It means the lowering of the physique of the nation, the poisoning of the people, the deterioration of our constitutions, and morally a fraud practised by the seller on the buyer – a cheating to which we have become so callous that it has hardened our conscience for honesty in bigger things.[27]

The view of some economists is that there is no absolute category that can be called 'adulteration', only deviations from norms established in laws that are socially constructed.[28] By the same token, ethical considerations aside, the skills deployed by all of those who manipulate food for gain deserve objective valuation. They are forms of 'substitutionism', by which actors in food systems seek to add value, through any means at their disposal.[29] Because milk production in the late nineteenth and early twentieth centuries was more seasonal than demand for drinking milk, there was a strong temptation for traders to stretch the supply in times of shortage. This was frequently the case in dry summers when forage was scarce.[30] Substituting milk and water or skim milk for whole milk was certainly profitable for those who were able to evade detection.

It could be argued that falsification was a technical accomplishment. Alessandro Stanziani makes this point when he asks the question

> why and by whom are particular technical processes and the ensuing products qualified as adulteration? How is innovation defined and perceived in a market economy? Why does one speak of 'innovation' for manufactures but evoke 'adulteration' when food and drinks are concerned?[31]

He goes on to argue that adulteration is best seen in terms of technical progress in food chemistry, the pressures of urbanization, and the internationalization of

---

26   Guillaume 2006.

27   *The Anti-Adulteration Review*, 1, 1871: 4, quoting Phillips Bevan at the Social Science Congress of 1870.

28   Fine and Leopold 1993: 158, Stanziani 2007a.

29   Goodman et al. 1987, Goodman and Redclift 1991.

30   Read Committee 1874: Q.3557, Farmer.

31   Stanziani 2007a: 376.

the economy. Together these were responsible for providing the opportunities and incentives to modify foods, but the development of rules to define 'adulteration' was a matter of contestation between the various actors and their interests. From this we can see that 'quality is not an objective, ahistorical category but one defined through competing notions ... which vary over time'.[32]

In the same spirit, Fine and Leopold contradict Filby's contention that adulteration peaked in the mid-nineteenth century. They argue instead that the Sale of Food and Drugs Acts can be seen as 'legislation *for* adulteration rather than its erosion'. This is because 'the law has itself laid down the terms and conditions under which one substance may displace another' but 'there is no right or wrong as such'.[33] Thus, 'adulteration' is a technical matter of whether the terms of the law are apparently met. In consequence, the law collapses time and space into a machinery that fabricates norms. Retrospective comparisons become difficult, most of all to the time before the first Act, but also from the most recent to earlier legislation. Judgements within a single jurisdiction ignore regional variations and force a consensus of natural circumstances. Those unable or unwilling to satisfy the court are penalized and, where open to manipulation, nature may, as a result, gradually take on a new identity.

But how can we research these legal ontologies? One approach, according to Tal Golan, is to fill the gap between the histories of science and the law. In particular his concern is with scientific expert testimony.

> Historians of science ignored it because they did not consider courts of law to be important sites of scientific activity before the twentieth century. Historians of law ignored it because they never considered science to be a significant factor in the development of judicial practices and jurisprudence related to evidence.[34]

Golan traces the history of scientific testimony in the common law back to the 1780s and analyses its chequered history in British and American courts in the nineteenth century. One problem over this period was its somewhat miscellaneous nature, an issue that was decisively tackled in the case of Daubert v Merrell Dow Pharmaceuticals Inc, in the American Supreme Court in 1993. The Justices' ruling held up for scrutiny the system of expert testimony in common law jurisdictions such as the United States and Great Britain, and found it wanting. Their conclusion was that insufficient quality control has been exercised over the years in who qualifies to be called to court as an expert, and that a more rigorous test should be applied in future.[35]

---

32 Stanziani 2007a: 377.
33 Fine and Leopold 1993: 158.
34 Golan 2004: 1.
35 Edmond and Mercer 2004.

Bruno Latour has explored this 'abyss that separates law from science'.[36] His intriguing comparative ethnology of a laboratory and the Conseil d'État in Paris highlighted the passion and energy of science as against the detached rules-based deportment of the law. Both claim a basis in fact but their epistemologies of facticity are profoundly at odds. It is not surprising that the intersections of these two circuits of knowledge and performance cause mutual incomprehension and at times friction. A court must pass judgement now. It cannot wait for the further accumulation of data and must limit time devoted to debates about interpretations. Despite its claims to the contrary, science exists in a fog of indeterminacy, whereas the law shares with the media and politics a carefully constructed fiction of certainty.

Latour's basic point is that 'society ... is the consequence of all the different types of association' and he explores the law as one of these.[37] He is very clear, however, that the social cannot be reduced to the legal or vice versa. Its ability to establish ties and to bind are based upon 'common sense reasoning, results from instruments, precedents, legal documents, signatures, etc.', thus sharing materials to a certain extent with science, but the authority of evidence is treated very differently and the consequence of argument is, of course, binding in a way that would be impossible for scientists to accept in their own sphere, since the very basis of science is refutation.[38]

Food Quality and Performative Law

Typically, Nature is spoken of in a manner that assumes the earth is ontologically unproblematic. Nature is cast as a given, as a discrete entity, as already 'there'. It is presented socially, legally and politically and something knowable and ultimately controllable in all its aspects. And it becomes the task (or, more accurately, the unfulfilled promise) of a 'value-free' science to equip administrators with the technologies for such control.[39]

Legal procedure under the Food and Drugs Acts from 1875 onwards was as follows. An inspector took a sample and divided it into three portions. One was the property of the trader for use in any possible defence. One was retained by the inspector and the third was sent to the analyst. If the analyst found a discrepancy, a certificate was made out and could be used as evidence in court. If the local authority decided to prosecute, the trader was summonsed to appear in the magistrate's court and could declare a warranty defence.[40] This was a means of showing that quality had

---

36 Latour 2004: 225.
37 Latour 2007: 6.
38 Latour 2007: 5.
39 Halsey 2006: 14.
40 Under the 1899 Act there were only seven days to declare a warranty.

been guaranteed by a supplier, thereby shifting responsibility. A conviction usually meant a fine, but the Acts allowed for the possibility of prison if there were repeat offences or a failure to pay the fine.[41] Appeals against conviction were possible, although the route depended upon technicalities. In England an appeal that the conviction was wrong in fact would go to the next Quarter Sessions (in Scotland to the Justiciary Court). An appeal on a point of law could go to the Quarter Sessions or direct to the High Court in London (in Scotland to the High Court of Justiciary). If the defendant was acquitted on a question of fact, the prosecution could not appeal, but on a point of law either side was entitled to go to a higher court.

Occasionally a local authority analyst's certificate was insufficient and evidence had to be given in court. Analysts therefore had to become authoritative and personable experts, willing to testify in court, behind whom there was an administrative and scientific weight that was beyond question. They would have been very busy indeed if the advice of one well-known writer had been followed:

> If it be possible, a public analyst should attend all adulteration prosecutions, not necessarily to give evidence, but to watch the case on the scientific side, to prompt the prosecuting solicitor, to advise the magistrates on technical points if asked, and generally, to prevent fairy tales being accepted as solid truth.[42]

Christopher Hamlin points out that the expert witness was often mistrusted by Victorian courts because 'courts have been particularly stubborn in believing that science should mean the straightforward application of general laws to particular circumstances'.[43] For Bauman, the sovereignty of the modern intellect 'is the power to define and to make the definitions stick – everything that eludes unequivocal allocation is an anomaly and a challenge'.[44] But science is in reality more complex and less certain than these expectations demand, with the result that 'the testimony of real living scientists often holds up rather badly in the adversarial courtroom situation' and 'research done according to the standards of scientists is often not impersonal and law-like enough to stand up to political and judicial scrutiny'. As a result, the science of food analysis had to adjust to the requirements of the law and lawyers if convictions were to be obtained and adulteration eliminated. Laboratories had to be run with reference to methods of analysis known to be acceptable to the courts, and at levels of efficiency in the processing of samples and the reporting of results that would stand up in court.

The incomprehension and scepticism of magistrates may be one explanation of their leniency towards those accused of adulteration, both on the part of the local authorities and by magistrates. Table 8.1, for instance, shows that the proportion of samples made the subject of a prosecution slipped from 61 per cent in 1891/2

---

41 Section 17 of the 1899 Act.
42 Liverseege 1932: 56.
43 Hamlin 1986, Porter 1995: 195.
44 Bauman 1991: 9.

to only 43 per cent in 1913/14. A decreasing share of these led to a fine being imposed, the average penalty being very small, on average little more than £2.[45] It seems that the gross fraud of the early period had, by the 1890s, been replaced by a subtler, more insidious adulteration.

Table 8.1   Legal action nationwide against the adulteration of milk, 1887–1913

	Samples reported against	Prosecuted (%)	Prosecutions leading to a fine (%)	Average fine (£)
1887–90	8,850	71.0	–	–
1891–95	9,659	64.2	85.9	1.73
1896–1900	10,880	62.8	81.8	1.91
1901–05	17,970	53.3	77.2	2.28
1905–10	24,505	45.2	70.3	2.20
1911–13	17,346	42.8	68.7	2.35

Source: Local Government Board, *Annual Reports*.

These small fines were taken in their stride by the more cynical dairymen. In Birmingham, adulteration at one point was so common that for consumers it was said to amount to 'an additional water rate', and for vendors the penalties were just 'a part of ordinary working expenses'.[46] Herbert Preston-Thomas, speaking to the Foster Committee in 1894 on behalf of the Local Government Board, commented on the 'extraordinary leniency' of magistrates with regard to fines for adulteration. He ascribed this to the poverty of many of the milk-sellers.[47] As a result, local authorities felt discouraged from investing in comprehensive systems of sampling and analysis.[48]

> For many years prior to 1879, [magistrates] discretion in matters of fines in Revenue cases was limited to reducing penalties to not less than one-fourth of the amount named in the Act. By the Summary Jurisdiction Act of 1879,

45   W.H. Grigg, Chairman of the Sanitary Inspectors Association, complained of 'the ridiculous fines that are imposed for adulteration ... [which] are no deterrent at all to adulterators'. Wenlock Committee 1901: Q.6662. A legal case in 1902 set an unfortunate precedent. By forming a company, an adulterator could escape personal liability for payment and imprisonment. By dissolving the company s/he could avoid the fine altogether. Paulus 1974: 112.

46   Liverseege 1932: 211.

47   Foster Committee 1894: Q.275.

48   Bell 1884: 136.

however, they were given full discretionary power in first offences, but the former restriction remains in force for second and subsequent offences.[49]

Fines were increased in the 1899 Act to a maximum of £50 for a second conviction and £100 subsequently. Yet adulteration remained a profitable option and fines a matter of routine.

> In 1919 a producer who sold milk by retail was convicted of a sixth offence. He employed no solicitor, and before sentence was passed he was asked whether he had anything to say in explanation. He evoked the Sherriff's astonishment by replying that he had nothing to say. The Sherriff inflicted a fine of £100; the defendant produced his pocket-book, enquired for the person to whom the money should be paid, and handed over the £100 on the spot.[50]

Contract, Warranty and the Network Concept of Quality

> The movement of the progressive societies has hitherto been a movement from status to contract.[51]

The literature on the history of food adulteration has taken a relatively unsophisticated view of agency. It sees only a fraudster and a victim, and little attention has been given to the food system context within which these manipulations took place. In this section we will argue that one of the reasons why the percentage of fraud differed so much from food to food was because the circumstances of each particular trade were divergent and the means by which the goods were brought to the consumer were also varied. A reference point here is the literature on conventions and the network definitions of food quality that we introduced in Chapter 1.

Let us start with the attribution of responsibility. Some historians of adulteration emphasize events such as the identification by Accum and Hassall of individual retailers who were accused of facing tea or substituting chicory for coffee.[52] One can imagine the dramatic impact this must have had upon customers' perceptions and upon the business of the traders singled out among the thousands who shared the guilt. But contemporary observers were less interested in the crimes of the individuals than in the dishonesty and sharp practices of groups of actors – farmers, railway employees, roundsmen, fixed-shop keepers – or in whole trades, such as bakers and dairymen. Stereotypes once formed, such as the baker adding alum to his bread as a whitener or the dairymen stretching his milk with water, were

49   Bell 1884.
50   Mackenzie Committee 1922: 870.
51   Maine 1861: 170.
52   Burnett 1966, Smith 2001.

difficult to shake, and by the mid-nineteenth century consumer pessimism about the state of the food supply generally had become a kind of complicit fatalism.

Accusations about true responsibility for milk fraud were very common but rarely based upon any factual knowledge other than unspecified 'inside information'. The evidence given to the various Select Committees is rich in accusations of the broadest kind, sometimes with guesses about motivation, but often this amounted to little more than mutual recriminations by different sections of the trade. The data in Table 8.2 are a rare example of a quantified comparison.

Table 8.2   Percentage of Birmingham milk adulterated at different points in the chain

	1889–93	1894–1903	1904–13	1914–23	1924–29
Farmers	14.1	12.0	3.7	3.4	3.3
Carts, creameries	23.4	15.0	6.5	4.0	2.0
Shops	–	–	7.2	7.0	4.6
Bottled milk	–	–	–	–	1.4

Source: Liverseege 1932, 209.

First, milk producers were blamed. George Barham of Express Dairies, who took up the milk trade in 1858, was surprised to find that some farmers could not be relied upon to deliver honest, whole milk.[53] This showed that the farmers were not in such a weak position as was sometimes asserted, although their actions were clearly under scrutiny. James Niven, Medical Officer of Health for Manchester, in his evidence to the Wenlock Committee, insisted that the adulteration by farmers was three times worse than that by dealers, a figure backed up by Dr A.K. Chalmers of Glasgow.[54]

Second, the guilt of the middlemen was asserted by the writer of the Local Government Board's Annual Report, who thought that fines for small retailers were probably low in the courts because magistrates believed that the particular retailer who is summoned 'is less culpable than the wholesale dealer who has been the actual adulterator, but who generally escapes scot-free'.[55]

The third group was identified by Richard Bannister, of the Inland Revenue laboratory. He told the Read Committee that milk 'is adulterated chiefly by the distributors, the retailers'.[56] Christopher Middleton, a dairy farmer, agreed: 'the greatest adulteration is practised by those irresponsible milk carriers, men not

---

53   Read Committee 1874: Q.2434.
54   Wenlock Committee 1901: QQ.3407, 9608 and Appendix XII.
55   *Twenty-Second Annual Report of the Local Government Board, 1892–93*, PP 1893–4 (C.7180) xliii.143.
56   Read Committee 1874: Q.1553.

representing large dairymen, but people who buy and sell a few gallons of milk in the day.'[57]

In the eyes of the law, this game of charge and counter-charge was of no interest or consequence. Unless s/he could prove otherwise, it was the retailer selling adulterated food who alone was responsible, irrespective of whether the milk purchased from a farmer or wholesaler had already been watered, or if it had been later tampered with by a servant, such as the milkman delivering to the doorstep. The Food and Drugs legislation was consistent on that point. Most prosecutions were of retailers and *mens rea*, or guilty knowledge, on their part was not necessary for a conviction. The only effective defence was in the law of contracts.[58]

Paul Mitchell has shown that the development of quality obligations in the sale of goods has a complex history in the nineteenth century. At the start of the century there was essentially no requirement for the seller of natural products to guarantee quality because defects were beyond their control. The balance of the law at this early stage was clearly towards *caveat emptor*, and there was a tacit assumption that the buyer would view the goods before purchase.[59] Manufactured goods were different since their very existence depended upon human intervention and ingenuity. They had to be 'merchantable' and fit for purpose.[60] A food such as milk was somewhere in between. It was a natural product but everyone knew that it was subject to manipulation by farmers and traders, and warranties were therefore offered and demanded at various stages of the purchase chain in order to attach some guarantee of quality to the highly variable product that went under the name of 'milk'.

By the mid-century, sale by description, such as 'new milk', amounted to a warranty of quality implied by law.[61] A warranty was an oral or, more usually, a written statement that the milk delivered would be whole and untampered with. It was not about adulteration as such but, rather, about the commercial transaction between two parties, the buyer and seller. It was not until Section 25 of the 1875 Act that this was strengthened into a legal assurance of quality, and for milk this remained until 1938 as the main defence against an accusation of watering. Warranties became especially common after the Food and Drugs Act of 1899, although in practice the courts required a carefully worded written document to guarantee protection for the recipient of milk delivered on contract.[62] The key

---

57   Foster Committee 1894: Q.2363.

58   The problem of responsibility for fraud was similar in the French legal system. Monier et al. 1909, Brousseau and Glachant 2002, Guillaume 2006.

59   Mitchell 2001, Barton 1994.

60   The word 'merchantable' came to the fore in the early nineteenth century as remained fundamental to sales law until 1987. Bridge 1991: 55.

61   Stoljar 1952.

62   Section 29 of the Food and Drugs (Adulteration) Act (1928) codified warranty. The defendant now only had seven days to produce the warranty or invoice, and the warrantor

warranty case was Pain v. Boughtwood (1890) where it was shown that a dealer entirely innocent of his supplier's adulteration was nevertheless liable under the law.[63] A written warranty would have overcome this and the supplier would have been held to account.

Warranties therefore became a foundation of the milk trade – a means of asserting quality and putting trust on a legal footing. It became so important in fact that the trade pulled together when the Milk and Dairies Bill (1912/13) sought to abolish the warranty defence. Together with the Board of Agriculture, they lobbied the Local Government Board, whose President, John Burns, was eventually forced to drop the relevant clauses.[64]

The law of warranties followed a long and winding path through many test cases in the High Court in the nineteenth and early twentieth centuries.[65] According to Bell, the question of what constituted an adequate written warranty was the most difficult aspect of the Food and Drugs Acts.[66] The following are just two examples of the contradictions.

First, in Harris v. May (1883),[67] Robertson v. Harris (1900),[68] and Watts v. Stevens (1906),[69] it was shown that a general warranty applying to a long-term contract was no defence unless it could be proved to apply to particular consignment. The best way of doing this was, for instance, by producing in court the paperwork that accompanied deliveries each day. However, the judges in Elliott v. Pilcher (1901),[70] Draper v. Newham (1910)[71] confusingly decided that a specific warranty with each delivery of milk was not necessary. Second, in Bacon v. Callow Park Dairy Co. (1902)[72] and Rees v. Davies (1908)[73] churns labelled with words such as 'warranted pure' were said to be covered in a contractual sense by a legal warranty. But the judgement in Dewey v. Faulkner (1923) directly contradicted this.[74]

Early debates about warranties under the Food and Drugs Acts were important in the formative period of a new concept of 'strict liability', which subsequently had a profound effect upon the part of the common law where, perhaps counter-intuitively, *mens rea* was not necessary for guilt to be established. Instead the test

---

had the option of appearing in court.

63   24 QB 353, 54 JP 469, 59 LJMC 45, 62 LT 284, 38 WR 428, 16 Cox CC 747, 6 TLR 167.

64   NA: MH 80/4. 'Note of an interview with Sir Thomas Elliott. Milk and Dairies Bill', 23 August 1912.

65   Such as Farmers' Dairy Co. v. Stevenson (1890).

66   Bell 1931: 158–72. See also Spencer 1904: 302–3.

67   12 QB 97, 48 JP 261, 53 LJMC 39.

68   2 QB 117, 64 JP 565, 69 LJQB 526, 82 LT 536, 48 WR 571, 19 Cox CC 495.

69   2 KB 323, 70 JP 418, 75 L.J.KB 828, 95 LT 200, 4 LGR 821.

70   2 KB 817, 65 JP 743, 70 L.J.KB 795, 85 LT 50, 20 Cox CC 18.

71   74 JP 124, 102 LT 280, 8 LGR 144.

72   66 JP 804, 87 LT 70, 18 TLR 573.

73   72 JP 375, 24 TLR 735, 6 LGR 1038.

74   39 TLR 130.

for the defence was one of 'reasonably foreseeable consequences' with regard to the condition of the product. The Sale of Goods Act (1893) was also important in introducing the idea that products sold had to be of sufficient quality to be 'merchantable'.

Not all authorities prosecuted according to strict liability. Charles Cameron, analyst for Dublin, for instance, 'always declined to put the Act into effect when I knew that there was no intentional adulteration; when I knew that the spirit of the Act was not violated, although the letter might be'.[75]

Evidence to the Read Committee in 1874 suggested that warranties were a means of protecting wholesalers and retailers from unscrupulous farmers.[76] But by the time of the Foster Committee the system of warranties was said to be unworkable because retailers declared their warranty at the very last minute, by which time the reserve milk samples had deteriorated beyond recall and the trader was therefore safe.[77] In other words, through the practice of the law this loophole had emerged. The 1899 Food and Drugs Act corrected this but a great deal of damage had already been done.

Twenty years later, the Mackenzie Committee (1922) found the warranty system to be failing once more. By then the mixing of milk had become commonplace and identifying those responsible for fraud had become more difficult as a result: 'after milk from two sources or more has been mixed, the separate warranties can no longer avail.'[78] The 1928 Food and Drugs (Adulteration) Act allowed a warranty defence only if a sample had been taken from unmixed milk, and if notice was given to the local authority within sixty hours of the sample being taken. Churns had to be labelled with the names and addresses of the consignor and consignee, and have a statement about it being 'pure and unadulterated new milk'.

On the subject of prosecutions, Mackenzie commented that 'it is no wonder if such a clumsy method of dealing with a very complex article of food should occasionally give rise to injustice'.[79] Wrongful convictions were possible and, according to the contemporary literature, common, but the law was not particularly helpful in tracing the real source of fraud.

Poor-Quality-but-Genuine versus Good-Quality-but-Adulterated

The gradual accumulation of case law after the Sale of Food and Drugs Acts of 1860, 1872, 1875, 1879, 1899 and 1928, and the issue by successive governments of regulations and explanatory circulars, fostered a changing understanding of the thresholds of legality with regard to food. However, the law was unable

75 Read Committee 1874: Q.4719.
76 For instance Read Committee 1874: QQ.2642–3, E.C. Tisdall.
77 A. de T. Egerton MP, Read Committee 1895: Q.7581.
78 Mackenzie Committee 1922: 889–90.
79 Mackenzie Committee 1922: 899.

'to suppress or eliminate everything that could not or would not be precisely defined'.[80] On the contrary, it revealed, in its pedantic reverence of the statutory text, uncertainties that no-one, from farmer to retailer to scientist, had foreseen. It created injustice by convicting innocent parties and pardoning the guilty; and it undermined informal trust that had existed in the trade for decades and encouraged the substitution for that of complex contractual obligations. Such paradoxical outcomes, and the attending inconsistencies of legal interpretation, would be no surprise to Valverde, whose sociology of legal knowledges has revealed a multitude of judicial standards and practices, preventing the fulfilment of the 'law's dream of a common knowledge'.[81]

The law, then, proved to be a blunt instrument in its inability to accommodate the slippery uncertainty of the question 'what is milk'? It demanded clarity, certainty, and resolution, but was met instead with confusion, inconsistency and the possibility of injustice. The best way to demonstrate this point is to explore a number of appeal cases that addressed the seemingly intractable problem of milk that was genuine but poor in quality. The Sale of Milk Regulations (1901) allowed for this by granting the accused a chance to prove her innocence, for instance by the so-called 'appeal to the cow'. The essence of this is that if the same cow was milked a second time and in the normal course of events produced a sample of similar composition, then this was said to be evidence for dismissing a case of adulteration.

If our definition of a genuine (non-adulterated) item of food is 'nothing added, nothing taken away', then genuine can be applied to milks that are 'as from the cow' but poor in quality. Here genuineness and naturalness part company with quality because it is perfectly possible to have watery, thin milk that has not been interfered with by a dishonest farmer or trader. The reasons may be to do with feeding regime, time of year, cattle breed and several other factors.

In one way it seems to be counter-intuitive that poor quality milk, which is below the official presumptive compositional limits, should be legal. This view was widely expressed at the beginning of the twentieth century but it caused controversy. It took a number of key legal cases to expose the unintended consequences of the logic underlying the regulations. The result was an extension for a further 70–80 years of the entrenched view of natural milk.

The case of Lane v. Collins (1884) is an example of a minor *cul de sac* of case law.[82] It was not cited subsequently but did, temporarily, create a bubble of 'truth' in the name of the law, which was very confusing for many observers. A milk seller who had been prosecuted for selling skim milk as milk was acquitted on appeal on the grounds that skim milk is indeed a type of milk.

---

80 Bauman 1991: 7–8.
81 Valverde 2003.
82 14 QB 193, 49 JP 89, 54 LJMC 76, 52 LT 257, 33 WR 365.

The first appeal case proper that we will look at is Smithies v. Bridge (1902).[83] Here milk had been sold that was low in fat after the cow had not been milked for 16 hours and much of the fat had been re-absorbed within the udder. Lord Alverstone CJ and Channell J held that milk should not have been sold from an animal that was improperly treated or diseased. In the words of Channell J, 'the cow was not in fact producing milk, but was producing another liquid which did not contain the constituent parts of milk'. Let us pause for a moment to consider this statement. The judge was, in effect, saying that there are many different types of milk but only one is legal in the sense that its constituents match with the expectations of the law. It is the farmer's responsibility to gauge whether the conditions under which the milk is produced are appropriate. On reflection, this is an intrusive judgement that holds a view even before the milk leaves the udder. The condition of the cow becomes crucial, including presumably her state of health and her feed, as well as the length of time between milkings. At a stroke it wrote off the custom in some regions of a single daily milking, in favour of two or three. Nowadays, highly bred animals produce so much milk that only one milking a day would be painful for the cow and a threat to animal welfare, but yields were much lower in 1902, and the judgement amounted to a rewriting of agronomic expertise. And where was this judicial commentary on farming practice to end? Might it extend to the practice, still prevalent as late as the 1950s, of farmers keeping back milk for their families and their labourers, usually richer milk from the top of the churn?[84] On small farms this was responsible for a substantial lowering of the average fat content but very few observers would have called this adulteration.

Smithies v. Bridge proved to be controversial. The opposite was decided in Scotland in Wolfenden v. McCulloch (1905),[85] and Scott v. Jack (1912) was another Scottish case that ignored Smithies v. Bridge and decided that milk as it came from the cow could not be adulterated.[86]

In England, Smithies v. Bridge was also criticized in our second landmark case, Hunt v. Richardson (1916).[87] This is probably the most famous milk case of all time, proof of which may be found in the presence of five appeal judges on the bench. They quashed a Cambridge magistrates' court conviction for adulteration.[88] The original decision had been that the deficiency of fat in the sample taken was due to the farmer, John Hunt, feeding his cattle on wet pastures and green maize, which was said to have affected the composition of the milk to the extent that it could not be regarded as 'of the nature, substance and quality demanded'. The

83   2 KB 13, 71 L.J.KB 555, 18 TLR 575, 66 JP 740, 87 LT 167, 50 WR 686.
84   Davis 1952.
85   92 LT 857, 69 JP 228, 21 TLR 411, 3 LGR 561.
86   SC(J) 87, 49 ScLR 989.
87   Buckley [1923]: 3.
88   2 KB 446, 115 LT 114, 32 TLR 560, 80 JP 305, 85 LJKB 1360, 25 Cox CC 441, 60 S.J. and WR 588, 14 LGR 854, *Analyst*, 41, 1916: 224–7, Anon. [1916a], Anon. 1916b, Anon. 1916c, Anon. 1916d.

defence of an 'appeal to the cow' was found to be admissible and it was said that 'a farmer is now to be tried by a jury of his own cows'.[89] This was because the Regulations made only a presumption of adulteration if a sample was not up to the standards set (*prima facie* evidence), and it was therefore possible to refute this with evidence that it was milk as it had come from the cow, with nothing added or taken away. However, this was a majority decision by three to two, so the case divided even the country's top judges. In dissenting from the majority decision, Bray and Scrutton JJ argued that the key to the case was not the farmer's cattle husbandry but whether the milk was, as demanded by the 1875 Act, of the nature, substance, and quality demanded. They suggested that, by falling outside the prescribed compositional guidelines, it could not be of merchantable quality and therefore was not legal.

The tension between the state, as represented through the courts, and the farming community, was in some ways the result of different interpretations of the rule of law. The state saw its role as the arbiter of positivist law, made by society for its mutual protection and based on morality only in as much as the standards were presumptive and defendants therefore had a chance to prove their innocence. Farmers, on the other hand, were proponents of natural law, the notion that justice is immanent in nature. They argued that compositional rules were unfair if they ignored the empirical experience of natural variation and imposed poorly researched and arbitrary standards. This was in effect the nature of rules versus the rules of nature.

The judgement in Hunt v. Richardson was later brought into disrepute when farmers partially milked their cattle and legally produced milk of a very low quality. To make matters worse, a complementary case, Williams v. Rees (1917), established that a cow kept on poor pasture could legally produce milk with fat 28 per cent below the limit simply because it was 'milk as it came from the cow'.[90]

Grigg v. Smith (1917) added further ammunition to farmers' defence when it established that there was no need for milk to be the outcome of an entire or uninterrupted milking.[91] This case was heard before Lord Reading CJ, Ridley J, and Atkin J. The magistrates in Stratford upon Avon had convicted farmer Smith under Section 6 of the 1875 Act for selling milk at 2.6 per cent fat, well under the official limit. He had only one cow and had not fully milked it, leaving some for the suckling calf. This meant in effect that he had delivered the thinner fore-milk to his customers and kept back the richer strippings. The appeal was dismissed and the original judgement upheld of justices in Stratford-upon-Avon. In dissent, Atkin J commented acidly that 'a farmer was now entitled by law to give preference to his own calves over the babies of his customers'.[92]

---

89   Liverseege 1932: 40, 58, 197.
90   82 JP 97, 87 L.J.KB 639, 118 LT 356, 26 Cox CC 173, 16 LGR 159.
91   117 LT 477, 33 TLR 541, 82 JP 2, 87 L.J.KB 488, 61 S.J. and WR 677, 26 Cox CC 26, 15 LGR 769, *Analyst*, 42, 1917: 323–4.
92   *The Times*, 27 July 1917: 4D.

Milk dealers did not receive judicial sympathy in the same way as farmers. In Dyke v. Gower (1892) a shopkeeper who served from the top of a vessel, without stirring from time to time, was at risk because, with the cream constantly rising, the milk at the bottom was likely to be deficient in fat.[93] It was said not to be a defence to claim that no adulteration had taken place because the milk was sold in an 'altered state' with notice being given to the customer. The same was decided in Knowles v. Scott (1918), where milk was served from a tap at the bottom of a can in which the cream had risen over the previous period.[94] An example of judicial inconsistency may be found in Bridges v. Griffin (1925),[95] however, where, under similar circumstances, a conviction was overruled.

Trader as Victim: A Discursive Construction

In 1886 the milk trade estimated that about half of those convicted of adulteration were innocent.[96] Their message, expressed in every trade paper, was that the working of the Food and Drugs Act was most unsatisfactory. We have no way of checking the figures, but the important point is that farmers, wholesalers and retailers all believed that they were hard done by. This was the birth of their self-image as victims of an unjust system. To some extent the battle between Somerset House and the public analysts was mirrored here by that between the public analysts and the milk trade. Some honest traders do indeed seem to have been caught in the net because supplies to them were of poor quality, and they also felt that the tests used by certain analysts were unreliable.

> Dairymen have no antipathy to public analysts as a class, but against certain individuals who are not quite fit for the position they occupy, whose certificates have been the means of ruining many honest traders.[97]

Evidence to the Read Committee established a new discourse, that of trader as victim, which lasted for over half a century. The Act was said to have 'inflicted considerable injury, and imposed heavy and undeserved penalties upon some respectable tradesmen'.[98] Augustus Voelcker was characteristically scathing about the efforts of his fellow analysts:

> I think the chemists have been very harsh upon the milk dealers lately; they have not taken sufficiently into consideration the natural variations to which milk is

---

93 1 QB 220, 56 JP 168, 61 LJMC 70, 65 LT 760, 17 Cox CC 421.
94 1918 JC 32, 55 ScLR 167.
95 2 KB 233, 89 JP 122, 133 LT 177, 23 LGR 564.
96 *Metropolitan Dairyman's Directory* 1886: 2a.
97 E.G. Easton speaking at a conference on adulteration. *Analyst*, 9, 1884: 158.
98 Read Committee 1874: 245.

liable, and therefore some of the decisions which have been given are rather harsh upon the individuals that sold the milk.[99]

Within a decade of the passing of the Sale of Milk Regulations in 1901 grave misgivings were expressed at the Board of Agriculture. The following quotation from a confidential memo of a senior civil servant captures this concern well and indicates a crossing of the psychological threshold from *caveat emptor* to *caveat venditor*.

> It is doubtful whether it is possible to administer the present law strictly enough to keep down adulteration without incurring great risk of instituting proceedings against sellers of milk which is genuine though poor ... The difficulty of distinguishing between the cases where poverty of milk is due to adulteration and those where it is due to natural or accidental causes is almost insuperable and great credit is due to officers of many local authorities, including Bristol, for the strenuous efforts which they make to avoid creating hardship by the execution of the law ... Numerous complaints are made by milk-sellers as to the operation of the present law as to the sale of milk. On the one hand it is stated that innocent persons are frequently prosecuted and even convicted for the sale of milk which has not been adulterated, or at any rate not by them ... The honest milk seller feels that he does not get enough protection against the unfair competition of those who adulterate milk or wilfully sell adulterated milk as milk.[100]

In February 1922 Colonel J.M. Rogers, a magistrate from Kent, wrote to the Minister of Health, Sir Alfred Mond. He had presided over the case of three farmers whose milk had been satisfactorily tested regularly for 11 years but who had recently fallen foul of the law by providing a sample with less than 3 per cent fat. He and his fellow magistrates in the Sevenoaks Police Court had dismissed the case because the herd had been tuberculin tested just two days before and the cows' metabolism was therefore thought to have been disturbed.[101] This view was supported by Robert Mond, who was a doctor with decided views about milk. Robert wrote to his brother, the Minister, saying that there should only be a prosecution where there is a repeated failure of the test. Sir Alfred listened and insisted on an official circular being issued in July to the effect that:

> The Minister is ... of opinion that it is extremely undesirable that a prosecution should be based upon the results of an isolated test when other tests of the particular milk supply have proved satisfactory, and I am to suggest for the

---

99   Read Committee 1874: Q.5527.

100   Haygarth-Brown, E.G. 1911. Memorandum on the law relating to the sale of milk, NA: MAF 52/10, File A/22021/1911.

101   NA: MH 56/110, *Kent Messenger and Sevenoaks Telegraph*, 4 March 1922.

consideration of the Local Authority, that in such cases prosecutions should be instituted only where a series of tests have shown repeated default.[102]

The Minister ignored a warning by his chief scientist that tuberculin tests were not thought to make any difference to milk composition. He ignored the discovery that the information provided to Col. Rogers about the regularity of sampling was inaccurate. And he also ignored the advice of his senior civil servants that any action might cause controversy and be embarrassing. They were correct and the Circular proved to be disastrous. About 80 Local Authorities wrote in protest, as did the British Medical Association and other health professionals, against what the journal *Public Health* called 'the Magna Charter of the fraudulent milk dealer'.[103] The circular was withdrawn after Mond was replaced by a new Minister the following year.[104]

Conclusion: Holmes and the Path of the Law

Although he was a Justice of the United States Supreme Court, Oliver Wendell Holmes Jr (1841–1935) is a key source in understanding the evolution of the common law in Britain.[105] As a proponent of the philosophy of pragmatism, practice was his main inspiration, particularly the precedence of past decisions but also the performative persuasion of counsel, experts and exhibits in court. In this way, judgements are not necessarily representative of infringements of the law but their significance is that they leave traces in the judicial record which influence subsequent outcomes. This is very much along the lines of Simondon's individuation, although here the skill of the design engineer in replaced by a combination of the legislative draughts(wo)man and the judge. Deakin claims that

> the juridical record is a trace of shifts in the wider social and economic environment. It is therefore open to an interpretation which can help to explain the nature and direction of historical change in the society of which it formed a part. It further follows that an evolutionary interpretation, by locating the influence of the past in the way suggested, can throw light on doctrinal disputes and tendencies in the modern law.[106]

---

102 Ministry of Health Circular 325, 17 July 1922.

103 *Public Health*, January 1923: 88, Wellcome Contemporary Medical Archives Centre: SA/BMA/F105.

104 Ministry of Health Circular 399, 16 May 1923.

105 For accounts of American food regulation and milk cases, see Law 2003, Wright and Huck 2002.

106 Deakin 2006a: 236.

The common law, with its reverence for precedence, has changed very slowly, and every decision has been contingent upon an historical databank of argument and wisdom. Edging forward, while at the same time looking back, has introduced a path-dependence to all of the judgements that we have studied in Chapter 8. Holmes called this 'the path of the law' and his most famous quotation is a pithy summation of this point: 'the life of the law has not been logic: it has been experience.'[107] The Holmes oeuvre is vast and his erudition enviable but for our purposes three aspects of his thinking were important, especially since he was an exact contemporary of many of the late nineteenth and early twentieth century legal debates that we have discussed.

First, Holmes was a member of the Metaphysical Club at Harvard, the famous philosophical circle centred on Charles Peirce and William James, whose members were participants in the development of American pragmatism. Although he seems to have had sharp disagreements with his friend James on many matters, nevertheless he was influenced by the ideas that were current and applied them, consciously or unconsciously, in his writings and judgements.[108] Holmes, for instance, was 'hostile to generalization and abstraction' and preferred 'situation sense' to principle.

Holmes was also a utilitarian and an historicist. The latter arose from his interest in Darwin and evolutionary biology and a conviction that much can be explained in terms of temporal context. But, in the common law, Holmes was not in awe of historical precedent for its own sake but rather he seems to have been grateful for the richness of argument and wealth of experience it represents.

Third, further to this point about Holmes and history, Clayton Gillette adds that he was convinced by what the modern literature calls 'path-dependence' and 'lock-in'. Once a technology or an institutional structure is established, it survives at least partly by inertia. At some point it may become suboptimal but may survive in preference to newcomers simply due to the economic, and maybe also the emotional, costs sunk in it. Holmes's celebrated article, *The Path of the Law*, sees precedent in these terms, as imparting rigour but also acting as a form of restraint. Since the facts of cases are rarely identical, this is not a deterministic process but preceding judgements are referred to by counsel and taken into account by judges. Even where the judgement is different, a reasoned argument is given to justify this, in order to help subsequent cases on similar topics.

Common law path dependence is in trouble where decisions become locked-in to a precedent that is out of line with the circumstances of the later period. A judgement from scratch might have come to a different conclusion but the burden of expectation, for the time being at least, lies with the precedent. There may even be a 'cost' to change in the sense that learning and adjusting to new rules is time-consuming and may cause confusion for a period in the relevant networks of interest or industries. This will be true where actors trim their activities to be just within the

---

107 Holmes 1881: 1, see Atiyah 1987: 8.
108 See Posner 1992, Burton 2000.

law. Even slight inconsistencies between judges, through time or between courts, may flush out those who were previously marginal, and everyone, including prosecuting authorities and court officials, are put on guard where previously their judgement would have been to do nothing. Under such circumstances of instability, legislation may be required to force a step-change in thinking.

Moving on from Holmes, Patrick Atiyah has argued that judges have subsequently absorbed ideologies of the late eighteenth and nineteenth centuries, such as *laissez-faire* and utilitarianism, and reproduced them in their judgements.[109] This is rarely explicit but Atiyah finds many clues in their rhetoric. An example is the gradual increase in stress, in the law of sales, upon *caveat emptor*, which was part of the moral discourse of self-reliance seen to be fundamental to free markets. Extending this argument is to imply a causal or, at least, a facilitating link between the structures of legal thinking and the evolving structures of the modern economy.[110]

Common law up to the eighteenth century took little interest in protection of the buyer from low quality goods. One reason was that the majority of deals were still made face to face and inspection was therefore possible before purchase; also, the price bargained would have incorporated a notion of quality. A second factor was that most transactions would have been, individually, of low value and therefore not worthy of legal action if they were unsatisfactory. Lawyers therefore had little experience of the workings of the food system by comparison with property or marriage settlements.

Ironically, in the early nineteenth century, contract law moved away from its specialization in mercantile deals and was merged with agrarian common law thinking. As a result, it tended to take a view favouring the seller or manufacturer at the expense of the buyer or consumer.[111] Atiyah identifies the 1870s as a turning point, exactly the time that warranty emerged in the milk industry.[112] Until then freedom of contract had been in full flood, and the period since has seen its modification in an increasingly regulated environment. The rise of egalitarianism since the late nineteenth century has played its part in terms of expectations. In sales law, *caveat emptor* was in retreat. Atiyah claims not to know the reason why, although he blames the authors of legal textbooks, who failed to spot the common law's return to its earlier ideals of fairness in contractual exchange.[113] We can speculate that the political and popular sentiment behind Sale of Food and Drugs Acts was influential across the spectrum of this type of law and that the passing of

---

109   One important intellectual well-spring, although from the French civil law tradition, was Pothier's *Traité des obligations*, which was translated into English in 1806.

110   Horwitz 1977: ch. 6.

111   Teeven 1990: 220.

112   As do Cornish and Clark 1989, and Teeven 1990. One reason for the 1870s being a threshold was the Judicature Acts of 1873 and 1875, which merged the courts of equity and common law and reformed what had become a dysfunctional process of justice.

113   Atiyah 1979: 479.

so much legislation on adulteration in the period 1860–1879 cleared a logjam of indignation that had existed for decades beforehand.

The law already had many case precedents on quality and fitness.[114] One group involved the issue of an 'express warranty', a piece of paper stating the nature of the goods. Another group relied on 'implied warranty', such as was used in the sale of horses.[115] Here a fair price was assumed to imply a sound commodity, although it was more likely in many instances that the exchange was based on trust in the seller's integrity. This is why so many food consumers returned, again and again, to the same market stall-holder or the same milk seller, once familiarity and trust had been established. Much in the understanding of quality flowed from this.

The Sale of Goods Act (1893) was another major event in the history of the common law, representing the first centralized, coordinated attempt to influence the law of contracts. The rights of the buyer were enhanced through the technical device of an implied warranty of merchantability. According to Teeven, 'this revision was necessary due to new techniques of marketing and financing through middlemen over great distances ... *Caveat emptor* was no longer an entirely workable rule since buyers were not relying on their own judgement as much as on the integrity of the seller'.[116] The Act was a codifying measure, designed to be used as a source of authority and it marked a new beginning for some aspects of the law of sales. It influenced the United States' Uniform Sales Act (1906) and was still in force in Britain many years later in the modified form of the Sale of Goods Act (1979).

The 1893 Sale of Goods Act stipulated that the buyer was entitled to reject goods on or after delivery if the conditions of a contract were breached. The breach did not have to be serious, so the seller had an incentive to deliver goods of the necessary standard.[117] Sections 13–14 of the Act spelled out three exceptions to *caveat emptor*. Goods had, first, to conform to their description; second they must be 'merchantable'; and, third, they had to be 'fit for purpose'.[118] The term merchantable simply meant 'an article that would be saleable in the market' and dates from the era when common law trials had juries, sometimes made up of merchants, who had the expertise to judge upon satisfactory quality for sale.[119]

How, then, to sum up this chapter? The most important lesson that we have learned is that the material quality of milk in our period has been closely, constitutively related to the individuation and practice of ontological devices in the common law. The law, despite the scientific uncertainties associated with the composition and definition of milk, has driven through its own vision and thereby retrofitted definitions of milk which thereby also changed our approach to nature. We will

---

114   Atiyah 1979: 472–5.
115   Llewellyn 1936, 1939.
116   Teeven 1990: 220.
117   Whittaker 2005: 251–2.
118   Whittaker 2005: 229.
119   Whittaker 2005: 235.

see something rather different in Chapters 9 and 10, where we will look at notions of clean milk. The relationship here between bacteriological and legal thinking about material quality was dissimilar because its methodology and quantification encouraged the development of clear standards, with allowable maxima. This was a metrological and classificatory solution, unlike our compositional view so far, which has been tied to ideas about cheating and therefore to a presumptive perspective in law.

PART IV
Impurity and Danger

Introduction to Part IV

In the 1850s the main public disquiet was about adulteration that might, in one way or another, introduce poisonous or contaminating material into milk. As we have seen, this anxiety continued into the first decades of the twentieth century, with revelations about chemicals such as preservatives and colorants. Dirt was the next source of alarm, from the 1880s to the First World War. The third fear, which ran parallel with these at the end of the nineteenth century, but which overwhelmed and outlasted them both, was the hazard of infectious disease. This was not a general alarm because there were germ theory deniers even fifty years after Pasteur's discoveries, but it was powerful for a couple of reasons. First, it seemed to match the in-depth epidemiology of epidemics, and, second, the influence of the science of bacteriology grew in the politics of public health and its interpretations became the ones acted upon. Dirt-based food hygiene eventually reinvented itself as applied bacteriology.

The focus of the following chapter and the next two is a history of milk quality as defined by the interests of public health in the period roughly 1880–1960. To begin at the beginning, the link between mother and baby, which is forged at the breast, has given milk a special quality which has a deep symbolic significance. It was not lightly, therefore, that Victorian and Edwardian commentators branded cow's milk a principal cause of ill-health, particularly for children under five. As we will see in Chapter 9, the material property of greatest interest initially was the absence of dirt. By the turn of the century this had changed and milk was thoroughly implicated in outbreaks of a range of diseases, from epidemic diarrhoea to bovine tuberculosis. Eventually concern about dirt faded in the 1930s and, as a result, 'clean milk' came to mean something rather different. We have here, then, changing notions of quality that were the result partly of social changes in health understandings, and partly due to ontological shifts with regard to the materiality of a food with an uncertain knowledge base.

Caveat emptor might have been successive governments' motto in their attitude to the dangers of drinking milk, and interference from the state was therefore minimal in most local authority areas until the 1930s. It was left largely to private and commercial initiatives to bring about a shift in attitudes among producers and traders.[1] As a background to developments in the culture of production, there were debates about the need for administrative regulation, about technological interventions such as pasteurization, and there were also doubts about the solidity of the science of dairy bacteriology.

It will become clear that the political context of the clean milk movement was especially important. At a general level the risks were well-known: everyone was

---

1   The political dynamic of the 'reluctant state' is explored in Atkins 2010.

aware that much milk was infected with tuberculosis and other diseases when it left the farm but no-one had a clear idea whether the particular pint that they had just purchased was safe.[2] Some retailers displayed warranties of quality but fraud was so extensive in the trade that the public was rightly wary.[3]

The grading of milk, when it came, was derived from two separate logics. One was the call to reward farmers for the quality of their products, whether milk, fruit and vegetables, or meat. In retrospect this might sound odd, but until the 1920s and 1930s the systems of provision for individual foods were not geared to paying for quality, with the result that farmers and growers did not prioritize it. For milk, quality usually meant butterfat content. The new approach gave some remuneration for the costs of production and also added an incentive against adulteration or the 'toning' of milk to a low standard. The second, and quite different, logic was the notion of milk quality defined as freedom from disease and dirt, where a consumer interest was at stake. It is the latter tradition which mainly concerns us in this chapter.

The framing of Victorian and Edwardian environmental ideas was partly driven by a fear of dirt and pollution, and Chapter 9 charts the way in which this drove certain early ideas about milk policy. Dirt in milk remained a major concern until the 1930s and a great deal of time and effort was devoted to monitoring and eliminating it. The National Clean Milk Society of Wilfred Buckley and Waldorf Astor lobbied heavily from a relatively privileged position and much of the political action with regard to legislation and regulation was played out in Whitehall rather than Westminster.

Cleaning up the milk supply chain was a slow process but there is evidence that hygienists, such as the team of researchers under Robert Stenhouse Williams at the National Institute for Research in Dairying, were having a significant impact by the later 1920s and the introduction of a National Milk Testing and Advisory Scheme in the war completed the process. These efforts received general approbation, although there was resistance among the dairy farming community, for whom cleanliness meant greater effort and expense. In the 1930s the issue of clean milk, which had been the focus of lively debate for decades, fell off the agenda. The National Clean Milk Society was disbanded in 1928, and the deaths of Williams (1932) and Buckley (1933) removed the principal activists.

---

2   Atkins 1992.
3   Hassard 1905, Atkins 1991.

Chapter 9
Dirty Milk and the Ontology of 'Clean'

Matter Out of Place

What about dirt and disease in food history? Their under-emphasis in the standard texts seems curious in view of their crucial role in the development of quality definitions, particularly in the nineteenth and early twentieth centuries when dirt was seen as one of nature's principal transgressions.[1] Dirty food, in particular, possessed the most visceral of nature's qualities and was increasingly associated with affective repulsion.

Before the age of great cities, farmers and agricultural labourers encountered nature in many ways. They made their living from the soil and many died young from the infectious diseases that spread quickly in cramped housing conditions, or as a result of eating contaminated food. 'Dirt' in its broadest sense, then, was implicated in both life and death. Organic matter, minerals, microbes were both friend and foe. Nature was simultaneously generative and destructive. By comparison, in the process of modernization, urban societies have experienced a number of ontological shifts, at first fearing dirt and eliminating it, then not seeing it, and, most recently, recognizing a category of 'good dirt' that is vital for the development and maintenance of the human body's immune system.[2]

The development of the concept of cleanliness has closely mirrored the civilizing process. Norbert Elias demonstrated this and related it to food through the concept of disgust. In late medieval Europe, table manners, and the use of implements such as the fork were both encouraged by increased feelings of repugnance at prospect of eating from a communal dish and the possible transfer of contamination from the fingers.[3] Ken Albala relates etiquette to class, as wealthy consumers, especially from the sixteenth century onwards, began to avoid food items and habits associated with the lower social orders. At the same time a strong cultural association developed between disease and a cluster of variables that included food hygiene, eating habits, and the humoral associations of individual foods and their combinations.[4]

---

1 An honourable exception is Kiple and Ornelas 2000, which devotes 19 of 169 chapters to disease, although only one is on food-borne infection.
2 Strachan 1989, Kline 2007.
3 Elias 1939, Mennell 1996, Romagnoli 1999. See also Miller 1997.
4 Albala 2002.

Another example of changing perceptions is Vigarello's account of personal hygiene in France.[5] In medieval times cleanliness was rarely mentioned and not associated with health. In the sixteenth and seventeenth centuries wiping the face was preferable to body washing and, for the wealthy at least, wearing white linen and perfume was greatly superior to having a bath. In fact bathing was dangerous because it opened the skin and thereby invited exogenous challenge. By the 1730s changes of attitude were noticeable. The advent of the bidet in aristocratic French households was followed by comments from the 1760s concerning the need to wash and the elimination of body odours. After about 1830 warm water was playing a part in bodily purification and was said to give some protection against the most frightening epidemic disease of the day: cholera. In wealthy homes, having specialized rooms for bathing and the necessary furniture became a means of establishing status and, at the same time, cleanliness came to be associated with morality.

Alain Corbin's history of the smellscape of Paris is a further elaboration of the same theme.[6] The idea of dangerous odours was nothing new but the miasmic theory of disease grew in popularity in the early phase of urbanization as a common sense correlation between illness and an increasingly dirty and disordered environment. At the end of the eighteenth century, enquiries into epidemics led to some speculation about the nature of contagion, principally in terms of climate but also increasingly in relation to smells and infectious gases rising from drains, graveyards, slaughterhouses, and city streets strewn with horse manure. By the turn of the nineteenth century public health was being rethought in terms of the geography of dirt and stench, and there were the first stirrings of an interventionist mentality.

Gradually from the 1870s, the discoveries of Pasteur, Koch and others rewrote popular environmental understandings in terms of the germ theory of disease. The porous skin that had been such a concern for hundreds of years was no longer a risk. Instead, cleanliness was now protective. In addition, bodily control became an expression of social control and hygienism was therefore an important marker of both biopolitics and governmentality.[7]

According to Mary Douglas, dirt in modernity has become 'matter out of place'. For her, symbolic concerns about dirt have been more important than worries about micro-organisms. This is because of implications of pollution and 'a polluting person is always in the wrong'.[8] Dirt, then, is a means of establishing the margins of the socially acceptable and this, of course, has varied according to social and cultural context.[9] By implication, dirt also has an epistemological history because the conceptual and practical tools of analysis have changed, so it is

---

5   Vigarello 1988.
6   Corbin 1986.
7   Lupton 1999: 40.
8   Douglas 1966: 113.
9   As Campkin 2007 notes, this is at odds with Douglas's structuralist universalism.

no surprise that specific commentaries, and even casual observations, have varied greatly in tone through time and across space.

Dirt and disease may also be interpreted in terms of what Bruce Braun calls the uncooperativeness of commodities.[10] By this he means that the physical qualities of material objects influence the ways in which they can be manipulated for commercial purposes and the modes of cultural and political regulation that can be deployed. Milk's susceptibility to contamination made it exceptionally difficult to handle and risky to consume throughout our period.

Dirty Milk: The Lost Discourse?

As I have shown elsewhere, cowsheds in both town and country were in poor condition, certainly up to the First World War.[11] Attempts had been made in the late nineteenth century to regulate the fabric of the buildings, the light and ventilation, drainage and the removal of manure. The Dairies, Cowsheds and Milkshops Orders of 1879, 1885 and 1899 were adopted enthusiastically by a few progressive authorities – mostly cities and larger urban districts – and, ironically, the result was that cowkeepers in those cities were bankrupted and an increased proportion of drinking milk was drawn from less well regulated, rural cowsheds. At the turn of the century, Leslie Mackenzie, Medical Officer of Health for Leith, published a colourful account of conditions in a typical cowshed.

> Whoever knows the meaning of aseptic surgery must feel his [sic] blood run cold when he watches, even in imagination, the thousand chances of germ inoculation. From cow to cow the milker goes, taking with her (or him) the stale epithelium of the last cow, the particles of dirt caught from the floor, the hairs, the dust, and the germs that adhere to them. Meanwhile, what with switching of cows' tails, what with stamping of feet, the byre is in a state of persistent agitation. The cows are feeding. The imprisoned dust of the dry fodder is scattered to the air currents. Meanwhile, too, the other milkers are collecting the milk. They perspire. They transfer the milk from pail to can. They leave the total to gather more germs and dust. Perhaps, the moderately careful dairyman sieves the milk roughly from the pail; but the sieve is not such as to enmesh any but the major particles. Everywhere, throughout the whole process of milking, the perishable, superbly nutrient liquid receives its repeated sowings of germinal and non-germinal dirt. In an hour or two, its populations of triumphant lives is a thing imagination boggles at. And this is in good dairies. What must it be where the cows are never groomed, where udders are never washed, where teats are never rubbed, and where the byres are never even approximately cleaned, where ventilators are never open, where the dung is a stale heap at the byre door, where pigs are

---

10   Braun 2006a. See also Bakker 2004.
11   Atkins 1992. For a French account of this period, see Rolet 1908.

a few feet away, where cobwebs are ancient and heavy, where the ammoniacal
emanations of decomposing urine nip the eyes, where hands are only by accident
all washed, where heads are only occasionally cleaned, where spittings (tobacco
or other) are not infrequent, where the milker may be a chance-comer from some
filthy slum, where in a word the various dirts of the civilized human are at hand,
reinforced by the inevitable dirts of the domesticated cow![12]

Mackenzie was no doubt exaggerating in the hope that his gothic account
would shock his councillors into action, but such descriptions were common
enough at the time and, as a result, the image of milk was undoubtedly tarnished in
the mind of the consumer. The perception had gradually accumulated of a product
that was adulterated with water, that had chemical preservatives and colorants
added, and which had a high risk of contamination with dirt and the germs of a
variety of diseases.

Comparing Mackenzie's and similar descriptions, we can identify a number
of dimensions of public disgust in the period roughly 1850–1930. The first, as
we have seen, was the conditions in which the animals were kept. There were
occasionally accusations of animal cruelty, and often representations that shared
much with contemporary surveys of slum housing. Physical and sensual attributes
were used to make the point: rickety buildings, poor drainage, strong smells.
Although most readers of these reports would never have visited a cowshed, the
graphic descriptions were enough to make them feel as if they were present and by
about 1900 revulsion was a common, learned response.

The second dimension was the visible evidence of dirt in milk. Consumers
may have experienced this as visible sediment at the bottom of a glass, but
more likely they would have read about something called creamery filtrate.
Contemporary observers were shocked to see this slimy mass of dung, hair, straw
and other, miscellaneous matter which was left after milk was processed. Their
outrage was usually directed at farmers whose contracts with dairy companies
usually required them to strain their milk before delivery. Yet separator slime,
as it was sometimes also called, received little scientific attention. On the rare
occasions that measurements were reported, there was no agreement as to method.
In Switzerland, Barthel found that

if this slime is examined microscopically it is seen that along with the mass of
bacteria, white blood corpuscles, and other cells, there are fragments of dung
and fodder, hair from animals and human beings, vegetable fibres, grains of
sand, parts of insects, in short, everything that can be found in the dust of the
stall. The amount of separator slime can vary from 0.03–0.25 per cent, of the
weight of the milk, but generally it does not exceed 0.1 per cent.[13]

---

12   Mackenzie 1899: 373–4.
13   Barthel 1910: 13–14.

The continental literature records laboratory tests where the sludge was dried and weighed, and then reported in parts per million by volume. Thus Jensen compared Hamburg, which had an average load of 13.5 ppm, with Oslo at 11, Helsinki 1.8, Berlin 10.3, Halle 14.9, Leipzig, 3.8, and Munich, 9.[14] British studies tended to report centrifuged, moist slime, as in Sheridan Delépine's detailed and long-running investigation of milk in Manchester, which in 1896 found 256.7 ppm, falling to about half that a decade later (Table 9.1). The situation in Yorkshire in 1908 was a little better (Table 9.2).

Table 9.1   Sediment in Manchester's milk supply, 1896–1906 (mm of sediment separated by centrifuge)

	Samples	Clean		Doubtful	Dirty
		0–6	7–8	9–10	>10
		%	%	%	%
1896–1900	454	4.2	35.3	48.0	12.5
1901–2	861	21.6	37.7	32.6	8.1
1903–4	857	19.1	34.5	38.0	8.4
1905	764	24.9	36.2	31.6	7.3
1906	704	26.3	42.0	26.7	5.0

Note: The author admitted that this classification was 'quite arbitrary'.

Source: Delépine 1909: 816.

Table 9.2   Percentage of Yorkshire milk samples containing sediment, 1908

	Parts per million					
	0–20	20–40	40–60	60–80	80–100	100–120
Cowshed	27.8	39.4	26.2	1.6	1.6	3.2
Retailer	52.4	39.3	6.5	1.6	–	–

Source: Orr 1908.

The third dimension of disgust was built from the many stories about the reluctance of farmers to implement improvements. Grooming and washing the flanks of milch cows was said to have been 'practically unheard of' in Cheshire, for instance, at the turn of the twentieth century, not out of laziness but because farmers firmly believed that washing a cow's udder would make her susceptible

---

14   Jensen 1909: 127.

to catching cold.[15] This popular veterinary theory of heat had physical form in the cowshed, with minimal ventilation and cows standing close together, and farmers claimed that yields were higher as a result. Any suggestions of the need for sanitary improvements were thought to be a 'doctor's fad'.[16] Common features of the public health literature of the day were a patronizing view of customary knowledge and a stated need for education, so that milkers could learn the 'true' path to hygiene in the cowshed. The commonest causes of contamination on the farm were said to be: (a) dirt falling into the open pail during milking; (b) the use of unsterilized milk pails, coolers, and churns; (c) straining a large amount of milk through just one strainer; and, (d) mixing fresh and stale milk.[17]

Most farmers felt they had done enough if they strained their milk. But, by the first decade of the twentieth century there were several portable devices on the market for visiting inspectors to use in the cowshed to convince such sceptics that more was needed.[18] The simplest was shaped in the form of a funnel with a detachable cotton pad at the apex, the colour of which changed as the milk was filtered through, according to the degree of contamination.[19] A problem with this method was that a third to a half of cowshed dirt was soluble and therefore quickly became invisible.[20] Also, Joseph Race worried that sanitarians might defeat their own object in this way if they emphasized dirt at the expense of bacteria. He suggested that using the absence of dirt as evidence of cleanliness was a fallacy.[21]

Fourth, the transport of milk was said to present problems of delay and contamination, particularly on the railway, which dominated milk traffic from the mid-nineteenth to the early twentieth century. Ventilated and sprung wagons were not always available and timetables required adjustment. The eight barn gallon churns used until the First World War were poorly designed.[22] One survey in 1908 found that over 70 per cent of the cans used for transport were of a type that admitted dirt and dust.[23] When it arrived in the big cities, the milk was sometimes decanted and mixed on the platform, which again increased the risk of contamination.

The lengthening of distance between producer and consumer certainly meant significant delays in delivery. To the average six hour rail journey to London was added the time needed by the wholesaler to supply the trade, and the retailers to supply their customers.[24] Milk was then stored in the home before consumption.

---

15 Laird 1905: 439.

16 Dorset County Council, Medical Officer of Health, *Annual Report* 1905: 25.

17 Milk Control Board 1919: 3.

18 According to Barthel 1910, a laboratory version of this approach was devised by Renk in 1891 and later modified by Stutzer and Gerber for general use.

19 Ministry of Agriculture and Fisheries 1924.

20 Larsen and White 1913: 124.

21 Race 1918: 183.

22 Eight barn gallons were the equivalent of 17 imperial gallons.

23 Orr 1908: 46.

24 The maximum journey times were: Great Eastern Railway four hours, Great Northern 5.5 hours, Great Western 11 hours, London, Brighton and South Coast 3.3 hours,

The interval between cowshed and teacup could therefore take over 24 hours, more than enough time in hot weather for the multiplication of pathogens.

The fifth dimension, as touched on in Chapter 8, was an accounting of responsibility in the various sections of the trade. In Yorkshire, for instance, Thomas Orr found that 44.4 per cent of the contamination of milk took place on the farm, 21.1 per cent on the rail journey, 18.5 per cent in the hands of the wholesaler and retailer, and 19 per cent in the consumer's home.[25] Others suggested that the small general shopkeepers and itinerant vendors known as 'purveyors' had the worst record (Table 9.3).[26] This was due to poor storage conditions and maybe some adulteration with less-than-pure water.[27] George Newman, then Medical Officer of Health for the Metropolitan Borough of Finsbury, London, found that all of the 221 milk shops inspected had sanitary defects, including 73 per cent who made no attempt to cover the counter pan used to ladle out the milk to customers.[28] J.F.J. Sykes noted that 'it is very difficult to get the keepers of small general shops to use any precaution to protect milk from contamination'.[29]

Table 9.3   Sediment in London's milk supply, 1904/5: percentage of samples in each class (classes in parts per million by volume, measured by centrifuge)

	Samples	0–24	25–49	50–74	75–99	100–149	>150
Cows	20	100	–	–	–	–	–
Rail stations	20	45	35	15	5	–	–
Wholesalers	20	60	35	5	–	–	–
Dairy shops	20	40	40	10	–	–	–
Purveyors	20	40	25	15	5	5	10

Note: Houston regarded 50 ppm as the upper limit for clean milk.

Source: Houston 1905: 17.

London and North West 8.75 hours, London and South West five hours, Midland Railway 12 hours. Swithinbank and Newman 1903: 25.

25   Orr 1908.

26   Under the Dairies, Cowsheds and Milkshops Order (1879), London's Metropolitan Board of Works could make regulations about basic standards of hygiene but they had no power to refuse registration to existing vendors. The LCC (General Powers) Act (1908) changed this, and in 1910–1911 alone 1,290 retailers were removed from the register in London. Sykes 1887/8, Renney 1906, *The Dairyman*, 33, 1911.

27   Niven 1896: 3.

28   Newman 1903: 23.

29   *Annual Report of the Medical Officer of Health to the Metropolitan Borough of St Pancras*, 1911.

These five dimensions were the collective perceptions that were acted upon and therefore made real. Disgust was a measure of risk, however, and we cannot say that the priority order of interventions – smells and visible dirt first, everything else second – was necessarily the best protection for the health of milk drinkers.[30]

Tellingly, the science of dirt in milk did not progress greatly in the early twentieth century. According to the Society of Public Analysts, as late as 1937 the most precise laboratory approach was still to set a sample of milk for 72 hours in a sedimentation vessel and then to centrifuge the deposit that settled. After that, the milk column was stirred and allowed to settle for a further 48 hours, thus freeing any remaining material attached to suspended fat globules.[31] However, the Society's statement was of little consequence because interest in dirt had by then long passed its peak. The milk-related public health literature had fewer references to dirt from the 1930s onwards because by then it had shifted over to a discussion of bacteria, although these were usually not identified and the discourse was therefore somewhat vague, merely changing from 'dirt is bad' to 'bacteria are bad'.

Early Clean Milk Initiatives

The earliest precursor of modern clean dairying was the philanthropic venture of William Harley, who established the Willowbank Dairy in Glasgow in 1810. By 1815 he had 260 cows and was producing milk with great attention to the hygiene of the buildings, the utensils and the milkers. The operation was so unusual for the day that it became a tourist attraction, with visitors including Russian and Austrian royalty, and part of the revenue generated was from entrance fees. Harley's example was copied by dairies in Edinburgh and London but they left no lasting legacy because the high cost of production made them vulnerable to the economic recession which followed the French Wars.[32]

It was in America that concern for milk hygiene was next manifest. The mid-nineteenth century scandal about swill milk in New York was a shock to many consumers and tainted the image of milk produced within the city's boundaries.[33] This milk came from cows fed on distillery waste, who were kept in foul conditions. Other cities recognized these descriptions of poor quality milk as being similar to their own supplies, and Boston was the first, in 1859, to employ an official Milk Inspector, whose task was to seek improvements. Other cities eventually followed, notably Providence, RI (1870) and Washington, DC (1871).

30 A point similar to this is made in Campkin and Cox 2007: 2.
31 Analytical Methods Committee, Society of Public Analysts 1937, Richmond 4e 1942: 344–51.
32 Harley 1828–29, 1829, Anon. 1863–65, Boutflour 1933.
33 Hartley 1842, Mullaly 1853.

These inspectors concentrated at first on sanitation, attempting to improve the conditions of production.[34] In addition, some local medical societies took a prominent role in an advisory capacity. The Essex County Medical Commission in New Jersey was an example, and in 1893, at the instigation of Dr Henry Coit, they arranged a supply milk of the very highest standard.[35] The criteria stipulated in their contract included the design of the stables, a pure water supply, the regular removal of manure, healthy cows and milkers, sanitary conditions of production, and the milk to be cooled and bottled with a label of certification. The aim was to secure special milk for infant feeding and as a clinical intervention for invalids.[36] In 1904 their term 'certified milk' was patented to stop it being used by dairymen who had no connexion with a medical commission.[37] The idea caught on and by 1921 there were 81 medical milk commissions in various parts of the USA, and an American Association of Medical Milk Commissions to promote best practice. There was also a Certified Milk Producers' Association of America, which held its first meeting in 1908.[38] What is often not mentioned is that such clean milk was expensive, on average double the cost of ordinary milk, and, as a result, model dairies supplied only 0.5 per cent of American urban milk demand in 1922. Its significance was, rather, in helping in the development of an intellectual climate of food hygiene.[39]

Sedgwick and Batchelder were among the first to count bacteria in milk.[40] Their pioneering work in Boston, from 1890 to 1892, was a revelation and their method was quickly followed by others. In 1900 New York was then the first city to use this approach as a basis for setting a bacteriological standard for ordinary market milk, at one million bacteria per cubic centimetre.[41] It was even more difficult to enforce than compositional standards because there was not necessarily any implication of cheating, just bad practice, and, anyway, who was responsible for the contamination? Boston followed, in 1905, with the tougher threshold of 500,000 bacteria per cc, and then Rochester in 1907 with 100,000. By the outbreak of the First World War over 20 American cities were counting bacteria in their milk supplies.

Setting a standard was not enough; farmers had to be encouraged or cajoled to change their practices. But which practices? One means of identifying the elements of milk production that were the least sanitary was to score each in turn. The first dairy score card was introduced in 1904 in the District of Columbia, borrowing a

---

34   We saw in Chapter 3 that sampling for fat content began in New York in the 1870s but this was not common elsewhere until the 1880s. MacNutt 1917: 65.
35   Coit 1909. For a copy of the first contract, see Heinemann 1921: 483–7.
36   North 1922: 134.
37   Erdman 1921: 37.
38   Rosenau 1912.
39   North 1922.
40   Sedgwick and Batchelder 1892.
41   Belcher 1903: 16, Rosenau 1912: 77–8.

point scoring method previously used to judge prize dairy cows.[42] The notion was popular, reaching 169 US cities by 1912, although a few critics noted problems, for instance the weak correlation between a high score for clean production methods and the bacteriological quality of the milk output.[43] The following quotation best sums up the pragmatic view taken by most local authorities.

> Without a score card it is exceedingly difficult to make a thorough and systematic inspection; in fact, it would be quite impossible to do so without long experience. The score card has an educational value for the inspector as well as for the farmer … At first blush the score-card method seems to be a kindergarten procedure. It is evidently impossible to express biological conditions and grades of sanitary excellence in exact mathematical figures. The advantages of the score card, however, far outweigh its limitations.[44]

After this introduction of quantification, the next logical step was to differentiate milk into market grades. New York was the first, in 1912, to introduce grade banding, with grades A, B and C, and the authorities there refined the grades in 1915 with various further combinations of bacteriological quality, score card results, and heat treatment. Once established, this interpretation of quality proved to be resilient despite much critical commentary. Interestingly, the objections made by various observers at this time were remarkably similar to those that surfaced in Britain about a decade later. First, the grades were said to be based on meagre experimental data and were therefore both tentative and arbitrary. Second, milk should not be on sale if it does not pass all reasonable tests of quality. Third, the public were said to be confused by the multiple grades and therefore unwilling to pay a premium. Finally, it was predicted that inferior milk would gravitate to poor areas.[45]

To give farmers an incentive to participate in the score card system, clean milk contests were held, the first being at the National Dairy Show in Chicago, February 1906.[46] From 1906 to 1915 there were 87 contests judged by the USDA's Dairy Division, Bureau of Animal Industry. The first city contest was in Cleveland, Ohio, March 1907.[47]

Dr Charles E. North of New York was not impressed by score cards and called them an 'utter failure'.[48] Instead, he proposed an alternative which has been described as the first 'rational system' of clean milk production.[49] In 1909 he was

---

42   Wing 1913: 44.
43   Buckley 1915: 2, Harris 1915, MacNutt 1917: 70.
44   Rosenau 1912: 164–5.
45   Parker 1917: 381.
46   Lane and Weld 1907, MacNutt 1917: 115.
47   Kelly, Cook and Gamble 1916: 1, Kelly and Taylor 1919, Kelly and Posson 1926.
48   North 1918: 117.
49   MacNutt 1917: 78.

hired by the New York Milk Committee, a group of self-appointed philanthropists, to find and supply clean milk to their welfare centres.[50] He did so at the outset by sourcing from the estates of country gentlemen nearby but found this to be too expensive. So in 1910 he began drawing over 5,000 gallons a day by rail from the village of Homer, 260 miles from the city. North's principles were as follows.[51] First, the farmer was given financial incentives for the extra work and the expense of meeting high bacteriological standards. At a premium of one cent per quart, this made the exercise profitable. Second, milking was done with clean, dry hands into covered pails. Third, milk pails and cans were thoroughly washed and sterilized. Fourth, the milk was cooled at a collection point before despatch to the city. Finally, samples were regularly taken and payments made on the basis of the laboratory tests of bacteriological and compositional quality. North's system was not as demanding for the milk producer as certified milk, and retailed for a great deal less, but the quality was regarded as being a significant improvement on ordinary market milk for only relatively modest changes in practice.

The final piece in the American clean milk jigsaw was the infant welfare centre. The first of these was established by Dr Henry Koplik in New York in 1889, but their most famous advocate was Nathan Straus. The latter came to the fore in 1892 through a public debate about pasteurization,[52] which he favoured, and his own centres began operation in the same city in 1897.[53] By 1922 there were 69 infant milk depots in New York feeding 25,000 babies daily. Straus encountered opposition from raw milk enthusiasts, at least until a devastating diphtheria epidemic in 1909 in which milk was implicated.[54] After that the argument was won. In 1911 pasteurization became standard in New York and by 1922 it was being used for 98 per cent of the milk supply of large American cities.[55] In 1936 the coverage was 73 per cent of the population even in small municipalities of 10,000 people, which was far ahead of Britain and most of Europe.[56] The US experience was well known in Britain, where the relevant literature was closely monitored.[57]

In Europe, Copenhagen was the city with the closest to an American-style emphasis upon clean milk. In the 1870s the circumstances there were rather similar to those in British cities, with urban and suburban producers gradually being ousted by a combination of tough hygiene regulations and competition from railway milk. The first large-scale organization was the Copenhagen Milk Supply

---

50   North 1918, 1922.
51   Rosenau 1912: 139–40, Parker 1917: 171–81, MacNutt 1917: 78–83 and 203–13.
52   Moore 1921, North 1922: 136.
53   MacNutt 1917: 87.
54   MacNutt 1917: 78, North 1922: 136.
55   North 1922: 137.
56   Frank and Moss 1932, Raw 1937, Wilson 1942: 72.
57   Eastwood 1909.

Company, established in 1879.[58] Its distinguishing feature was the voluntary adoption of comprehensive standards with regard to the veterinary health, the cleanliness and the feeding of the herds that supplied it, along with the health of the milkers and its own 400 employees.[59] In 1911 Rider Haggard reported it as a philanthropic enterprise, never taking more than a 5 per cent dividend and using surplus profits for investment.[60] The milk came from forty selected farms and was brought by train in special vans owned by the company, 6,000 gallons a day in sealed cans. It was cooled to 5°C by ice and filtered through fine-meshed cloth and sterilized gravel.[61] The company was opposed to pasteurization.

The experience of this company influenced the general development of milk provision in the Danish capital and further afield.[62] In 1922 there were three large depots, two of which were cooperatives, serving most of the city's needs. All of their milk was delivered in bottles, much earlier than was the case in British cities.[63] In 1927 Copenhagen, a city of 700,000, was dominated by the Copenhagen City Milk Distributive Co., which got its milk from 3,500 farms with a total of 55,000 cows.

Stockholm, somewhat smaller at 450,000 people, was similarly dominated by a large cooperative known as 'Milk Central'. This started in 1915 with 1,523 farmers and 33,500 cows and by 1927 had expanded to 7,260 suppliers with 87,000 cows. Each farmer invested in shares proportional to the size of their milking herd and interest was paid annually *pro rata*. All of the milk was machine bottled.[64]

As we have seen, the intellectual roots of concern for milk quality in America were a matter of moral outrage at swill-milk, coupled with philanthropic attempts to provide safe milk for infants and public health initiatives to manage the risk of infectious disease. These motivations were also present in Britain but later and on a lesser scale. An early example of a modern commercial experiment was the Manchester Pure Milk Company, which traded from 1898 to 1902 using Copenhagen as its model. Nine farms in Cheshire and Derbyshire supplied the milk, which was ice-cooled and tuberculosis-free.[65] The basic principles were sound but the company collapsed financially because it did not find the necessary demand. It started with £15,000 of capital but served only 1,000 families.[66] Using the same Danish consultant as Manchester, the York Citizens' Public Health

---

58   Smith-Gordon 1919: 11.

59   Macgregor 1890, Anon. 1900: 316, Swithinbank and Newman 1903: 484–96, Jensen 1909: 157–9.

60   Haggard 1911: 86–96.

61   Anon. 1900: 316.

62   For instance Dresden's Milchversorgungsanstalt, founded in 1909. Wolff 1912: 101–03.

63   Murray 1922.

64   Hobday 1927: 739.

65   C.W. Sorensen, consultant, in evidence. Maxwell Committee 1902: Q. 7048.

66   Anon. 1900: 316.

and Housing Committee organized a similar supply of pure milk in 1901. And a third example at the turn of the century was the milk brought in bottles by rail to Newcastle and County Durham from Northallerton. This operated from 1905 until the company was dissolved in the First World War, by which time it had supplied 624,273 gallons.[67]

One of the earliest calls in Britain for graded milk, on American lines, came in 1900 from Nevil Story-Maskelyne, who explained to the Wenlock Committee that ordinary milk should be distinguished from 'prime' milk by the regulation of different proportions of fat content. He suggested 3.25 and 3.5 per cent as appropriate thresholds. At the same hearings F.E. Walker preferred three grades, again differentiated by fat, at 2.4, 3.4 and 4.4 per cent. But Wenlock concluded that

> all who have experience of the milk trade concur in the statement that the public as yet will not buy milk on this basis. Attempts have been and are being made to offer milk of a superior quality, or prepared in a particular way, to the public at a higher price, but the demand is very limited, and so far as it exists it is a demand for purity, i.e., freedom from adulteration or contamination, rather than for extra richness.[68]

Despite this, there *was* a market for high grade milk in wealthy areas. In 1910, for instance, E.J. Walker, a cowkeeper who had a dairy in Sloane Street, Chelsea, was selling bottled milk from tuberculin tested cows at 6d a quart, a 50 per cent mark-up.[69] However, without such a guaranteed market and favourable location in the West End of London, most small dairymen avoided selling graded milk because it would have been difficult for them to guarantee the conditions in which the milk was produced.[70]

The certification of milk was also rare in Britain. The only example known to the Local Government Board in 1907 was that of East Sussex County Council, which had encouraged an experiment in the Eastbourne area, where the local authority issued certificates in recognition that a particular milk supply was pure.[71] An earlier attempt had been made in Sunderland in 1896 to certify tuberculosis-free free herds but only five farmers had volunteered for the scheme.[72] In the first decade of the new century the Borough of Plymouth also introduced a system of identifying the dairy farms that complied with their regulations and issued certificates to that effect. In 1914 it mounted a poster campaign which listed

---

67   NA: FS 17/72, 'Wensleydale Pure Milk Society Ltd'.
68   Wenlock Committee 1900: 389.
69   Anon. 1910: 1501.
70   Letter from D.R. Hughes, Secretary of the Retail Dairymen's Mutual Supply Ltd: to Waldorf Astor, 25 June 1914. Reading University Archives, Astor Papers: MS1066/1/1019.
71   NA: MH/80/3, 'Milk supply: dairies, cowsheds and milkshops', 29 October 1907.
72   Scurfield 1923: 37, Dodd 1905: 13.

twenty such farms, about one-fifth of the town's supply.[73] These certificates could be displayed in shop windows but we have no record of what consumers thought of the experiment. There was no price premium but the incentive was that large firms, hospitals, and clubs sourced bulk supplies with them and that doctors would recommend their milk to patients.

Predictably, it was the large dairy companies with prosperous consumers in cities such as London who were the most enthusiastic trade proponents of certification:[74]

> At the present time there is no protection for dealers who endeavour to sell to the public a clean product; on the contrary, those who desire to do so find it most difficult to persuade purchasers to buy such milk, as careless competitors too frequently argue that their product is at least as clean and wholesome.[75]

These companies no doubt hoped that this would give them some competitive advantage because their smaller rivals would struggle to find high quality milk and the customers to purchase it. On behalf of the smaller retailers, *The Dairyman* tended to be either cynical or hostile:

> We are disposed to think that the great public will regard the efforts of the certified milk cult with sympathetic consideration, which is invariably extended to those humorous little cliques and associations whose object is to amend or benefit the community by amending some trifling unit in our welfare ....[76]

The Reading Vision of Clean Milk

Non-commercial dairy research in Britain was on a small scale at the end of the nineteenth century. The British Dairy Institute, which had been founded by the British Dairy Farmers' Association at Aylesbury in 1888, found itself isolated from a scientific environment and only eight years later moved to University College Reading, where a Department of Agriculture had recently been established with support from the Board of Agriculture.[77] In 1910 the Institute was expanded with better teaching equipment and a larger student body, consolidating its position as the country's leading centre of dairy instruction.[78] Other dairy education at this time

---

73   Anon. 1911, University of Reading Archives, Astor Papers: MS1066/1/1020.

74   *The Dairyman, The Cowkeeper and Dairyman's Journal*, July 1914: 481–2.

75   *The Times* printed a letter applauding the notion of certification signed by The Aylesbury Dairy Co., Express Dairies, Lord Rayleigh's Dairies, Ben Davies and Son, and Welford and Son. *The Times*, 25 June 1914: 10b.

76   *The Dairyman, The Cowkeeper and Dairyman's Journal*, August 1914: 529.

77   Russell 1966: 412, Holt 1977.

78   Mackintosh 1939: 399.

was given mainly by itinerant instructors and at centres such as the Midland Dairy Institute (later renamed the Midland Agricultural and Dairy College) at Kingston-upon-Soar in Nottinghamshire, the Royal Agricultural College at Cirencester, and the Eastern Counties Dairy Institute, Ipswich. These were teaching ventures only and relatively little dairy research was undertaken, apart from work on cattle at Leeds University and Armstrong College, Newcastle.[79] The Liberal Government's new Development Commission sought to change this situation just before the First World War by recommending investment in agricultural education and research.[80] Seventeen agricultural colleges and research institutes were to be given support for teaching and most also became technical advisory centres, with practical support to farmers from county organizers.

Dairying was identified as an important field of agricultural research in the Commissioners' 1911 report and Reading was chosen as an appropriate centre of excellence, on the initiative of Daniel Hall.[81] A Research Institute of Dairying was established there in 1912, with two academic posts and two assistants. This was the modest staff complement of the Departments of Chemistry and Bacteriology soon to be led respectively by John Golding and Robert Stenhouse Williams.[82] In 1919 a Dairy Husbandry Department was added, under James Mackintosh. The Shinfield Manor Estate near Reading, complete with dairy farm and buildings, was purchased in 1921 with the help of Rupert Guinness, Viscount Elveden,[83] who guaranteed a loan. The Institute moved there in 1924, now known as the National Institute for Research in Dairying (NIRD).[84]

The historiography that exists on clean milk has made Robert Stenhouse Williams (1871–1932) a pivotal figure. He had been a hospital bacteriologist and Lecturer in Public Health at Liverpool University before arriving at Reading in 1913.[85] He seems to have been an unusual character. For years he refused to take the position of Director of the Institute because of his egalitarian views, but his stubborn independence and aggressively expansionist tactics made him unpopular in Whitehall.[86] The NIRD struggled financially throughout the interwar period, partly it seems because dairy research was regarded somewhat dismissively as a pseudo-science by the funding establishment.[87] One participant later recalled that many of the bacteriological discoveries at the NIRD in the 1920s and 1930s were never published 'because they were thought not to be sufficiently important or

---

79   Mackintosh 1939: 399.

80   The Development Commission was established by Asquith in 1909 under the Development and Road Improvement Funds Act.

81   For more on Hall and his influence, see Dale 1956, Brassley 2004.

82   *The Times*, 2 September 1921: 5a, Russell 1966.

83   Later the Earl of Iveagh.

84   *The Times*, 21 July 1924: 18c–d, NIRD 1924, Kay 1968, Holt 1977: 12.

85   Burgess 1947.

86   Vernon 1997.

87   Vernon 1997.

generally known, and partly because there was at that time no suitable journal'.[88] Nevertheless, Williams and Mackintosh were influential. Williams, in particular, was consulted by Medical Officers of Health and graded milk producers about bacteriologcal standards.[89] His main interest was 'clean milk' and he frequently spoke out against the need for pasteurization if milk production conditions were properly controlled.[90] This is why Reading was said to be the spiritual home of Certified and Grade A milks.[91] The constituency of Williams and Mackintosh is, nevertheless, difficult to define.

The milk trade had an obvious interest in clean milk because of its losses through souring. Williams estimated that 11 per cent of the value of liquid milk was wasted: 1 per cent of milk went sour, and 9.7 per cent of milk needed to be pasteurized because it was old and near to going sour.[92] Even condensed milk, after its manufacturing process, was prone to the blowing of tins due to bacteria-induced fermentation.[93] But the relationship between the NIRD and the industry remained at arm's length because of Williams' opposition to heat treatment, which by the 1920s had become indispensible to the logistics of the drinking-milk system. For this reason, United Dairies deliberately withheld any financial support.[94] After Williams' death in 1933, the NIRD became more embedded in the industry. Funding started coming in from the Milk Marketing Board and other, commercial sources now that the position on pasteurization had softened.[95] In 1939 one staff member, A.T.R. Mattick, even felt able publicly to advocate the compulsory pasteurization of milk for communities of over 20,000 people.[96]

As early as 1917, Williams had spoken of a Reading vision of clean milk that included freedom from microbes and visible dirt.[97] He did realize, however, that this would be difficult to achieve for two reasons. First, in his opinion, 'the consumer has had no real understanding of the difference between clean and dirty milk and has therefore been quite unwilling to pay more than a flat rate for his milk supply'.[98] Second, many farmers were thought to have no knowledge of contamination with microscopic particles. But Williams was convinced that the

---

88   Davis 1983: 4.

89   Williams and Mattick 1922.

90   *The Dairyman, the Cowkeeper and Dairyman's Journal*, August 1922: 538–9, October 1924: 78–9, August 1925: 630–33, Enock 1943: 32.

91   Orr 1925: 33, Maddock 1926: 537.

92   Williams 1917: 25.

93   Hiscox 1923.

94   Davis 1983: 3.

95   Kay 1944: 123.

96   The Division for the Social and International Relations of Science of the British Association in Reading. See Mattick 1944: 142.

97   Williams and Cornish 1917. Much of the Reading work was prefigured in the USA. For a much earlier discussion of clean milk, aimed at the farmer, see Belcher 1903.

98   Williams and Cornish 1917: 6.

blame for this was collective: 'if the farmer is lacking in education on this point, then the fault lies with the nation.'[99]

Williams and Mackintosh soon turned their attention to clean production methods. The old cattle sheds at Shinfield were converted by laying concrete floors, and improving the lighting, drainage and ventilation.[100] Mackintosh manipulated these variables, along with trials on methods of stock husbandry and milking methods, to see whether a Reading model of optimum clean milk production could be established. Specifically he looked at:

- the hygienic condition of the cowshed, including the cleanliness of the stall and cowshed at milking time;
- the animals, including their dry grooming and washing before milking;
- milking methods;
- the importance of sufficient light in the farm dairy;
- provision of clean water;
- temperatures for cooling, storage and transport;
- the cleanliness and design of dairy equipment such as milking pails, milking machines, strainers and straining materials, milk coolers, thermometers, and milk churns;
- the sterilization of utensils by scalding, steam or chemicals.

Already by 1920, Williams and his research group claimed to have discovered five basic principles.[101] First, they recommended that milk should be cooled within three hours of milking as a means of slowing the growth of bacteria and the souring process. Since many farmers lacked a water supply, this was not a practical proposition for them, although there was the alternative of investment in mains water and machinery at small creameries near to any railway stations used by groups of farms. Second, they confirmed the findings of the Barlow Committee that the covered type of milking pail made a significant difference because dust and dung was less likely to find access. Third, steam and the sterilization of cans and dairy utensils reduced contamination greatly.[102] Cumming and Mattick had found that bacteria were generally present in milk churns, and later a comparison of the costs of the sterilization of utensils versus cooling came down in favour of the former as a means of increasing the shelf-life of milk, for cost reasons.[103] Fourth, grooming and washing cows diminished the dirt at source; and, fifth, a trained and motivated work force was also identified as a key factor. In sum, the

99   Williams and Cornish 1917: 8.

100   Russell 1966.

101   Freear, Buckley and Williams 1919, Knight, Freear and Williams 1920. See also Mackintosh 1922, Mattick 1927.

102   On steam, see also Mattick 1921, Hoy and Williams 1921, Hoy 1923, 1924, Hoy and Hickson 1924, Proctor and Hoy 1925.

103   Cumming and Mattick 1920, Hall 1951: 4.

NIRD discovered that 'adequate light and water, and clean utensils were found to be essential, and the required degree of cleanliness could very rarely be attained without steam sterilization'.[104] In addition, clean milking required 'workers who possess not only intelligence and enthusiasm but also that kind of vision which is not dependent upon the eye for its conception of dirt'.[105] The encouragement of clean milk also succeeded best where there were financial returns for farmers.

The NIRD was instrumental in the standardization of laboratory methods in the bacteriological examination of milk, and it helped with the establishment of the Advisory Bacteriological Service.[106] From 1924 the Ministry of Agriculture set up a number of advisory bacteriological posts in various cities to test milk for farmers and dairymen, and these were funded 40 per cent by the County Councils and 60 per cent by the government.[107] The first dairy bacteriologist appointed was in Bristol, to serve Gloucestershire, Herefordshire, Somerset, Wiltshire and Worcestershire. He was followed in the same year by others at Cambridge and Wye (Kent), and by 1939 there were 13 at various regional centres.[108] These appointments seemed to have been hastened by the fact that Reading was physically unable to provide a testing service for the whole country. Practices varied from region to region:

> In most provinces the advisory officer [by 1939 had] established relationships with producers and with local authorities, whereby he examines milk samples at regular intervals or as required, and makes use of the results for advisory purposes. At some centres samples from official samples are not received at all, at others the adviser accepts samples submitted by the authority supervising milk supplies, but only in an advisory capacity; no official action is taken on the results of the tests beyond suggesting to the producer that advice is needed and can be obtained. These methods have on the whole been very successful.[109]

---

104   Mackintosh 1939: 415.

105   National Institute for Research in Dairying 1924: 42, National Institute for Research in Dairying 1929: 7.

106   Mackintosh 1939: 416.

107   Hanley 1939: 121. In addition, the 1925 Tuberculosis Order provided for the submission of samples of milk or faeces to approved pathological laboratories to assist with diagnosis. One hundred and four Local Authorities applied and in 1925/6 the Ministry of Agriculture approved 41 laboratories in 27 counties. Ministry of Agriculture and Fisheries 1927. *Report of Proceedings under the Diseases of Animals Acts* London: HMSO.

108   In addition to Bristol, there were Aberystwyth (Mid Wales), Bangor (North Wales), Cambridge (Bedforshire, Cambridgeshire, Essex, Huntingdonshire, Norfolk, Suffolk), Harper Adams College (Shropshire, Staffordshire, Warwickshire), Leeds (Yorkshire), Manchester (Cheshire and Lancashire), Midland College (Derbyshire, Leicestershire, Lindsey, Kesteven, Nottinghamshire, Rutland), Newcastle (Northumderland, Durham, Cumberland, Westmorland), Reading (Berkshire, Buckinghamshire, Dorset, Hampshire, Middlesex, Northamptonshire, Oxforshire, Isle of Wight), Seale-Hayne College (Cornwall, Devon, Isles of Scilly), South Eastern Agricultural College (Kent, Surrey, Sussex).

109   Hanley 1939: 119.

Another element of the NIRD's outreach programme was the organization of demonstrations at agricultural shows and the provision of advice on the planning of clean milk competitions. The first large-scale competition in Britain was held in Essex in 1920, with both Williams and Mackintosh as judges.[110] The Ministry was slow to respond to this initiative. Its 1919 instruction leaflet *The Dairy Cow and Milk Selling* makes little mention of clean milk.[111] But in 1924 it did, at last, give its *imprimatur* by publishing a pamphlet entitled *Guide to the Conduct of Clean, Milk Competitions*.[112] The same year, some of its county organizers began competitions around the country under standardized rules,[113] creating what one sceptical journal called 'a positive epidemic' of clean milk competitions.[114] In 1925 the county organizers arranged lectures and demonstrations attended by a total of 18,620 farmers, but there was no consistency because their duties under the graded milk schemes varied from county to county.[115] In some they acted in an advisory and non-official capacity, while in others they reported in detail to the local authority on the suitability of buildings, equipment, and milking methods.[116] By 1927/8 there were competitions in 42 counties involving 1,144 farms and 31,727 cows.[117] There were even inter-county contests, with cups given by the National Milk Publicity Campaign, and the whole project was judged a success in raising consciousness among agriculturalists.[118] These competitions continued until 1936, when the Milk Marketing Board's Accredited Milk scheme started and financial incentives took over from evangelism.[119]

The NIRD was undoubtedly the intellectual focus of the clean milk movement between the wars and of dairy bacteriology generally.[120] In 1929 Williams inaugurated an annual series of academic conferences, held at Reading University, which laid the foundations of what in 1931 became the Society for Agricultural Bacteriologists (later renamed the Society for Applied Bacteriology, and most recently the Society for Applied Microbiology).[121] Gradually the NIRD's practical ideas also diffused out into the farming community. Clean milk practice does not seem to have been taught systematically in agricultural colleges until 1913, and

---

110   *The Dairyman, The Cowkeeper and Dairyman's Journal*, August 1920: 460, Williams and Mattick 1922: 389, Billington [1954]: 1, Russell 1966: 419.
111   Board of Agriculture and Fisheries 1919.
112   Ministry of Agriculture 1924.
113   Blackshaw 1926.
114   *The Dairyman, The Cowkeeper and Dairyman's Journal*, December 1924: 197.
115   For more on the county organizers, see Blackshaw 1924.
116   Hanley 1939: 106.
117   National Institute for Research in Dairying 1929. In 1920 there was one competition, two in 1921, three in 1922 and 1923, 19 in 1924 and 33 in 1925. See also Hole 1924, Williams 1925.
118   Harvey and Hill 1936.
119   Billington [1954]: 2.
120   Kay 1950.
121   Thomas 1963.

not even at Reading until 1916,[122] but in 1920s Williams hit upon several means of publicity. The first was to use the new farm as a demonstration unit and it seems that Shinfield gradually became a place of pilgrimage.[123] Short courses were mounted there for county dairy instructors, sanitary inspectors, farmers, milk dealers, veterinary surgeons, and Medical Officers of Health. Second, Williams and Mackintosh arranged for demonstrations of the principles of clean milking, for instance at agricultural shows, and frequently attended themselves to give lectures. Third, the NIRD achieved official status for its work through MAF bulletins on methods of clean milk production. As a result of the NIRD's efforts, six firms in Reading were selling 'Grade A' milk by 1923, and four years later one-sixth of the town's milk was graded. There was early enthusiasm for graded milk also in Leicester, where the pioneer was Dr John Donald, and in Cardiff, Newcastle and Scarborough.[124]

Williams' successor in 1933 was Herbert Davenport Kay (1893–1976), a biochemist, who was Director until 1958.[125] The NIRD received research money under the Milk Act of 1934 and Kay came into contact with John Boyd Orr, then a powerful figure capable of opening doors in Whitehall. In 1936 the function of Imperial Bureau of Dairy Science was added to its portfolio, and in 1938 the inaugural Chair of Dairying was Edward Capstick. The NIRD had 50 staff in 1929, 68 in 1933/4, and by 1946 this had grown to 1,007, an astonishing rate of growth from the modest beginnings of 30 years earlier.[126]

Bottling Up Purity

The Express Dairy experimented with bottling retail milk in 1884, using a ginger beer type of bottle secured by wire, and in 1894 James Niven, Medical Officer of Health for Manchester, suggested compulsory bottling as a means of excluding dirt.[127] The disc-stoppered glass bottle was tried by several companies in 1897/8 but the charge of 1d extra per quart was too much for most consumers.[128] The first large-scale bottling in Britain started in 1905–1906 with bottles imported at first from Europe and America until the first Owens automatic rotatory glass container-making machine was first imported in 1912. Fibre discs were the most common

---

122   National Institute for Research in Dairying 1924. In 1917 courses began for dairy teachers.
123   Enock 1943: 62. The British Dairy Farmers Association had started the British Dairy Institute in Aylesbury in 1888. It moved to Reading in 1896 in association with University College, Reading, and in 1910 became the NIRD.
124   Williams and Mattick 1923: 11.
125   Blaxter 1977.
126   Crossley 1958.
127   Morgan 1964: 31.
128   Astor Committee 1919: 683.

means of closure but were replaced in the 1930s by aluminium.[129] Early bottles had grooves to hold the stopper disc and this made them difficult to clean.[130]

London had about 10 per cent of its milk bottled by 1914, and the proportion grew rapidly from the early 1920s.[131] By 1927 there were 50 million bottles in circulation and large companies such as United Dairies quickly adopted this container for all of their supply.[132] By 1930 70 per cent of Glasgow milk was bottled but it remained a phenomenon of the cities and larger towns until after the Second World War.[133]

In 1921 the Standing Committee on Trusts in 1921 favoured bottling because 'there can be little doubt that the most hygienic method of distribution would be for all milk to be delivered in bottles'. A key advantage was in-bottle pasteurization.

Opinion in the trade was split between the larger firms, who were in favour, and the small producer-retailers, who opposed bottling with the same vigour that they resisted pasteurization.[134] To them, it was an additional expense that could not be justified by their small turnover. They saw it as a technology benefiting the larger dairies and so further concentrating the trade in fewer hands.[135] The argument was finally settled in 1954 when new Special Designations regulations required all pasteurized milk to be sold in bottles as a guarantee of quality.

Conclusion

As we will see in Chapter 10, the clean milk movement was successful in persuading the milk industry, the public and politicians of the need for hygiene. But in purveying a strong, simple message, it tended to neglect the hazard of specific diseases. This led Graham Wilson to complain in the mid-1930s that more attention was being paid to the cleanliness than the safety of milk, which after all should have been the object of public measures.[136] Also, the early interest in visible dirt meant a long period of concentration on clean conditions of production rather than upon hygiene at later stages of the system, including the home, and the under-emphasis of keeping quality, a key variable for both the milk industry and the consumer.

---

129   Enock 1943: 194.

130   Hobday 1927: 739. See also Gibbs-Smith and Hobday 1928.

131   Letter from C.F. Green, manager of a large milk depot in Stamford Hill, dated 7 April 1914. University of Reading Archives, Astor Papers: MS1066/1/1018. See also Forrester 1927: 91.

132   *Parliamentary Debates*, 211, 1927: 234–5. All United Dairies milk was bottled by 1929. Davies 1930: 8.

133   Macgregor 1930.

134   *The Field*, 137, 3551, 15 January: 62.

135   Anon. 1921a.

136   Wilson 1936: 494.

Chapter 10

The Material Politics of Milk

Counting Bacteria

We started Chapter 9 with a discussion of dirty milk. The measurement of sediment was a central concern of milk hygiene at the turn of the twentieth century but, as we saw, the definition of clean milk gradually changed in the following decades to become principally bacteriological. In this chapter we find that bacteria gained a curious kind of power, to the extent that the regulation of cleanliness in the milk industry from the First World War onwards was, and still is, dominated by counting these organisms. The microscopic world gained a privileged status and what had been called 'visible dirt' became invisible again. Bacteria were dutifully cultured on Petri dishes, like crops in miniature fields, and politicians then set limits of the number that were acceptable in any particular sample. The bacteria in these statements were usually undifferentiated, although by the First World War increasingly sophisticated science had become available to comment on the significance of the individual species. The underlying assumption of what we might call bureaucratic bacteriology was that the more bacteria there were, the worse it was for the consumer. Interestingly, this is still one of the quality criteria for milk in the milk regulations of the present day, now known as the Total Bacterial Count.

As early as 1828, the power of the compound microscope had revealed to the popular imagination the liveliness of the bacterial world of water, with William Heath's satirical print of the 'monster soup' that was London's supply. After that, the self-made identity and professional status of the discipline of bacteriology depended upon there being micro-environmental issues in the food and water supply, although much of the microscopic population of milk was shown to be made up of non-pathogenic micro-flora and fauna (Table 10.1). It was the seeming lack of human control at this scale that was the challenge.

The science of bacteria in milk was stimulated by Pasteur and Koch, especially the latter's discovery of the tubercle bacillus in 1882. Bactériologie, as it was classified in the *Bibliographia Lactaria*, was accelerating as a science in the last two decades of the nineteenth century.[1] It was responsible for 43 papers on milk worldwide in the 1880s and 332 in the 1890s. From there it grew further as one of the 'useful', applied divisions of microbiology.

In the late nineteenth century, bacteria were discussed in books on dairy manufacture mainly because of the role of lactic acid streptococci in cheese-

---

1   Rothschild 1901, 1902.

Table 10.1   The principal bacteria found in milk

Non-pathogenic
- Acid-forming bacteria, such as *Streptococci lactis*, *Streptococci foecalis*, and lactobacilli. They ferment the lactose in milk and cause souring.
- Gas-forming bacteria, for example coliform bacilli
- Proteolytic bacteria
- Alkali-forming bacteria
- Inert bacteria.

The main pathogens
- *Mycobacterium tuberculosis*
- *Brucella abortus*
- Streptococci causing mastitis in cattle, and scarlet fever in humans (*Streptococcus pyogenes*)
- *Staphylococcus aureus*, may cause gastro enteritis
- *Corynebacterium diphtheriae*
- Typhoid, paratyphoid, food-poisoning, dysentery
- Q fever (*Rickettsia burneti*).

Source: Wilson and Miles 1975: 2669–72.

and butter-making and in the fermentation process of products such as kefir and koumiss. This was the positive side of bacteria, and the negative aspects – souring, slimy or ropy milk, red milk and disease – received less attention at first. In 1895 Aikman's book, which had the word bacteriology in its subtitle, only about 40 per cent of the text was about bacteria, and the following year the English edition of Fleischmann's internationally famous textbook on milk, had only 16 pages devoted to bacteria out of a total of 330. We have to wait for the textbooks on dairy bacteriology by Freudenreich (1893) and Russell (1894), a series by Herbert Conn (1901, 1903, 1907), and Swithinbank and Newman's excellent study (1903), to see the specialism beginning to mature. Certainly by the time of the First World War it was no longer possible to publish a book on milk, academic or popular, which did not have a substantial discussion of bacteria and their consequences. By that time it was known that large numbers of lactic acid bacteria were common when milk was not properly cooled but that they were not as harmful in the same way as the faecal bacteria associated with manure or the pathogenic mycobacteria of tuberculosis. Buttermilk commonly contained tens of millions of bacteria but was generally considered safe to drink.[2] It was also known that any pathogenic bacteria in warm milk multiplied quickly and that this is where the principal hazard lay.

Early work on the bacteriology of milk fell into two streams. On the one hand there was the exploration of a new microscopic environment, with the

---

2 Heinemann 1921: 410.

discovery that milk shared many of the bacteria previously found in water. The way that these were reported suggested the extraordinary exoticism of a coral reef. Discovering whether bacteria were active and harmful was the priority, along with the temperature at which they could, if necessary, be de-activated. On the other hand, an interest quickly developed on the origins of bacteria and their potential for contamination. From the mid-1890s, papers began appearing that compared the bacterial load of milks from different situations and different cities in order to identify the attendant risk factors. Samples were taken in the cowshed, in shops, and in the domestic setting in order to establish the circumstances of, and the responsibility for, contamination. In Britain, Sheridan Delépine, the Professor of Bacteriology at Owens College in Manchester, was a leading figure in both the academic and public health versions of this type of study.[3] He pioneered laboratory methodologies and published widely on behalf of the Manchester Corporation and the Local Government Board.

The counting of bacteria for the detective work of public health was not straightforward. Somehow the organisms had to be grown in a culture medium before they could be counted. But which culture medium and which method of counting? Only by standardizing procedures was it possible to compare results, but progress was slow at first. Koch's agar plate method was refined first by Petri and later by others.[4] A problem for public health control, though, was the lack of precision in the early bacteriological work. At least with the chemical analysis of milk there was, by the time of the First World War, a consensus about methods and replicability of results. But the results of bacterial examinations varied sometimes within wide limits and it was impossible to say that a specific disease germ was absent just because it could not be found in a particular sample.[5]

By the 1920s there were two methods of counting bacteria in milk: by cultivating and counting colonies on agar plates or by the direct microscopic examination of milk stained on a slide. The former was more common but involved a time delay for the incubation of the bacteria and the results were therefore never ready until after the milk had been consumed and had possibly done its damage.[6] It was an analytical system performing perpetually in retrospect. The means of preparing the agar plates varied from laboratory to laboratory and this was the main reason for some extraordinary inconsistencies in results, as illustrated by Table 10.2. The method of counting of colonies was another source of frequent disagreement. It was not until 1929 that the Ministry of Health's Memo 139/Foods set out the techniques for the bacteriological examination of milk, as later codified by the British Standard Methods for the Sampling of Dairy Products, BSI No. 809 (1938).[7]

---

3   Worboys 2004.
4   For an account of the early growth media, see Swithinbank and Newman 1903: 35–41, Bulloch 1938, Davis 1950: 459–69.
5   Heinemann 1921: 412.
6   Parker 1917: 439–47, Heinemann 1921: 413.
7   Howell 1934.

Table 10.2   **Variations in bacterial counts according to plate media**

	Bacteria observed per cc using different media		
Sample	Ordinary agar	Litmus-lactose agar	Ministry of Health agar
1	20,000,000	24,000,000	5,000
2	8,000,000	36,000,000	76,000

Source: McIntosh and Whitby 1931: 148–9.

Use of the plate method produced results that astonished and alarmed contemporary commentators. George Newman, as Medical Officer of Health for Finsbury in central London, commissioned Alexander Foulerton, a bacteriologist at the Middlesex Hospital, to look at a number of samples from his district. Foulerton found life in every one, including staphylococci, streptococci, diplococci, bacillary forms, yeasts, sarcinae, epithelium, acid-fast organisms, pus and dirt. These were treated as equally serious forms of contamination, although without any indication of the specific consequences for human health.[8] Table 10.3 is another early example, showing the results of a survey of retail outlets in different parts of central London, particularly indicating the hazard of buying milk from street vendors, whose milk was often of poor quality in several ways. Table 10.4 compiles Manchester data from a decade later and suggests that the milk provided in hospitals there also left a great deal to be desired. Delépine performed these analyses and found that the souring time of 96 per cent of the samples was less than 20 hours.

One risk factor identified by many Medical Officers of Health was the sale of milk in small quantities by purveyors, who were poorly regulated and rarely inspected. Their premises were often unsanitary. In St Pancras, for instance, a survey just before the First World War of 389 shops selling milk showed that 52 per cent dealt in less than two gallons daily, of which 63 per cent were 'general' shops.[9] In this last category, half had floors that were not kept properly clean, and three-quarters did not have covered milk vessels. These small purveyors were concentrated in poor areas, and it was also here that the risk of domestic contamination of milk was greatest. Working class housing at this time rarely had facilities for the hygienic storage and preparation of food.[10]

Popular though it was with most practical bacteriologists, the plate count method came under critical scrutiny in the 1930s. According to Wilson, it was unsatisfactory for several reasons. The technique was complex, difficult to standardize and required highly skilled workers. It also had a very large experimental error of up to ±90 per cent but, ironically, had a spurious aura of accuracy. Irregularities in the

---

8   Newman 1903: 44.

9   *Annual Report of the Medical Officer of Health to the Metropolitan Borough of St Pancras*, 1913. For a very similar account of Finsbury, see Newman 1903.

10   Of 2,669 houses inspected in Colchester 1905–8, for instance, 92.8 per cent were without any larder accommodation. Savage 1912: 2.

Table 10.3   The contamination of milk retailed in London, 1903–4 (million bacteria per cubic centimetre of milk sampled)

	Good class shop	Poor class shop	Street vendors
Lambeth			
Inner wards	1.9	4.7	5.2
Outer wards	1.7	4.2	3.2
Finsbury	1.8	2.1	–
City of London	4.8	–	–
Islington	1.6	–	–
Westminster	1.6	–	–
Holborn	4.8	–	–

Note: According to Dodd 1905, 'good' milk contained 50,000 to 500,000 bacteria per cc.

Sources: Newman 1903: 40; *Annual Report of the Medical Officer of Health to the Metropolitan Borough of Lambeth*, 1904.

Table 10.4   Bacteriological results of 70 samples in Manchester, 1916–17

	Bacteria per cc			
	<10,000	10,000–100,000	100,000–1,000,000	>1,000,000
Dairies, winter 1916/17	1	6	0	0
Hospitals, winter 1916/17	0	8	19	11
Hospitals, summer 1916	0	0	2	8
Schools for Mothers, 1916/17	3	2	0	0
Shops, spring 1917	3	2	2	3
Total	7	18	23	22

Source: Delépine 1921: 7.

clumping of organisms made counting problematic because the average number of bacteria was variable from one milk to another and from time to time.[11] What was needed was a means of classifying the bacteriological quality of milk without having to count bacteria under the microscope or on agar plates. This needed to be quicker, more accurate and less labour-intensive than the plate method.

Methylene blue was one possible answer. It had been known since 1900 as a test for the bacterial loading, and in milk its change from a blue oxidized to

11   Wilson 1935.

a colourless, reduced state was known to be related to the quantity and type of organisms present, which, as they grew, used up the oxygen.[12] In its unimproved state the test was a rather crude and only picked out gross contamination, but in the mid-1930s a modification by Wilson made it suitable for all types of milk and propelled it into the front line of public health as a possible universal test for the bacteriological quality of raw market milk. Wilson's innovation was in inverting the test tube every 30 minutes in order to keep the fat and any attached bacteria more or less uniformly distributed throughout the milk. This greatly improved the test's accuracy and made it quicker.[13] Unlike the plate method, the methylene blue test was cheap, simple and subject to only small variations in experimental results. Wilson's links with the Ministry of Health and their high regard for him meant that the methylene blue test was officially adopted for raw, graded milks in the Milk (Special Designations) Order, 1936.[14]

The coliform count was another widely used test in the early twentieth century. It was adapted from the water industry and involved the incubation of organisms in MacConkey broth. These bacteria were from the coli-aerogenes group, which are associated with faecal contamination and with souring.[15] They produce acid and gas from lactose in the presence of bile salt. In water these *Bacteria coli* were of human origin but in milk they were from cattle, so comparability was limited. Coliform bacteria do not grow in direct proportion to the original contamination and, in short, the test had 'little advantage over a standard plate count'.[16] It was, however, used extensively from the first to the fourth decades of the century.

During the Second World War the resazurin test was introduced as an alternative to methylene blue in the testing of the bacteriological keeping quality of milk.[17] Farmers referred to it as the 'reassuring' test.[18] There were two versions: a 10-minute test used with any milk that smelled off and a one hour test for routine grading. Resazurin worked well with churn milk that had been cooled with water but for bulk tanker milk it was less effective than the plate count method. In addition, there was the Phenolphthalein test for sourness to check the opinion of the 'sniffer' who was employed by every milk collection centre to detect 'off' aromas.[19]

There were many other tests and variations on the main bacteriological themes.[20] This is not the place to describe them but an important further point to

---

12   Arup 1917: 20.

13   Wilson 1936: 496.

14   Ministry of Health, *On the State of the Public Health: Annual Report of the Chief Medical Officer for 1935*. London: HMSO: 136.

15   Davis 1950: 125–9.

16   Wilson 1936: 495.

17   Chalmers 1945: 34–46.

18   Davis 1983: 7.

19   Jenkins 1970: 43.

20   Two fast and straightforward tests of keeping quality used in creameries were the 'clot on boiling' test and 'titratable acidity' test.

make here is that their cost effectiveness was a key consideration. In a sense, that is directly comparable with the chemical tests that we discussed in Chapter 3; those using bacteriological methods had to decide on a balance between precision and cost. In an industry that dealt every day with a highly perishable product, speed was essential and even the incubation period of just a few days using the plate count method seemed intolerably slow. Table 10.5 neatly summarizes these factors.

Table 10.5   The accuracy and speed of bacteriological tests

Test	Time required	Accuracy in relation to keeping quality	Advantages	Disadvantages
Smell	3 seconds	*	Immediate	Subjective
Sediment	30 minutes	*	Best psychologically. Very simple. Immediate result.	Weak correlation with bacteriological quality
Resazurin	10 minutes	***	Simple, inexpensive, quicker than MB, accurate enough for grading	Less accurate than MB
Methylene blue	2–8 hours	*****	Simple, inexpensive, very accurate measure of keeping quality	Cannot be used for pasteurized milk
Plate count	2 days	****	Good psychological effect	Fairly expensive
Coli	2 days	****	Measures both keeping quality and hygienic aspects	Expensive microscope needed, fatiguing for worker, not accurate for clean milk
Keeping quality	2–3 days	*******		

Source: Modified from Davis 1951: 37, 563.

Despite the technical issues that we have identified with the use of bacteriological techniques from their inception in the 1880s through to their maturity in the 1930s, the idea of bacteriological quality gradually began to dominate thinking, along with fat content in the compositional sense. But, while the chemistry may be thought of as static, this was certainly not true of bacteria, which were dynamic, and it was their rapid action that made milk so perishable and so potentially dangerous for consumers. In the next section we will build on this foundation by looking at how the concept of bacteria came to dominate milk politics after the First World War.

Clean Milk: Politics and Policy

Wilfred Buckley (1873–1933), a shipping magnate turned country estate owner, was an early leader of the British clean milk movement.[21] He had started milk production on his farm in 1907 using methods that he had observed in America.[22] One factor seems to have been that his daughter had earlier contracted bovine tuberculosis and he wanted to eliminate this disease from the milk supply. He later recalled that 1910 was the year he realized that certification was the only way forward to general improvements. This was when he helped to convene a committee to work on the idea of clean milk.[23] It was chaired by Sir Thomas Barlow, President of the Royal College of Physicians, and undertook practical experiments at Buckley's farm at Moundsmere, near Basingstoke from November 1910 to September 1911, with bacteriological analysis provided by the Lister Institute in London.[24] The first results were disappointing because, despite Buckley's best efforts, his milk contained 35,000–75,000 bacteria per cc. But the introduction of covered pails reduced this to an average of 9,144 per cc and it soon became obvious that even relatively small changes in practice could make a difference. The Barlow Committee sent its report to the Local Government Board in October 1911, advocating the adoption in Britain of an American-style system of milk certification for cleanliness. This drew approbation from *The Times* and support also came from the leading dairy companies in London, mainly those that saw a market for a high quality product.[25]

Another player at this time was the Pure Food and Health Society of Great Britain, which established a Central Milk Certification Committee. In December 1913 it invited the British Medical Association to send representatives, and Dr T. Barrett Heggs of Sittingbourne was delegated to attend.[26] The Society was considering the possibility of voluntary milk certification and was looking at the New York experience. They also discussed the possible promotion of milk

---

21   Sheail 2003. Obituaries: *The Times*, 28 October 1933: 7, 1 November 1933: 7, *The Milk Industry*, 14, 1933: 84.

22   Buckley's mother and wife were American and he lived for ten years in New York from 1895, no doubt aware of the work of Nathan Strauss. Various newspaper articles on his milk enterprise mention the American inspiration. See, for instance, *The Times*, 14 November 1910: 4a–b. He also publicly advocated using the American system of certification. *The Spectator*, 14 March 1914: 433–4, *Mark Lane Express*, 20 July 1914, *Farmer and Stockbreeder*, 20 July 1914.

23   Buckley 1922. It is not clear whether the committee was the precursor of the National Clean Milk Society. *The Times*, 25 September 1912: 3f.

24   *The Times*, 12 October 1912: 11f, 14 October 1912: 4a, 14 November 1912: 4a–b, 6 December 1912: 15f. The Lister Institute was reorganized in 1906 and began taking on routine public health analysis work in order to boost income. See Chick et al. 1971: 77.

25   A certified milk supply, *The Times*, 12 October 1912: 11G.

26   Wellcome Contemporary Medical Archives Centre: SA/BMA/F105.

certification centres throughout the country. However, in March 1914 Heggs reported back negatively. It was his opinion that the Society was

> not prepared to deal with the subject in a sufficiently comprehensive and detailed manner to warrant the interest of the BMA, but was chiefly concerned with preparing for the society a scheme whereby as soon as possible the Society could undertake the granting of their authority to dairymen who are prepared to conform to certain sanitary and other requirements.[27]

Lobbying of government on the subject of certification came in 1914 from the Agricultural Organization Society and Barlow's informal committee.[28] Buckley was a member of both delegations and he suggested three milk grades: certified, inspected, and ordinary. The Barlow Committee also wanted legal protection for the word 'certified' and the introduction of official machinery for carrying out certification.

Despite the outbreak of war in 1914, the clean milk movement continued to gather momentum in Britain. A National Milk Hostels Committee of the Pure Food Society was formed after hostilities began 'for the purposes of securing and supplying fresh, pure and wholesome milk to members of those poor families who would be most likely to suffer poverty and loss in whose case a supply of milk was an urgent necessity'.[29] The Chair was Muriel, Viscountess Helmsley, and there were 25 Vice Presidents, including names in milk politics such as Christopher Addison, Waldorf Astor, and Charles Bathurst. But this proved to be largesse on a miniature scale, supplying only 200 gallons a week to 2,200 recipients.[30] Astor soon withdrew from such a feeble effort, but he retained his interest in the subject. Already in 1912 he had chaired the Local Government Board's Departmental Committee on Tuberculosis, no doubt motivated by his own personal experience of having contracted the bovine variant of this disease.[31] In April 1914 he advocated the principle of grading in the House of Commons during a debate on Milk (Supply and Sale).[32] This was the same debate in which Jack Hills, MP for Durham, argued for the use of the American score card system as a means of allocating points during an inspection for each aspect of production: the cow's health, the location and construction of the cowshed, cleanliness of the cowshed and cleanliness of the utensils.[33] This was also advocated outside parliament by Buckley, who, with the

---

27   Wellcome Contemporary Medical Archives Centre: SA/BMA/F105.
28   NA: MH/80/5, 'Milk grading. Notes of interview with representatives of Agricultural Organization Society at Local Government Board on Friday, 13 March 1914'.
29   Pure Food Society [1915].
30   University of Reading Archives, Astor Papers: MS1066/1/1021.
31   Astor was MP for Plymouth from 1911 to 1919. He then moved to the House of Lords as Viscount Astor.
32   Astor 1931: 771, *Parliamentary Debates*, 61, 1914: 1023–67.
33   *Parliamentary Debates*, 61, 1914: 1023–67.

assistance of James Mackintosh of Reading University College and Ben Davies of United Dairies, devised a score card suitable for British conditions. By 1916 this had been adopted by the city authorities in Birmingham and Bradford, and was experimented with by the Agricultural Organization Society.[34]

As a result of the efforts of Astor and others in parliament, provisions were made in the Milk and Dairies Act of 1914 for the grading of milk by quality. But the postponement of this legislation in 1915 due to the war was a set back.[35] Astor's private correspondence shows that he received support for his ideas on clean milk from a wide range of individuals and organizations and in March of 1914 it seems that he began to cooperate with Buckley.[36] Buckley's letters begin with the rather formal salutation of 'Dear Mr Astor', then after six letters or so it becomes the more familiar 'Dear Astor' and for many years subsequently they worked closely.[37] It seems that Buckley approved of Astor's plan of asking Parliamentary questions on milk but wanted to talk to him before he proceeded any further. Buckley was having bacterial counts carried out by the Lister Institute on milk that he had purchased in London shops and he wanted the results to be widely known. The average in the West End, for instance, was astonishingly high at three million bacteria per cc. Buckley suggested possible parliamentary questions for Astor to table, such as whether the Local Government Board was proposing to grade milk; the need to control descriptions such as 'nursery milk'; and whether any action would be taken to reduce railway rates for bottled milk in order to encourage a cleaner supply. He fed Astor with information from America, such as the contact details of Charles North in New York.[38] And he also reported to him on friends and opponents of the 1914 Milk and Dairies Bill and recommended courses of action, for example writing to newspapers or sampling the milk used in the Palace of Westminster.[39] The latter proved to be especially useful propaganda because the sample taken of House of Commons milk yielded an unimaginable 73,080,000 bacteria per cc.[40]

Both Buckley and Astor continued to be active during the war and expanded their arena of milk politics. Although not a Member of Parliament, Buckley was asked to become Director of Milk Supplies at the Ministry of Food in August 1917, and he managed to persuade his patron to allow licences and higher prices for hygienic milk producers. Lord Rhondda unexpectedly died before the regulations came into force but the new Food Controller, John Clynes, a prominent Labour

34   Anon. [1916a]: 31–3.
35   Certified milk was mentioned in Section 1 of the Milk and Dairies (Consolidation) Act (1915).
36   Astor Papers: MS1066/1/1018.
37   Astor Papers: MS1066/1/1020.
38   Astor Papers: Buckley to Astor, 5 and 11 April 1914.
39   Astor Papers: numerous letters from Buckley from April to July 1914.
40   Letter from C.J. Martin, Director of the Lister, 20 April 1914. Astor Papers: MS 1066/1/1018.

MP, was also in favour and, as a result, grading was established on a small scale in that year.

Meanwhile, Astor was moving in higher political circles, first as a Parliamentary Secretary to Lloyd George, and later in the Ministry of Food. Then in March 1917 he was appointed to chair an official enquiry into the production and distribution of milk. This sat for two years and gave him ample time for reflection about grading and other milk matters.

A significant moment in our story was the foundation in 1915 of the National Clean Milk Society by Wilfred Buckley, who acted as its Chairman for the next thirteen years.[41] The Society's stated aim was 'to raise the hygienic standard of milk and milk products and to educate the public as to the importance of a clean and wholesome milk supply'.[42] See Table 10.6 for its objectives. Its performance was varied. Over the years it produced about 40 pamphlets, 20 leaflets, some 14 articles in the *Observer* newspaper, and numerous other forms of publicity about clean milk.[43] Buckley, who was said to be 'an inspiring leader in the clean milk campaign', toured the country giving lantern slide lectures, stressing simple messages.[44] His mantra was that 'milk must be kept clean from the beginning'.[45]

Support for certified milk came in principle from the Consumers' Council, the Central Agricultural Advisory Council (Ministry of Food/Board of Agriculture), the Travelling Milk Commission (Ministry of Food), and the Associated Milk Producers' Council. Some members of these quangos had the interests of consumers at heart but the Astor committee made it clear that their advocacy for grading was really a covert means of supporting the dairy industry:

> It has become generally recognized that if there is to be a real improvement in the cleanliness and hygienic quality of the milk supply of this country, it is essential to abandon the theory that all milk is of one quality, or that it is possible to raise the whole of the milk supply rapidly to the highest grade. The public realise that the eradication of disease or the use of more skilled labour to produce cleaner milk necessitates additional cost to the farmer, which he will only and can only be expected to bear if he gets in return either some official certificate of the better quality of his milk, which enables him to charge a compensating price, or to receive a higher price under an official system of fixed prices. Differentiation between different grades of milk so that producers and distributors of clean milk may be remunerated for the additional labour and expense incurred is essential.[46]

---

41 *The Times*, 3 June 1915: 5c, Buckley 1931.
42 Anon. [1916a].
43 Waldorf Astor was proprietor of the *Observer*.
44 Crichton-Browne 1918: 179.
45 *The Times*, 9 May 1919: 14e.
46 Astor Committee 1919: 665.

Table 10.6   The objectives of the National Clean Milk Society

---

- To advocate the score-card system.
- To advocate assistance being given by the Board of Agriculture and other authorities to farmers and dairy owners who wish to free their herds from tuberculosis, and to urge that such authorities should grant renewable certificates to the owners of herds which have been found by the tuberculin and other tests to be free from the disease, so long as such herds remain under the supervision of such authorities.
- To induce railway companies to provide refrigerated cars.
- To induce railway companies to offer reasonable rates for the carriage of milk in bottles.
- To promote and assist in prompting legislation making it illegal for any person to open churns or other vessels containing milk or milk products whilst in transit.
- Score-card system of inspection of milk shops.
- Compulsory annual licensing of all dairy farms, milk shops and places where milk and milk products sold either wholesale or retail.
- To promote improvements in methods of distribution to consumers of milk and milk products.
- To call the attention of consumers to the necessity and means of guarding milk and milk products from contamination in the home.
- To promote clean milk competitions.
- To advocate the authorisation by Act of Parliament of a term or designation defining a grade of milk inferior to 'certified milk', but which shall be produced from herds free from TB and shall contain at the time of its delivery to the consumer not more than 60,000 bacteria per cc.
- To consider and define matters which it is advisable should be incorporated in the Milk and Dairies Orders to be issued by the Local Government Board under Section 2 of the Milk and Dairies Act 1915.

Source: Anon. [1916a].

Advertisements had already appeared in 1912 in *The Dairyman* claiming that 'Clean milk is the present demand'.[47] The Dairy Supply Co. promoted itself on the following basis:

> Safety in the milk supply. Making nearly 4,000 analyses every month of the milk purchased and sold by them, the Dairy Supply Co., Ltd., are able to guarantee with confidence based upon knowledge the absolute purity and the high standard of quality of all milk sold by them. Any milk which shows the slightest impurity is immediately rejected.[48]

---

47 *The Dairyman, The Cowkeeper and Dairyman's Journal*, December 1912, advertisement for the Dairy Outfit Co.

48 *The Dairyman, The Cowkeeper and Dairyman's Journal*, July 1912: 407.

There was some competitive lobbying by the different sections of the trade, with separate delegations attending the Ministry of Health to put their case.[49] While the Astor Committee was sitting, the Ministries of Food and Health unilaterally introduced grading in the Milk (Special Designation) Order (1917) and the Milk (Prices) Order (1918) also recognized it; and from September 1918 the Food Controller granted some licences based on the following Local Government Board standards:

> *Grade A*: Milk produced under exceptionally clean and hygienic conditions from a herd certified by a vet as free from tuberculosis and immediately bottled on the premises where it was produced. These premises had to score a minimum of 80 per cent for cleanliness of buildings, equipment and methods of production on the Local Government Board's score card for the inspection of dairy farms. There was a premium of 3d per pint retail.

> *Grade B*: Milk produced under clean conditions, at least 60 per cent on the score card, again from a tuberculosis-free herd, for sale in bulk. The premium here was 4d per gallon.[50]

The official scheme made a very slow start: in 1918–1919 only 38 licences were issued nationwide (Table 10.7). In 1919 the new Ministry of Health, created out of the Local Government Board in 1919, took over responsibility. The Ministry of Agriculture was not involved because the policy was intended to protect the public rather than to introduce a quality classification for marketing purposes.[51] Successive governments imposed fees for the licences they granted and consistently refused to waive the cost of the tuberculin test necessary for the higher grades.[52]

The existing licences all lapsed in March 1920. In the meantime, there was discussion in government, particularly about changes to the classification. Buckley had suggested A1, A2, B and C.[53] But the new scheme was organized by the Food Controller on the advice of the Fisher Committee on the cost of milk

---

49　The larger dairymen and farmers were represented by an alliance of the National Famers' Union and the National Federation of Dairymen's Associations, the smaller enterprises by the Farmers' Cooperative Societies, the Agricultural Organization Society, the London and Provincial Master Dairymen's Association, the Retail Dairymen's Mutual Supply, and the London Retail Dairymen's Association. *The Milk Industry*, 2(7) 1922: 27, *The Milk Industry*, 3(10): 22.

50　Hole 1924: 534.

51　People's League of Health 1932.

52　*Parliamentary Debates*, 191, 1926: 1260, *Parliamentary Debates*, 248, 1931: 2271.

53　NA: MH 56/92, 9 November 1919: 'Suggestions re grading.'

Table 10.7   Licences for designated grades, England

1918   16 Special Licences
1919   22 Grade A and Grade B
1920   9 Grade A (Certified) and 11 Grade A
1921   23 Grade A (Certified) and 15 Grade A
1922   63 Grade A (Certified) and Grade A

Year	Certified	Grade A (TT)		Grade A	
		Producer bottling	Production only	Producer bottling	Production only
1923	47	31	–	–	9
1924	63	45	–	52	60
1925	73	9	62	89	110
1926	94	9	90	179	203
1927	118	17	121	245	266
1928	136	137	127	342	361
1929	144	40	144	436	462
1930	152	43	145	498	524
1931	165	44	151	547	578
1932	188	60	152	620	–
1933	208	65	142	641	–
1934	228	92	172	692	–
1935	275	144	208	1094	–
1936	331	333	217	14,353	–

Year		Tuberculin Tested		Accredited	
		Producer bottling	Production only	Producer bottling	Production only
1937	–	731	812	1610	15,411
1938	–	812	1216	1808	18,016
1939	–	1128	1994	3651	19,345
1947	–	1268	8240	1065	18,739
1948	–	1543	10,470	908	17,744
1949	–	1834	13,278	814	17,390

Sources: Williams and Mattick 1922; Hole 1924; Hole 1926; National Institute for Research in Dairying 1929: 31; Ministry of Health, *Annual Reports*.

production.[54] Grade A became Grade A (Certified) and Grade A replaced Grade B. A more detailed score card was introduced and a more stringent tuberculin test.[55] Making the grades appear as if they were at the top level was the result of lobbying by the milk industry, which had included a threat to find a way to block the package in the Commons if necessary.[56] The most significant change in the proposed replacement scheme was the introduction of bacteriological standards for the quality of Grade A (Certified) milk, and the compulsory sterilization of bottles with steam. For Grade A there was no test, but the milk had to be bottled on licensed retail premises. The Ministry of Health now assumed full control of both producers' and distributors' certification.

Later in 1920 the lack of safeguards for the milk supply as a whole was revealed in an appeal case (*Kenny v. Cox*) concerning a prosecution for dirty milk. The case had been based upon the assertion that dirty milk was 'not of the nature, substance and quality demanded'.[57] But the Lord Chief Justice pointed out that dirty milk was not illegal in the current state of the law, and he also commented, with an obvious sideswipe at clean milk campaigners, that

> I tremble to think what would be the price of milk if we gave effect to your views
> ... [W]hat would the price be if farmers had to take such precautions that there
> were absolutely no foreign substances in their milk?[58]

The Milk and Dairies (Amendment) Act (1922) and its accompanying Order made specific provision for certified milk.[59] These gave detailed definitions of graded milk (Table 10.8), introduced bacteriological counts, and also defined pasteurized milk for the first time.[60] Despite this legislative activity, the government did not actually expect graded milk to become popular. According to the Chief Medical Officer, 'the system has been introduced and maintained mainly to afford

---

54   NA: MH 56/92, Fisher, C.B. (Chairman), 'Report of the Travelling Commission of enquiry into the cost of production of milk'. *The Dairyman, The Cowkeeper and Dairyman's Journal*, March 1920: 233–5.

55   Farmers aspiring to both grades had to achieve a minimum of 300 on the score card, with 250 for the milking methods practised. This was out of a total of 400 points, of which 100 were for the health of animals, 100 for equipment and 200 for methods used.

56   NA: CAB 58/186, Economic Advisory Council, Report, Proceedings and Memoranda of Committee on Cattle Diseases (EAC (CD) Series, 88–109A), 1932–34, vol. 3, Memorandum no. 99: Stenographic notes of the evidence of Wilfred Buckley.

57   *The Times*, 23 October 1920: 4a.

58   *The Milk Industry*, 1(6), 1920.

59   The Milk (England and Wales) Order 1921 and the Local Authorities (Milk) Order 1921 continued until 1st January, 1923, when Section 3 of the new Act came into operation. The Act was explained in the Ministry of Health Circulars 335 and 356 and the new grades set out in the Milk (Special Designations) Order (1922).

60   Certified milk replaced Grade A (Certified) and Grade A became Grade A (TT). Hole 1924, Savage 1929: 78–9.

a working example of how milk can be made reasonably safe and attractive' but the Earl of Onslow, then Parliamentary Secretary in the Ministry of Health, considered it 'possible that the amount of this very high grade milk produced and sold may never be very large compared with the total volume of all supplies'.[61] Partly this was due to some reluctance by farmers to become involved.[62]

In a very rare burst of enthusiasm for regulation, *The Dairyman* declared itself to be content with the Ministry's attitude to grading, which 'marks the beginning of happier relations between Whitehall and the dairy industry ...'.[63] The 1922 Order did face certain technical problems with the way it was drafted, however, and was soon replaced by an Amendment Order and then again by a fresh Order in 1923.[64] The latter came after much industry pressure and relaxed the bacteriological standards that had been set in 1922. The maximum plate count was changed from 50,000 to 200,000 bacteria per cc up to the end of 1923, and to 100,000 thereafter.[65]

Criticisms of graded milk followed a number of lines of argument. One group complained that it was too expensive to produce and for the customer to buy. It could therefore only be consumed by the rich, and in future tuberculous milk would be more concentrated at the lower end of the market.[66] The retail price differential between ordinary milk and Grade A was 1d per pint, which hardly made it worth the while of hard-pressed retailers whose margins were anyway low. Grade A for them was little more than an advertisement of quality which they hoped would win them new customers for their main business.[67]

Second, there were opponents who wanted the quality of *all* milk to be improved. Thus Alderman Shelmerdine, Chairman of the Maternity and Infant Welfare Committee of Liverpool Corporation commented that since 'only one grade of water is admissible to the home; why should milk, which is much more to be feared, be admissible in three grades?'

> I confess that I have little sympathy with the American principle of grading milk into three classes, and then selling the higher grades as certified milk at

---

61   Ministry of Health, *On the State of the Public Health: Annual Report of the Chief Medical Officer for 1923* London: HMSO: 139, Onslow 1922: 21.

62   Williams and Mattick 1922.

63   *The Dairyman, The Cowkeeper and Dairyman's Journal*, September 1922: 9.

64   The first amendment, made only 10 days after the Milk (Special Designations) Order (1922), postponed the application of regulations about pasteurization by six months to allow the upgrading of equipment. *The Dairyman, The Cowkeeper and Dairyman's Journal*, September 1922: 11, January 1923: 217, 220–24.

65   *The Dairyman, The Cowkeeper and Dairyman's Journal*, May 1923: 432, 442, June: 504, 507–8, 511.

66   *The Dairyman, The Cowkeeper and Dairyman's Journal*, August 1918: 281, *Public Health*, 34, 1922: 118, *Public Health*, 36, 1922/3: 23, *Parliamentary Debates*, 165: 1923, 2635.

67   *The Dairyman, The Cowkeeper and Dairyman's Journal*, June 1924: 525.

Table 10.8   The definitions of graded milk under the 1923 Order

	Certified	Grade A (TT)	Grade A	Grade A (Pasteurized)	Pasteurized
Licence to produce	Ministry of Health	Ministry of Health	County Council or County Borough Council	County Council or County Borough Council	County Council or County Borough Council
Licence to distribute	Local Sanitary Authority	Local Sanitary Authority	Local Sanitary Authority	Local Sanitary Authority	
Cows	Must be TT and free from tuberculosis, inspected every 6 months	Must be TT and free from tuberculosis, inspected every 6 months	Must be inspected every 3 months by vet and certified free of clinical tuberculosis	Must be inspected every 3 months by vet and certified free of clinical tuberculosis	–
Milk containers	Bottled on farm with special tops	Unventilated, sealed, labelled containers or bottled on farm with special tops	Unventilated, sealed, labelled containers or bottled on farm with special tops	Unventilated, sealed, labelled containers or bottled on farm with special tops	Must be labelled
Bacteriological standards	Not more than 30,000 bacteria per cc; no B. coli in 0.1 cc	Not more than 200,000 bacteria per cc; no B. coli in 0.1 cc	Not more than 200,000 bacteria per cc; no B. coli in 0.1 cc	Not more than 30,000 bacteria per cc; no B. coli in 0.1 cc	Not more than 100,000 bacteria per cc
Heat conditions	No heat treatment allowed	No heat treatment allowed	No heat treatment allowed	145–150°F for at least 30 minutes, then cooled to 55°F; apparatus to be approved.	145–150°F for at least 30 minutes, then cooled to 55°F; apparatus to be approved.

Source: Savage 1929: 79; Davis 1950: 405–6.

an almost prohibitive price because it has a smaller bacterial content. All milk for sale should be free from pathogenic bacteria whether it is sold as marketing milk, cooking milk, or certified milk. As a matter of fact most of the poorer and more ignorant classes look upon all milk, simply as milk, disregarding dangers therefrom which they cannot see, and in actual practice the needy housewife buys the cheapest grade rather than pay the higher price and runs the risk, which she neither understands or appreciates; only one grade of water is admissible to the home, why should milk, which is much more to be feared, be admissible in three grades?[68]

Others complained that there was no minimum quality standard for milk.[69] Ben Davies argued that, rather than paying a bonus for certified milk, fines should be imposed on farmers who consistently produced dirty milk. The latter he called 'discredited milk', to mirror the new official designation of 'accredited milk' that was eventually introduced in 1935.[70]

Third, criticisms came from those who argued that much so-called graded milk was not safe to drink. Of Grade A school milk samples investigated in one survey between 1930 and 1935, for instance, 5.6 per cent contained *M. bovis*, scarcely less than the 7.4 per cent for ungraded samples.[71] In 1937 the London County Council even reported a higher rate of infection (10.6 per cent) for its Accredited than for its ungraded milk (6.6 per cent).[72]

Fourth, some argued that graded milk did not take off because doctors were reluctant to recommend it. Earl de la Warr, then Parliamentary Secretary to the Ministry of Agriculture, complained in the House of Lords that:

If the medical profession would talk a little less about dirty milk, and do a little more to encourage the buying of clean milk which was already in the market, they would be of more use to the nation as a whole and to the agricultural industry. They would also help the government if they would make up their minds as to what they really felt about milk. Before the medical profession come down on the farming industry for not taking certain steps about milk, they should really make up their minds what they want the farmers to do.[73]

Finally, many farmers were suspicious of the motives of the elite of influential and wealthy clean milk campaigners. They were worried that they might be expected to invest heavily in special equipment and new buildings which, as predominantly

68   Shelmerdine 1921: 300.
69   *The Dairyman, The Cowkeeper and Dairyman's Journal*, July 1920: 411, *Public Health*, 36, 1922/3: 267.
70   Davies 1933: 6–9.
71   People's League of Health 1936.
72   Wilson 1942: 15.
73   *The Times*, February 11 1931: 14C, Anon. 1931.

small operators in terms of their herd size, they could not have afforded. But how could they resist a movement that was picking up momentum and attracting attention in Parliament? The answer was through a combination of action and inaction, including minor forms of everyday resistance that Michel de Certeau called tactics. When James Mackintosh came to Chesterfield in 1920 on behalf of the National Clean Milk Society to give lecture on clean milk production, for instance, it was boycotted by local farmers and cancelled.[74] Some Medical Officers despaired of this negative attitude:

> The fact is that the farmer is inert, and in spite of prolonged agitation and discussion the general public a little less so. The necessary measures of reform will have to be formulated and imposed by government if clean milk is to be obtained.[75]

This defiance by farmers gradually changed to cooperation as it became obvious that:

> the standard for graded milk is well within the compass of an average farmer who insists on his milkers and dairy servants carrying out the ordinary rules of cleanliness, even although his byres may not be too good structurally.[76]

According to Mattick, clean milk did 'not mean milk from which visible dirt is absent or has been extracted, but rather raw milk from healthy cows, which has been produced and handled under hygienic conditions, which contains only a small number of bacteria and which is capable of remaining sweet from delivery to delivery in summer and for five or more days in winter'.[77] To him, well-trained and motivated labour was more important than expensive buildings or equipment.[78]

But some farmers had difficulty in believing in what they could not see – microscopic bacteria. Charles North addressed this by magnifying an image cheese cloth with a very fine mesh and showing that at 75x it was easy to see how '400 bacteria could be placed in a row in the space between two threads of the finest cheese cloth, and through one of the square openings of such a cloth a regiment of 160,000 bacteria could march abreast'.[79]

---

74 *The Dairyman, The Cowkeeper and Dairyman's Journal*, August 1920: 462.
75 Niven 1923: 140.
76 *The Dairyman, The Cowkeeper and Dairyman's Journal*, October 1923: 65. This quotation is from a Dairy Steward at Royal Agricultural Society.
77 Mattick 1927.
78 But Mattick's pamphlet does set out the required physical conditions in milking shed (light, floors and walls, water supply, disposal of manure, control of flies, ventilation), utensils (sterilizing), cleaning cow before milking, covered pails, straining and cooling milk, also clean bottling and distribution.
79 North 1918: 19.

In 1933 Ben Davies, former Director of Laboratories, Processing and Inspection for United Dairies, gave an insight into the thinking of dairy farmers:

> In gatherings of milk producers discussions on the cleanly and healthful character of the milk supply are likely to prove contentious and provocative no matter from what aspect they are approached. This is perhaps due in part to the hard-working practical farmer's distrust of propaganda for what he looks on as fanciful fads and theories, and also to the fact that he has not been made to realize that everything that can be done to assure the cleanly purity of the milk supply, and the health of his herds is not so much a tax on his precarious profits as a definite contribution to his own prosperity.[80]

In the 1920s the market for graded milk slowly expanded. Complaints from producers that they could not find markets faded as national advertising of quality milk was mounted by the new National Milk Publicity Campaign. In 1924 1,400,000 gallons of Certified milk were produced and 1,200,000 gallons of Grade A (TT).[81] Given the high elasticity of demand for milk, price was the main basis of competition and, at a retail premium of 28–93 per cent, it was only affluent consumers who were engaged at this time. But at least this top end of the market had discovered a desirable product which transcended the previous image of milk and the additional demand was helpful to a branch of farming that was going through a difficult time, with oversupply and falling real prices.[82]

The Milk and Dairy (Consolidation) Act (1915), which had been postponed during the war, finally came into operation in 1925. The Milk and Dairy Order followed soon after, in 1926, and was an important threshold in thinking about the milk industry, including the topic of clean milk. First of all, it updated the regulations about buildings and equipment, in the hope of modernizing the conditions of production.[83] The sterilization of milk vessels and appliances, for instance, was required, despite chemical sterilization probably being of greater value to small farmers.[84] Second, there were provisions about animal health, allowing a potentially diseased source of milk to be stopped. Third, all producers and retailers of milk now had to be registered, which made it easier for the authority to monitor their activities. Fourth, the cooling of milk was made compulsory.

Despite its initial misgivings, the industry gradually realized that grading meant the commodification of quality and therefore additional profit for those with

---

80   Davies 1933: 5.

81   *Annual Report of the Chief Medical Officer of the Ministry of Health, 1924.*

82   Liversage 1926. One could argue that there are parallels here with the creation of an organic food market in the 1990s.

83   It revoked and replaced the Dairies, Cowsheds and Milkshops Order of 1885, 1886 and 1899.

84   Sodium hypochlorite was allowed in 1943 under wartime emergency conditions but it was not formally recognized by the 1949. Hall 1951: 3, Rowlands and Hoy 1949.

the capability to meet the specifications. There was resonance here with the mood of the times because quality had generally not previously been rewarded in British agriculture. The grading of fruit and vegetables, for instance, was in response to serious structural problems in the sector that contemporary politicians thought could only be solved by marketing initiatives. Under the Agricultural Produce (Grading and Marking) Act (1928), the Ministry of Agriculture sought powers to licence the use of a National Mark on graded goods. The first commodities to be dealt with in this scheme were apples, pears, eggs, broccoli, and beef.[85] Beef producers were somewhat reluctant to join this voluntary scheme, but there was greater enthusiasm from egg producers and vegetable growers.[86]

The élite producers of graded milk soon formed themselves into Associations to protect their mutual interests,[87] and after ten years of rivalry the Certified and Grade A (TT) Milk Producers' Associations eventually joined forces when they realized that the National Farmers' Union meant to have them excluded from the Milk Marketing System. In March 1933 they lobbied the Ministry of Agriculture and complained that:

> There is not a single member of the FU who is producing graded milk, and it is rather like a red rag to a bull to mention it to them ... They have never had the vision to see that a clean milk supply means bigger sales ... we have no friends amongst the trade and amongst the members of the NFU.[88]

In 1926 graded milk still accounted for less than 1 per cent of consumption and in 1933 it was still produced by less than 1 per cent of cows.[89] It was not until the introduction in 1935 by the Milk Marketing Board of a 1d premium per gallon that the situation was revolutionized by a supply-side surge.

Hope in the 1920s

In 1923 Dr Harold Scurfield, Medical Officer of Health for Sheffield, reminisced that there had been little change to milk quality during his career: 'I do not think experienced Medical Officers of Health note much improvement in our milk

85 Whetham 1978: 223.
86 Brown 1987: 115.
87 *The Dairyman, The Cowkeeper and Dairyman's Journal*, July 1923: 569. An advertisement for The Certified Milk Producers' Association, 11 St Jas Square, London. There was also a Grade A (TT) Milk Producers' Association, which held a conference in 1924 to discuss graded milk. *The Dairyman, The Cowkeeper and Dairyman's Journal*, May 1924: 473, June: 524–38.
88 NA: MH 56/100, 'Ministry of Agriculture and Fisheries. Notes of a deputation to the Parliamentary Secretary from Special Milk Producers, Thursday, 9 March, 1933'.
89 Newman 1926, Harvey and Hill 1936: 279.

supply during these thirty years.'[90] But in the early 1920s there was a tipping point. By then some city authorities had already persuaded or coerced cowkeepers within their jurisdictions to address the issue of hygiene. Alderman Shelmerdine of Liverpool commented in 1921, for instance, that 'generally it may be said that the quality of milk, owing to the force of public opinion, has been sensibly improved'.[91] A few years later it was possible for one prominent Medical Officer to say in retrospect:

> From my study of milk, which has extended over 25 years, the position in England was that there was a very little improvement in the quality of the milk resulting from the first 20 years of those 25, but during the last five or six years, there has been an enormous improvement in the cleanliness and quality of the milk supplied in this country.[92]

Also in 1928, two bacteriologists working at the National Institute for Research in Dairying were even more upbeat:

> It seems that the outlook in the milk industry was never better. On the one hand there is the vast amount of work which is being done to improve the cleanliness of the general supply, and on the other there are the licensed milk producers who are scattered all over the country, and are not only setting an example to their fellows but are themselves acquiring new knowledge and becoming more and more efficient…There is no doubt that an improvement is taking place in the cleanliness of ordinary milk, and in that of milk produced under licence.[93]

Similar remarks were made in the late 1920s and 1930s by other commentators, although their estimated date of the start of the improvement varied.[94] The most important was the opinion of the Chief Medical Officer that 'here has been steady, and in some ways exceptional, advance. The ordinary common milk supply has been enormously improved in a single generation'.[95]

According to the informal Inter-Departmental Committee on the Law Relating to Milk sitting in 1931, this 'considerable improvement' was due to a number of factors.[96] First, there was the operation of the Milk and Dairy Act 1915 (from

90   Scurfield 1923: 28.
91   Shelmerdine 1921: 297. See also Anon. 1921b.
92   White 1928, comment in discussion by William Savage: 329.
93   Williams and Hoy 1928: 76.
94   Williams 1928, Wood 1928, Williams and Mattick 1931, Waley-Cohen 1933, Brockington 1937, McHugh 1943.
95   Ministry of Health, *On the State of the Public Health: Annual Report of the Chief Medical Officer for 1932*. London: HMSO: 147.
96   NA: MH/56/88. Second Interim Report of the Inter-Departmental Committee on the Law Relating to Milk, 31 March, 1931.

1925), the Amendment Act (1922) and the Milk and Dairy Orders (1922, 1923 and 1926), although in all cases their impact was gradual. Second, there were the various educational measures of the agricultural colleges and local authorities, and other services provided regionally, such as the veterinary inspection of cattle and bacteriological analysis of milk samples on behalf of farmers. Third, cow-sheds were increasingly being brought up to date in the post-war period and milking methods were more hygienic.

Such qualitative assessments were based upon expertise and observation. From the 1920s it was possible to add scientific weight through the large-scale, systematic testing of cleanliness under laboratory conditions. One example is the time series compiled by the Leicestershire County Council monitoring scheme. Table 10.9 shows a bacteriological classification of samples that was an extension of ideas of the day about designated milks. It is interesting to note that there was little change in the proportions in the middling categories over the 10 years from 1925, but a steady decline in the numbers producing the worst milk.

Table 10.9   Leicestershire County Council's monitoring scheme, percentage of milk samples in each category

	Good	Fair	Moderate	Bad
1925	54.3	19.0	2.4	24.3
1926	46.8	18.9	4.1	30.3
1927	40.3	24.3	4.3	31.1
1928	58.8	22.7	2.6	15.9
1929	62.0	24.4	1.7	11.9
1930	61.7	27.1	1.2	10.1
1931	62.1	19.1	1.5	17.3
1932	65.7	18.7	1.5	14.2
1933	69.4	16.0	1.1	13.5
1934	67.0	19.1	1.2	12.7

Good (Grade 'A') = less than 500,000 per cc, and no B. Coli in 1/100 cc.
Fair = 500,000–1,000,000 per cc, and no B. Coli in 1/100 cc; or, less than 500,000 per cc and B. Coli present in 1/100 cc.
Moderate = 500,000–1,000,000 per cc and B. Coli present in 1/100 cc.
Bad = over 1,000,000 per cc, or B. Coli in 1/1000 cc, or both.

Source: Fairer 1930: 125–6; Graham 1935.

Also in central England, Edwin White of the Midland Counties Dairy Ltd was the instigator of a scheme to pay bonuses to any of his 140 suppliers who could

supply clean milk.[97] Prizes of £100 and £50 were offered for the best two suppliers
in the period June to September 1922 and, at the same time, the company installed
their own laboratory to aid testing, using the plate count method.[98] Monthly lists
of test results were published, and the top 12 farmers were paid a bonus of 1d per
gallon for milk with a low bacterial count and the next 24 received 0.5d per gallon.
At the bottom of the list, those who persisted in producing unclean milk had their
contracts terminated at the end of the season.[99] In 1923 payments were added for
fat content.[100] The standard price was for 3.5 per cent fat, with a bonus of 0.1d per
gallon for every 0.1 per cent of extra fat above that.[101] The farmers' attitude to the
scheme gradually changed:

> At first most of them regarded the movement as a fad, and not many took the
> matter seriously. After a time, when certain farmers' names were regularly near
> the top of the list and receiving extra payment, more interest was taken, and the
> quality of our milk supply began to improve.[102]

White's experiment seems to have paid off with year-on-year improvements
in milk quality across the full range of his suppliers (Tables 10.10 and 10.11).
By 1933 65 per cent of its milk was coming in at the certified milk standard or
above.[103] By 1931 29 other dairy companies were emulating White's scheme,[104]
and many had also established their own laboratories for monitoring quality.[105]
However, the experiment with premium payments seems to have ended with the
outbreak of the Second World War.[106]

Although the Midland Counties Dairy was renowned for the quality of its
milk, there were powerful forces elsewhere in the industry resisting the payment
of premia. Ben Davies opposed them as a Director of United Dairies, the largest
dairy company in the country, his argument being that high standards should be
expected of all and that policy should be shifted instead to penalties for poor
quality.[107] For him, the certification of milk was the legitimation of 'grades of
dirt'.[108] Eventually, however, United Dairies joined the trend and their suppliers

97   Jones [1924]: 8, White 1928: 322, National Institute for Research in Dairying
1931, Enock 1943: 45, Davis 1983: 5.
98   White 1928: 322.
99   White 1928: 322.
100   Cook Committee 1960: 6.
101   White 1928: 328.
102   White 1928: 322.
103   Raison 1933: 87. The quality of Midland Counties Dairy supplies slipped in the
1930s as farmers preferred the easier premia of accredited milk.
104   Jeffcock 1937: 65, Atkins 2007a.
105   Williams and Mattick 1931: 142.
106   Cook Committee 1960: 6.
107   Enock 1943: 48.
108   Davies 1926: 71.

Table 10.10   Average bacterial count per cc of Midland Counties Dairy Ltd milk

	1922	1923	1924	1925	1926	1927
1st farmer	4,566	4,286	933	611	591	354
36th	365,535	211,960	73,751	20,347	7,342	3,647
60th	3,665,311	796,951	390,708	107,609	18,621	7,773
100th	–	–	5,843,740	851,540	130,479	38,966

Source: White 1928: 326.

Table 10.11   The percentage of farmers supplying milk in each of three bacteriological standards to the Midland Counties Dairy, Birmingham

	Thousand bacteria per cc		
	<10	10–30	30–200
1922	3.2	16.1	80.6
1923	10.8	31.7	57.5
1924	12.4	28.3	59.3
1925	16.4	32.3	51.2
1926	23.1	32.7	44.5

Source: White 1928.

began receiving payments for low bacterial counts and high butterfat.[109] By 1929 the company's laboratories were conducting the following tests as a matter of course:

- sediment tests to ensure that their supplies were produced in a clean manner;
- acidity tests to guarantee that the farmer despatched soon after milking;
- chemical tests to satisfy themselves that the full quantity of cream was present and that no water had been added;
- bacteriological tests to detect disease.

109   Tustin 1929: 313.

Policy in the 1930s

Proof of the consolidation of ideas about milk quality came in the years 1933/4/5, which were a major threshold in our story. The Milk Marketing Board of England and Wales was formed in 1933 and the new Milk Marketing Scheme inaugurated the following year.[110] Then in 1935 a new grade of Accredited Milk was started as part of a narrative of a new start for cleanliness: '"clean milk" is a term which admits the existence of dirty milk; the reproach is long standing, and must be removed from the minds of the consuming public at the earliest possible moment.'[111] To qualify for the accredited roll, farmers had to submit their herds to six monthly veterinary inspections, with the removal of any tuberculous cattle. In addition, their milk was tested at least three times a year and had to have less than 200,000 bacteria per cc, and *B. coli* absent in 1/100 cc. These were the criteria and there was no mention of dirt. In order to achieve the hoped for uplift in cleanliness, it was obvious that there would need to be a substantial expansion of existing laboratory facilities. The Milk Boards were keen to have uniformity in methods of sampling and analysis, and they specified a protocol of times within which the analysis must take place, the temperature to which samples must be cooled and even the types of analysis to be performed.[112]

The Boards were in effect monopoly cooperatives run by farmers for farmers and it should therefore be no surprise that substantial financial incentives, 1d per gallon, were offered for producers to join the accredited milk scheme. But there were also politics. The Ministry of Health vigorously opposed the Boards' plans from the outset, arguing that the new grade would not necessarily clean or safe.[113] Their verdict was that it would just be Grade A milk by another name. Ben Davies, well-known in the milk trade, said that 'it is nonsense. It is impracticable'.[114] The People's League of Health objected, as did the Society of Medical Officers of Health, because the milk would neither be from tuberculin tested herds nor pasteurized: 'it might be clean but it cannot claim to be safe.'[115] The influential Hopkins Committee also withheld its approval because 'all milk sold for liquid consumption should reach a fixed standard of cleanliness at the farm and ... it is

110   The Scottish MMB, the North of Scotland MMB and the Aberdeen and District MMB all date from 1934. The Northern Ireland MB was founded in 1955.

111   Milk Marketing Board 1934a.

112   NA: JV 6/83, Milk Marketing Board, Joint Quality Control Committee, 13 November 1934, 'Register of accredited milk producers. Memorandum on the bacteriological examination of milk samples'.

113   NA: MH/56/85. Minute by J.W. Hamill, 14 May, 1934.

114   NA: CAB 58/186, Economic Advisory Council, Report, Proceedings and Memoranda of Committee on Cattle Diseases (EAC (CD) Series, 88–109A), 1932–34, vol. 3, Memorandum no. 97: Stenographic notes of the evidence of Mr Ben Davies, 6 February, 1933.

115   People's League of Health 1934.

wrong to pay a bonus to certain producers for attaining a standard which is within the reach of all producers'.[116]

County Councils objected to paying the bill for the increased veterinary inspections, agricultural extension and bacteriological testing that were the cornerstones of the Scheme. The majority refused, delaying the start from October 1934 to May 1935.[117] Only 800 producers signed up in the first year, but by 1937/8 there were 22,711 licensees, producing about 40 per cent of the Board's supply.[118]

Meanwhile, the Ministry of Agriculture was making further moves on milk quality. In February 1935 it introduced an Attested Herds Scheme, which paid a bonus to owners of tuberculosis-free herds. Although this was really a producer-centred scheme to encourage the elimination of disease, it did introduce the notion of another quality of milk. The consumer knew nothing of this, however, because the milk was mixed by the Milk Marketing Board.[119] In 1936 a new Milk (Special Designations) Order recognized three new grades: Tuberculin Tested, Accredited, and Pasteurized.[120] The first two grades were raw milks that corresponded the previous Grade A (Tuberculin Tested) and Grade A milks respectively.[121]

The 1930s, then, saw a restructuring of the production and wholesale parts of the milk industry in Britain. Cleanliness of the supply was improving, although there were complex politics about the balance between the interests of public health and those of the farmers. The latter policy network was strong in this period and there was support in Whitehall from the Ministry of Agriculture, which meant that it was their initiatives that went ahead despite the protests of the weaker Ministry of Health.

There are two further points to be made. First, the beginnings of school milk in the 1920s were planned by the National Milk Publicity Campaign, and the national scheme of 1934 came with the blessing of the government. This is not part of my account of material politics of cleanliness here because there is little evidence that the milk sold to children was any different from the mainstream supply.[122] Second, a great deal of political energy in the 1930s was devoted to the issue of bovine tuberculosis in the cattle herd and in the milk supply. Views differed on whether this was a serious hazard to human health and what should be done in policy terms.

---

116   NA: MH 56/101.

117   NA: MH 56/101, 1 August 1934, copy of letter from Elliot to Baxter, 'Report on proceedings at a conference on the Accredited Producers Scheme held between representatives of the MMB, the County Councils' Association, the AMC, the MAF, the MH, on Tuesday, 4 December, 1934', 21 December 1934, A.N. Rucker, 'Milk. Deputation to the Ministers of Health and Agriculture'.

118   Wilson 1942: 171–2, Whetham 1978: 257.

119   NA: MH/56/85. Undated document 'Improvement of the milk supply'.

120   In addition, if the first of these was bottled at the farm it could be described it as Tuberculin Tested Milk (Certified).

121   Ministry of Health, *On the State of the Public Health: Annual Report of the Chief Medical Officer for 1935*. London: HMSO: 134.

122   See Atkins 2005a, 2005b, 2007b.

The Second World War and After

Deborah Dwork has argued that the special conditions of the First World War allowed the implementation of milk policies that had seemed impossible just a few years earlier.[123] We might argue the same for the Second World War because there were similarly radical shifts with regard to welfare milk, school milk, and also milk quality. In 1942, for instance, a National Milk Testing and Advisory Scheme was established, operated by the Ministry of Agriculture through an Advisory Committee, on which the Ministries of Food and Health were represented, as well as producers' and distributors' organizations.[124] By the end of the war 83 per cent of farmers were having their milk tested for keeping quality on a fortnightly basis and losses from souring had been halved.[125] The 1943 White Paper 'Measures to Improve the Quality of the Nation's Milk Supply' proposed shifting responsibility for cleanliness to the Ministry of Agriculture because of evidence of significant variations in performance between Local Authorities. This was enforced by Food and Drugs (Milk and Dairies) Act (1944). The Ministry appointed a Chief Milk Officer, with Regional Milk Officers to inspect farms and give advice on buildings, equipment and production methods. There was also a system of Regional Advisory Bacteriologists. The milk hygiene part of these arrangements were changed again in 1948 with the formation of a Joint Milk Quality Control Committee, comprised of members from the producers and buyers of milk, with observers from the government.

In 1949 the Milk (Special Designations) Act and Orders insisted for the first time that all milk had to be graded. By 1954 there were only three categories: Tuberculin Tested raw milk, which came from herds attested free from tuberculosis; Pasteurized milk, which had to be sold in bottles; and Sterilized milk. Once the national herd was declared free from tuberculosis, there was no further need the TT designation from 1963, but technical advances enabled a new one of Ultra Heat Treated in 1965 (Table 10.12).

From October 1982 a new test for the hygienic quality of milk was adopted, the Total Bacterial Count or Standard Plate Count, using a 72-hour incubation at 30°C and expressed as colonies per millilitre of milk.[126] This seldom exceeds 10,000 per ml in healthy cows kept under hygienic conditions. The cooling of milk to 4.5°C is a contractual requirement in Britain but this may delay bacterial growth for only a day or so if the initial loading is high. In 1985 the Health and Hygiene Directive 85/397/EEC set out Europe-wide standards, and these were replaced shortly after in Directive 92/46/EEC, as enacted in Britain in the Dairy Products

---

123 Dwork 1987.

124 This was encouraged by the hot summer of 1941 when souring losses were high.

125 Ministry of Health, *On the State of the Public Health: Annual Report of the Chief Medical Officer for 1939–45* London: HMSO: 122.

126 A popular modern, automated measurement technique, which counts individual bacteria, is the Bactoscan.

Table 10.12 Grades of milk specified in the Milk (Special Designation) Orders and Drinking-Milk Regulations

Date	Grades
1917	Grade A, Grade B
1923	Certified, Grade A (Tuberculin Tested), Grade A, Grade A (Pasteurized), Pasteurized
1936	Tuberculin Tested (Certified), Tuberculin Tested, Tuberculin Tested (Pasteurized), Accredited, Pasteurized
1949	Tuberculin Tested, Accredited, Pasteurized, Sterilized
1954	Tuberculin Tested, Pasteurized, Sterilized
1963	Untreated, Pasteurized, Sterilized
1965	Untreated, Pasteurized, Sterilized, Ultra Heat Treated
1998	Raw, Whole (Standardized), Whole (Non-standardized), Semi-skimmed, Skimmed

Source: Various, including *Dairy Facts and Figures*.

(Hygiene) Regulations (1995). The present limit for the plate count is 100,000 colonies per ml. Another quality measure is the presence of mastitis in dairy herds, which has serious economic consequences for the farmer and can cause illness in humans. The current maximum for this is 400,000 somatic cells per ml. Finally, there are also measures for a range of other issues, such as antibiotic residues and the presence of Staphylococcus aureus, and creameries also monitor the presence of a number of infectious diseases that are dangerous for humans.

Conclusion

The politics of clean milk were, from about the time of the First World War onwards, dependent upon bacteriology. As we have seen, the conceptual foundation of thinking moved in this direction, and statistics on cleanliness published in public health reports and even newspaper articles were increasingly in bacteria per cc. It was the visual image of microscopic fauna and flora, translated into a mental image of teeming, unseen life, that caught the imagination of everyone, from consumers to politicians. But it is rarely pointed out that all of this was dependent upon the material capacity of bacteriological laboratories. Without the plate count and coliform tests, for instance, the 1923 regulations on graded milk would not have been possible in the form set out in Table 10.8. Later critiques of the technology brought into question whether these tests were accurate, and therefore whether counting bacteria served any real purpose. The methylene blue test used officially from 1937 onwards was different. It was a chemical test of bacterial activity and gave only a yes or no answer to questions about the marketable quality of milk. In that sense it was a retreat from the felt need for close observation and precise quantification.

Chapter 11

Conclusion[1]

This book has been about policing the boundaries of material quality using the moral authority of nature.[2] It has addressed the means by which cow's milk has been manipulated in its composition over a period of 200 years, 1800 to date. Although naturalized to the point of being taken for granted, this organic fluid in reality has been a coeval participant and powerfully influential in the scientific, commercial and political interests that have emerged to shape the modern food system. Building an understanding of such commodities helps us grasp our mutually-constituting relationship with food, which is among our immanent and most intimate contacts with nature.

But such ontological work is not necessarily visible or continuous. Husserl's concept of 'sedimentation', for instance, sees objectivities (in our case the definition of a key food commodity) as gradually accumulating authority while their origins may be forgotten, and Jonathan Rée warns lest, with the ageing of concepts and facts, we 'take them to be infinitely old and preternaturally wise'.[3] Distance in time or space in this way may lend perceived qualities of naturalness, set apart from the hybridities of immediate human impact.

I have attempted, by a process of questioning and unsettling the taken-for-granted, to show that the material quality and composition of all liquid milk retailed in the United Kingdom bears the traces of scientific, technological, commercial and legal influences over a period of 200 years. Milk's apparently timeless qualities have hidden, beneath a blanket of innocent whiteness, the significant variations of composition in time and space that are palpable upon close inspection. Only since 1993 have the many milks become standardized into a single, stabilized Euro-commodity. Our ontogenetics of this single commodity has sought to reveal its trajectory and the dynamic of its evolution. In doing so, the book has tried also to redress the 'neglect of "food*stuffs*"' identified by Stassart and Whatmore as a reason for 'the lack of analytical purchase' in the literature on food.[4]

By investigating the emergence of a consensus of what 'milk' is and the policing of material boundaries of that foodstuff, this book has attempted a number of insights that have wider significance than the mundane trope of the daily diet.

---

1   This chapter relies heavily on Atkins 2007c.
2   Daston and Vidal 2004.
3   Laclau 1990; Rée cited in Hacking 2002: 6; see also Daston 2000.
4   Stassart and Whatmore 2003: 450.

First, the mass marketing of milk is transductive in one sense intended by Simondon.[5] It facilitates the transformation of a variable, perishable, organic fluid, produced by the cow for her calf, into a commercial product loaded with technicity – its standardized constituents, its artificially lengthened shelf life, its purification from micro-organisms, and a quality that is reliable over successive iterations of demand and supply. This technicity is itself the result of other transductions, for instance the invention and evolution of the technology of pasteurization, which in turn depends upon many other inter-relations between ideas, speech acts and materials. No one single act of forming matter is sufficient to comprehend this process of transduction and the resulting individuation. In the case of milk, its organic components continue to interact with each other and with the environment right up to the moment of consumption (and beyond, in the gut, until metabolized) and 'processing' is therefore largely a means of inhibiting and redirecting the fluid's inherent energies so that its transduction is into a product acceptable to the consumer and not into a degraded or poisonous one.

The boundary work between 'milk' and the various fluids sold as milk, with more or less butterfat and milk solids content, involved the refinement of a significant new area of science, organic chemistry, and the establishment of an innovative strand of governmentality – food law and regulation. As I have argued elsewhere, legal geographies can teach us about the ethical 'proximity' between parties in a chain of service provision and, in the late nineteenth and twentieth centuries, nature was made less mysterious as a combination of laboratory science and legal standards sought to encompass its compositional variations and bring it into the modern realm of the observed, the regulated, the trusted.[6] In codifying and enforcing the limits of nature, the law was a plane of transcendence, ultimately creating a basis for the discussion of human behaviour and guiding both thought and action in a direction that in reality was at times tangential to the interests of all of the stakeholders. The reduction of risk was a project in governmentality, 'a strategy of regulatory power by which populations and individuals are monitored and managed'.[7] There was a shift from a collective form of discursive hygienism that arose from Victorian debates about urban dirt and 'the great unwashed', towards control of the production process of individual farmers.

Second, analysing a commodity, not as a point in space-time but as a series of events, benefits from Simondon's notion of transduction, that 'a diversity of actors, interests, institutions and practices are articulated together through specific technologies' into 'collectives' and the present book has given an account of this in the historical setting of the interactive and mutually constituting flows between various sources of capacity: biological, commercial, legislative, legal, scientific, technical, and consumer politics.[8] The very complexity and astonishing intensity

---

5   Mackenzie 2005: 395.

6   Atkins et al. 2006.

7   Lupton 1999: 87.

8   Mackenzie 2002: 118, DeLanda 2000.

of activity associated with milk made it a locus of controversy and acrimony, not least because of the indeterminacy of its natural material form and also due to the irresolvable 'vitality' it was assumed to possess, even after the discovery of vitamins.[9] It achieved the status of a technical ensemble at a time when most other foods were simply processed, and it was therefore exemplary from an early date of the regulated and, in Foucault's terminology, 'normalized', foodstuffs that were a major element in the normative thrust of modernity – the production of truth through power. This provided raw material for the transformation and naturalization of society itself. Commercial interests were well served but the large dairy enterprises had to sacrifice some of their freedom of action as they became hybrid creatures, gorging themselves on a regular diet of scientific samples and motivated by the need to be more hygienic than their competitors. The rhetoric of trust had been initially fostered by the National Milk Publicity Campaign and commercial advertising but eventually also Medical Officers of Health and doctors joined in the chorus backing the 'drink more milk' marketing message. Interestingly, the large dairy corporations were selling better quality milk than their small-scale competitors. In the first half of the twentieth century their resources for and commitment to quality monitoring were second to none, seemingly the reverse of expectations in the risk society, where risks are said to proliferate as a result of corporatism and commodification.[10]

Finally, I conclude that histories and historical geographies of food, and of commodities generally, deserve greater attention than they currently attract.[11] Danny Miller understood this when he observed that 'objects are important not because they are evident and physically constrain or enable, but often precisely because we do not "see" them'.[12] The everyday material of our lives, including food and drink, because it is unconsidered, because it is unchallenged in its significance, is a powerful means of guiding our expectations – in the case of food, our habituated, embodied norms of nutritional sufficiency and bodily reproduction.

9 Atkins 2000a.

10 Dean, 1999; Lupton, 1999.

11 I am intrigued by the possibilities of applying the methodology of this book to other commodities. For the beginnings of a technical history of wheat quality, for instance, see Lásztity and Abonyi 2009.

12 Miller 2005: 5.

References

Adams, M.A. 1885. On a new method for the analysis of milk. *Analyst*, 10: 46–54.

Adshead, S.A.M. 1992. *Salt and Civilization*. Basingstoke: Palgrave Macmillan.

Aikman, C.M. 1899. *Milk: Its Nature and Composition*. London: A. and C. Black.

Albala, K. 2002. *Eating Right in the Renaissance*. Berkeley: University of California Press.

Albala, K. 2007. *Beans: A History*. Oxford: Berg.

Alexander, A. and Nicholls, A. 2006. Rediscovering consumer-producer involvement: a network perspective on fair trade marketing. *European Journal of Marketing*, 40: 1236–53.

Allaire, G. 2004. Quality in economics: a cognitive perspective, in *Qualities of Food*, edited by M. Harvey, A. McMeekin and A. Warde. Manchester: Manchester University Press, 61–93.

Allen, A.H. and Chattaway, W. 1886. Suggestions for the more ready employment of Adams' method of determining fat in milk. *Analyst*, 11: 71–3.

Analytical Methods Committee, Society of Public Analysts and Other Analytical Chemists 1937. Report of the Sub-Committee on dirt in milk. *Analyst*, 62: 287–301.

Anderson, B. and Wylie, J. 2009. On geography and materiality. *Environment and Planning A*, 41: 318–35.

Andrew, R.L. 1929. The cryoscopic method for the detection of added water in milk. *Analyst*, 54: 210–16.

Anon. 1799. Adulteration of milk. *The Times*, 11 January: 4B.

Anon. 1800. A brief enquiry into the true causes of the present high price of provisions in the metropolis, and the means of applying an adequate remedy. *The Times*, 10 November: 3B.

Anon. 1894. Three months in the London milk trade. *Economic Review*, 4: 177–88.

Anon. 1897. Special Sanitary Commission on the use of antiseptics in food. *Lancet*, i: 56–60.

Anon. 1817. Description of a lactometer. *Quarterly Journal of Science, Literature and the Arts*, 3: 393–4.

Anon. 1821. A treatise on adulterations of food and culinary poisons, *Quarterly Review*, 24(48): 341–52.

Anon. 1830. Lait falsifié vendu à Paris, *Moniteur de l'Industrie*, February: 53.

Anon. 1845. Falsification du lait: condamnation à la prison. *Journal de Chimie Médicale*, 3(1): 429.

Anon. 1846. Falsification du lait et de la crème par la fécule et de l'amidon. *Journal de Chimie Médicale*, 3(2): 286.

Anon. 1853. Falsification du lait. *Journal de Chimie Médicale*, 3(9): 188–9.

Anon. 1855. Falsification du lait soumise aux lois des falsifications alimentaires. *Revue de Thérapeutique Medico-Chirurgicale*, 3: 278.

Anon 1856a. Metropolitan Milk Company. Lancet, 67, 674.

Anon. 1856b. *The Tricks of Trade in the Adulterations of Food and Physic.* London: David Bogue.

Anon. 1860. Sur la falsification du lait. *Journal de Chimie Médicale*, 4(6): 225–6.

Anon. 1863–65. William Harley and his dairy system. *Quarterly Journal of Agriculture*, 24: 207–19.

Anon. 1886. Abstract of the work of the Milk Committee. *Analyst*, 11: 3–11.

Anon. 1894. Three months in the London milk trade. *Economic Review*, 4: 177–88.

Anon. 1900. The Copenhagen control-system of milk supply. *British Food Journal*, 2: 316.

Anon. 1910. London cows for London nurseries. *Lancet*, ii: 1501.

Anon. 1911. Voluntary control of milk supply. *Public Health*, 24: 266.

Anon. 1914. Memoir of the late Sir George Barham. *Journal of the British Dairy Farmers' Association*, 28: 9–14.

Anon. [1916a]. *Campaign for Clean Milk: A Series of Articles that Have Appeared in the 'Observer'*. London: St Catherine's Press for National Clean Milk Society.

Anon. 1916b. Pure milk problems. *Truth*, 80, 2067, Special Supplement no. 33.

Anon. 1916c. Regulation of prices and distribution of milk. *The Producer*, 1: 10–11.

Anon. 1916d. Quality of milk: the feeding of cows. Hunt v. Richardson. *Analyst*, 41: 224–7.

Anon. 1921a. Bottled milk and the small retailer. *The Milk Industry*, 1(8): 19–20.

Anon. 1921b. Improving the milk supply. *The Field*, 137(3567): 560.

Anon. 1931. Pure milk. *Lancet*, i: 387–8.

Appadurai, A. 1986. Introduction: commodities and the politics of value, in *The Social Life of Things: Commodities in Cultural Perspective*, edited by A. Appadurai. Cambridge: Cambridge University Press, 3–63.

Arthur, W.B. 1989. Competing technologies, increasing returns, and lock-in by historical events. *Economic Journal*, 99: 116–31.

Arup, P.S. 1917. The reductase test for milk. *Analyst* 42: 20–31.

Ashworth, W.J. 2001. 'Between the trader and the public': British alcohol standards and the proof of good governance. *Technology and Culture*, 42: 27–50.

Astor, Viscount 1931. The problem of the milk supply: revision of designations. *Lancet*, i: 771–3.

Athey, L.L. 1978. From social conscience to social action: the Consumers' League in Europe, 1900–1914. *Social Science Review*, 52: 362–82.

Atiyah, P.S. 1979. *The Rise and Fall of Freedom of Contract*. Oxford: Clarendon Press.

Atiyah, P.S. 1987. *Pragmatism and Theory in English Law*. London: Stevens and Sons.

Atkins, P.J. 1978. The growth of London's railway milk trade, c.1845–1914. *Journal of Transport History*, new series 4, 208–26.

Atkins, P.J. 1984a. Sir George Barham (1836–1913) milk wholesaler and retailer, in *Dictionary of Business Biography*, edited by D.J. Jeremy. London: Butterworths, volume 1: 157–61.

Atkins, P.J. 1984b. Arthur Saxby Barham (1869–1952) milk wholesaler and retailer, in *Dictionary of Business Biography*, edited by D.J. Jeremy. London: Butterworths, volume 1: 156–7.

Atkins, P.J. 1984c. George Titus Barham (1860–1937) milk wholesaler and retailer, in *Dictionary of Business Biography*, edited by D.J. Jeremy. London: Butterworths, volume 1: 161–3.

Atkins, P.J. 1984d. Sir Robert Butler (1866–1933) milk and dairy products wholesaler and retailer, in *Dictionary of Business Biography*, edited by D.J. Jeremy. London: Butterworths, volume 1: 531–3.

Atkins, P.J. 1985a. Joseph Herbert Maggs (1890–1964) dairy company chairman, in *Dictionary of Business Biography*, edited by D.J. Jeremy. London: Butterworths, volume 4, 79–80

Atkins, P.J. 1985b. Leonard Maggs (1890–1959) dairy company chairman, in *Dictionary of Business Biography*, edited by D.J. Jeremy. London: Butterworths, volume 4: 81–2.

Atkins, P.J. 1985c. Sir William Price (1865–1938) milk retailer and wholesaler, in *Dictionary of Business Biography*, edited by D.J. Jeremy. London: Butterworths, volume 4: 769–71.

Atkins, P.J. 1986a. Edwin White (1873–1965) milk wholesaler and retailer, in *Dictionary of Business Biography*, edited by D.J. Jeremy. London: Butterworths, volume 5: 774–6.

Atkins, P.J. 1986b. Edmund Charles Tisdall (1824–1892) milk retailer and wholesaler, in *Dictionary of Business Biography*, edited by D.J. Jeremy. London: Butterworths, volume 5: 534–6.

Atkins, P.J. 1991. Sophistication detected: or, the adulteration of the milk supply, 1850–1914. *Social History*, 16: 317–39.

Atkins, P.J. 1992. White poison: the health consequences of milk consumption, 1850–1930. *Social History of Medicine*, 5: 207–27.

Atkins, P.J. 2000a. The pasteurization of England: the science, culture and health implications of milk processing, 1900–1950, in *Food, Science, Policy and Regulation in the Twentieth Century: International and Comparative Perspectives*, edited by D.F. Smith and J. Phillips. London: Routledge, 37–51.

Atkins, P.J. 2000b. Milk consumption and tuberculosis in Britain, 1850–1950, in *Order and Disorder: The Health Implications of Eating and Drinking in the Nineteenth and Twentieth Centuries*, edited by A. Fenton. East Linton: Tuckwell Press, 83–95.

Atkins, P.J. 2004a. The Glasgow case: meat hygiene and the foundations of state food policy in the 1890s. *Agricultural History Review*, 52: 161–82.

Atkins, P.J. 2004b. Edmund Charles Tisdall (1824–1892), in *Oxford Dictionary of National Biography*. Oxford: Oxford University Press. Available at: http://www.oxforddnb.com/view/article/48346 [accessed 14 May 2009].

Atkins, P.J. 2004c. Sir William Price (1865–1938), in *Oxford Dictionary of National* Biography. Oxford: Oxford University Press. Available at: http://www.oxforddnb.com/view/article/48123 [accessed 14 May 2009].

Atkins, P.J. 2005a. The Milk in Schools Scheme, 1934–40: 'nationalization' and resistance. *History of Education*. 34: 1–21.

Atkins, P.J. 2005b. Fattening children or fattening farmers? School milk in Britain, 1900–1950. *Economic History Review*, 58: 57–78.

Atkins, P.J. 2007a. Le concept de lait sain en Grande-Bretagne et sa mise en oeuvre, 1900–1960, in *Un Aliment Sain dans un Corps Sain*, edited by F. Audoin-Rouzeau and F. Sabban. Tours: Presses Universitaires François-Rabelais, 273–87.

Atkins, P.J. 2007b. School milk in Britain, 1900–34. *Journal of Policy History*, 19: 395–427.

Atkins, P.J. 2007c. Laboratories, laws and the career of a commodity. *Environment and Planning D: Society and Space*, 25: 967–89.

Atkins, P.J. 2008. Fear of animal foods: a century of zoonotics. *Appetite*, 51: 18–21.

Atkins, P.J. 2010. *The Reluctant State? Scientific Uncertainty, Policy Making and Bovine Tuberculosis, 1880–1960*. Forthcoming.

Atkins, P.J. and Bowler, I.R. 2001. *Food in Society: Economy, Culture, Geography*. London: Arnold.

Atkins, P.J., Hassan, M.M. and Dunn, C.E. 2006. Toxic torts: arsenic poisoning in Bangladesh and the legal geographies of responsibility. *Transactions of the Institute of British Geographers*, 31: 272–85.

Atkins, P.J., Hassan, M.M. and Dunn, C.E. 2007a. Environmental irony: summoning death in Bangladesh. *Environment and Planning A*, 39: 2699–714.

Atkins, P.J., Hassan, M.M. and Dunn, C.E. 2007b. Poisons, pragmatic governance and deliberative democracy: the arsenic crisis in Bangladesh, *Geoforum*, 38: 155–70.

Atkins, P.J., Lummel, P. and Oddy, D.J. (eds) 2007. *Food and the City in Europe since 1800*. Aldershot: Ashgate.

Atkins, P.J. and Stanziani, A. 2008. From laboratory expertise to litigation. The municipal laboratory of Paris and the Inland Revenue laboratory in London, 1870–1914: a comparative analysis, in *Fields of Expertise: Experts, Knowledge and Powers in European Modern History*, edited by C. Rabier. Newcastle: Cambridge Scholars Press, 317–39.

Austrin, T. and Farnsworth, J. 2005. Hybrid genres: fieldwork, detection and the method of Bruno Latour. *Qualitative Research*, 5: 147–65.

Babcock, S.M. 1890. A new method for the estimation of fat in milk, especially adapted to creameries and cheese factories, *Bulletin* 24. Madison: Wisconsin Experimental Station.

Babcock, S.M. 1892. The estimation of the total solids in milk from the percent of fat and the specific gravity of the milk. *Wisconsin Experimental Station, Annual Report, 1891*: 292–307.

Bachelard, G. 1998 (original 1949). *Le Rationalisme Appliqué*. Paris: Presses Universitaires de France.

Bacon, C. 2005. Confronting the coffee crisis: can fair trade, organic and specialty coffees reduce small-scale farmer vulnerability in northern Nicaragua? *World Development*, 33: 497–511.

Baines, A. 1991. How the National Food Survey began, in *Fifty Years of the National Food Survey 1940–1990*, edited by J.M. Slater. London: HMSO, 17–23.

Baird, T. 1793. *General View of the Agriculture of Middlesex*. London: J. Nichols.

Bakker, K. 2004. *An Uncooperative Commodity: Privatizing Water in England and Wales*. Oxford: Oxford University Press.

Bakker, K. and Bridge, G. 2006. Material worlds? Resource geographies and the 'matter of nature'. *Progress in Human Geography*, 30: 5–27.

Balland, A. 1902. *La Chimie Alimentaire dans l'Oeuvre de Parmentier*. Paris: Baillière.

Bannington, B.G. 1915. *English Public Health Administration*. London: King.

Barad, K. 2003. Posthumanist performativity: toward an understanding of how matter comes to matter. *Signs*, 28: 801–31.

Barham, E. 2003. Translating *terroir:* the global challenge of French AOC labelling. *Journal of Rural Studies*, 19: 127–38.

Barnes, F.A. 1958. The evolution of the salient patterns of milk production and distribution in England and Wales. *Transactions of the Institute of British Geographers*, 25: 167–95.

Barnett, C. and Land, D. 2007. Geographies of generosity: beyond the 'moral turn'. *Geoforum*, 38: 1065–75.

Barrientos, S. and Dolan, C. (eds) 2006. *Ethical Sourcing in the Global Food System*. London: Earthscan.

Barruel, M. 1829. Considérations hygiéniques sur le lait vendu à Paris comme substance alimentaire, *Annales d'Hygiène Publique et de Médecine Légale*, 1(1): 404–19.

Barry, A. 2002. The anti-political economy. *Economy and Society* 31: 268–84.

Barry, A. 2005. Pharmaceutical matters: the invention of informed materials. *Theory Culture Society*, 22: 51–69.

Barry, A. 2006. Technological zones. *European Journal of Social Theory*, 9: 239–53.

Barry, A. and Slater, D. 2002. The technological economy, *Economy and* Society, 31: 175–93.

Barthel, J.G.C. 1910. *Methods Used in the Examination of Milk and Dairy Products*. London: Macmillan.

Barthélémy, J.-H. 2005. *Penser l'Individuation: Simondon et la Philosphie de la Nature*. Paris: L'Harmattan.

Barthélémy, J.-H. (ed.) 2006. Gilbert Simondon. *Revue Philosophique de la France et de l'Étranger*, 196(3).

Barthes, R. 1972. *Mythologies*. London: Cape.

Bartlett, S. and Kay, H.D. 1950. Milk quality. *Journal of the Royal Agricultural Society of England*, 111: 87–98.

Barton, J.L. 1994. Redhibition, error and implied warranty in English law. *Tijdschrift voor Rechtsgeschiedenis*, 62: 317–29.

Bauman, Z. 1991. *Modernity and Ambivalence*. Cambridge: Polity Press.

B.D. [Bernard Dyer]' 1908. Obituary: James Bell CB, DSc, FRS, *Analyst*, 33: 157–9.

Beck, U. 1992. *Risk Society*. London: Sage.

Beck, U. 1999. *World Risk Society*. Cambridge: Polity Press.

Beckman, E. 1894. Beitrag zur Milch Analyse. *Milch Zeitung*, 23: 702.

Behrend, P. and Morgen, A. 1879. Ueber die bestimmung der trockensubstanz in der milch nach dem specifischen gewicht derselben. *Journal für Landwirtschaft*, 27: 249–59.

Belcher, S.D. 1903. *Clean Milk*. New York: Hardy.

Bell, J. 1883. *Analysis and Adulteration of Foods. Part II: Milk, Butter, Cheese, Cereal Foods, Prepared Starches, etc.* London: Chapman and Hall.

Bell, J. 1884. Food adulteration and analysis. *Analyst*, 9: 133–47.

Bell, J. 1887. Somerset House and milk analysis. *Analyst*, 12: 221–2.

Bell, W.J. 1931. *Bell's Sale of Food and Drugs Acts*. London: Butterworths.

Bennett, J. 2004. The force of things: steps toward an ecology of matter. *Political Theory*, 32: 347–72.

Bennett, J. 2005. The agency of assemblages and the North American blackout. *Public Culture*, 17: 445–65.

Bensaude-Vincent, B. 2000. 'The chemist's balance for fluids': hydrometers and their multiple identities, 1770–1810, in *Instruments and Experimentation in the History of Chemistry*, edited by F.L. Holmes and T.H. Levere. Cambridge, MA: MIT Press, 153–83.

Bensaude-Vincent, B. and Stengers, I. 1996. *A History of Chemistry*. Cambridge, MA: Harvard University Press.

Berglund, E.K. 1998. *Knowing Nature, Knowing Science: An Ethnography of Environmental Action*. Knapwell: White Horse Press.

Bergson, H. 2001 (original 1889). *Time and Free Will: An Essay on the Immediate Data of Consciousness*. New York: Dover.

Berman, A. 1966. The Cadet circle: representatives of an era in French pharmacy. *Bulletin of the History of Medicine*, 40: 101–11.

Berry, M. 1993. Tempest in a milk pail: the lactometer and 19th century dairy disputes. *Rittenhouse*, 7(27): 76–83.

Berzelius, J.J. 1814–15. Experiments to determine the definite proportions in which the elements of organic nature are combined. *Annals of Philosophy*, 4: 323–31, 401–9; 5: 93–101, 174–84, 260–75.

Berzelius, J.J. 1841 *Rapport Annuel sur les Progrès de la Chimie*. Paris: Fortin, Masson.

Billington, T. [1954] *The Campaign for Better Milk*. Benfleet: Naish.

Bingen, J. and Busch, L. (eds) 2006. *Agricultural Standards: The Shape of the Global Food and Fibre System*. Dordrecht: Springer.

Binnie, J., Holloway, J., Millington, J. and Young, C. 2007. Mundane geographies: alienation, potentialities, and practice, *Environment and Planning A*, 39: 515–20.

Blackshaw, J.F. 1924. Methods adopted in England and Wales to convey dairy education and the principles of cooperation to the farmer, in *Proceedings of the World's Dairy Congress, Washington, DC, October 2, 3, Philadelphia PA, October 4, Syracuse NY, October 5, 6, 8, 9, 10*, 1923, edited by L.A. Rogers and K.D. Lenoir. Washington, DC: Government Printing Office, volume I, 366–73.

Blackshaw, J.F. 1926. Steps taken by the Ministry to promote the production of graded milk. *The Dairyman, The Cowkeeper and Dairyman's Journal*, May: 523–8.

Blaxter, K. 1977. Herbert Davenport Kay: 9 September 1893–24 November 1976. *Biographical Memoirs of Fellows of the Royal Society*, 23: 283–310.

Block, D. 2005. Saving milk through masculinity: public health officers and pure milk, 1880–1930. *Food and Foodways*, 13: 115–34.

Blyth, A.W. 1927. *Foods: Their Composition and Analysis*. London: Griffin.

Blyth, A.W. and Cox, H. 1927. *Foods: Their Composition and Analysis*. London: Griffin.

Board of Agriculture 1919. The dairy cow and milk selling. *Guides to Smallholders*, No. 8. London: Board of Agriculture and Fisheries

Bobrow-Strain, A. 2008. White bread bio-politics: purity, health, and the triumph of industrial baking. *Cultural Geographies*, 15: 19–40.

Bokma, B.H. 2006. Role of import and export regulatory animal health officials in international control and surveillance for animal diseases. *Annals of the New York Academy of Sciences*, 1081: 84–9.

Bolduc, J.-S. and Chazal, G. 2005. The Bachelardian tradition in the philosophy of science. *Angelaki*, 10: 79–87.

Boltanski, L. and Thévenot, L. 2006. *On Justification: Economies of Worth*. Princeton: Princeton University Press.

Bondeson, G. 1983. *The Growth of a Global Enterprise: Alfa-Laval 100 years*. Stockholm: Alfa-Laval.

Boschma, R. and Martin, R. (eds) 2007. *Handbook of Evolutionary Economic Geography*. Cheltenham: Edward Elgar.

Bourdieu, J., Bruegel, M. and Atkins, P.J. 2009. 'That elusive feature of food consumption': historical perspectives on quality in Europe, *Food and History*, 5, 247–67.

Boussingault, J.-B. and Le Bel, J.A. 1839. Recherches sur l'influence de la nourriture des vaches sur la quantité et le constitution chimique du lait, *Annales de Chimie et de Physique*, 71: 65–79.

Boutflour, R. 1933. The Harleian dairy system. *Journal of the Ministry of Agriculture*, May: 119–23.

Bové, J. and Dufour, F. 2001. *The World is Not for Sale: Farmers against Junkfood*. London: Verso.

Bowker, G.C. and Leigh Star, S. 1999. *Sorting Things Out: Classification and its Consequences*. Cambridge, MA: MIT Press.

Braconnot, H. 1830. Mémoire sur le caséum et sur le lait: nouvelles ressources qu'ils peuvent offrir à la société. *Annales de Chimie et de Physique*, 43: 337–51.

Bradbury, S. and Turner, G.L'E. (eds) 1967. *Historical Aspects of Microscopy*. Cambridge: Heffer.

Brassley, P. 2000. Livestock breeds, in *The Agrarian History of England and Wales, VII: 1850–1914*, edited by E.J.T. Collins. Cambridge: Cambridge University Press, 555–9.

Brassley, P. 2004. Hall, Sir (Alfred) Daniel (1864–1942), in *Oxford Dictionary of National Biography*. Oxford: Oxford University Press. Available at: http://www.oxforddnb.com/view/article/33647 [accessed 11 May 2009].

Braudel, F. 1982. *The Wheels of Commerce*. New York: Harper and Row, 555–9.

Braun, B. 2006a. Environmental issues: global natures in the space of assemblage. *Progress in Human Geography* 30: 644–54.

Braun, B. 2006b. Towards a new earth and a new humanity: nature, ontology, politics, in *David Harvey: A Critical Reader*, edited by N. Castree and D. Gregory. Malden, MA: Blackwell, 191–222.

Braun, B. 2007. Biopolitics and the molecularization of life. *Cultural Geographies*, 14: 6–28.

Bridge, M.G. 1991. The evolution of modern sales law. *Lloyd's Maritime and Commercial Law Quarterly*: 53–69.

Brissenden, C.H. and Shepheard, G.E. 1965–66. Refrigeration in the dairy industry. *Proceedings of the Institute of Refrigeration*, 62: 60–69.

Brisson, M.J. 1787. *Traité de la Pesanteur Spécifique des Corps*. Paris: Impr. Royale.

British Friesian Cattle Society 1930. *History of British Friesian Cattle, their Performance, their Pedigrees, and History of the British Friesian Cattle Society 1909 to 1930*. Lewes: BFCS.

British Standards Institution 1959. Horvet test: BS3095. London: BSI.

Brock, W.H. 1997. *Justus von Liebig: The Chemical Gatekeeper*. Cambridge: Cambridge University Press.

Brockington, C.F. 1937. The evidence for compulsory pasteurisation of milk. *British Medical Journal*, i: 667–8.

Brousseau, E. and Glachant, J.-M. 2002. Contract economics and the renewal of economics, in *The Economics of Contracts*, edited by E. Brousseau E. and J.M. Galchant. Cambridge: Cambridge University Press.

Brown, B. 2001. Thing theory. *Critical Inquiry*, 28: 1–22.

Brown, B. 2003. *A Sense of Things: The Object Matter of American Literature*. Chicago: Chicago University Press.

Brown, B. (ed.) 2004. *Things*. Chicago: Chicago University Press.

Brown, C.A. 1925. The life and chemical services of Frederick Accum. *Journal of Chemical Education*, 2: 829–51, 1008–35, 1140–86.

Brown, J. 1987. *Agriculture in England: A Survey of Farming, 1870–1947*. Manchester: Manchester University Press.

Buckley, W. 1915. *The Score Card System of Dairy Farm Inspection*. London: National Clean Milk Society.

Buckley, W. 1922. The history of graded milk. *The Dairyman, The Cowkeeper and Dairyman's Journal*, July: 494, 497.

Buckley, W. [1923]. *Some Observations on the Butter Fat in Cow's Milk*. London: Printed for Private Circulation.

Buckley, W. 1931. Milk and the public health. *Journal of State Medicine*, 39: 28–31.

Bulloch, W. 1938. *The History of Bacteriology*. London: Oxford University Press.

Burger, J., Kirchner, M., Bramanti, B., Haak, W. and Thomas, M.G. 2007. Absence of the lactase-persistence-associated allele in early Neolithic Europeans. *Proceedings of the National Academy of Sciences*, 104: 3736–41.

Burgess, H.F. 1947. Robert Williams and the early development of the National Institute for Research in Dairying. *British Medical Bulletin*, 5: 222–5.

Burnett, D.G. 2007. Cross-examination? *Isis*, 98: 310–14.

Burnett, J. 1958. History of food adulteration in Great Britain in the nineteenth century, with special reference to bread, tea and beer, unpublished PhD thesis, University of London.

Burnett, J. 1966 *Plenty and Want: A Social History of Diet in England from 1815 to the Present Day*. London: Nelson.

Burrows, G.T. 1950. *History of the Dairy Shorthorn Cattle*. [London]: Vinton.

Burton, S.J. (ed.) 2000. *The Path of the Law and its Influence: The Legacy of Oliver Wendell Holmes, Jr.* Cambridge: Cambridge University Press.

Busch, L. and Juska, A. 1997. Beyond political economy: actor networks and the globalization of agriculture. *Review of International Political Economy*, 4: 688–708.

Busch, L. and Tanaka, K. 1996. Rites of passage: constructing quality in a commodity subsector. *Science, Technology, and Human Values*, 21: 3–27.

Butler, G., Nielsen, J.H., Slots, T., Seal, C., Eyre, M.D., Sanderson, R. and Leifert, C. 2008. Fatty acid and fat-soluble antioxidant concentrations in milk from high- and low-input conventional and organic systems: seasonal variation. *Journal of the Science of Food and Agriculture*, 88: 1431–41.

Callon, M. 1998. An essay on framing and overflowing: economic externalities revisited by sociology, in *The Laws of the Market*, edited by M. Callon. Oxford: Blackwell, 244–69.

Callon, M. 2005a. Let's put an end on uncertainties, in *Quality: A Debate*, edited by C. Musselin and C. Paradeise. *Sociologie du Travail*, 47: S94–100.

Callon, M. 2005b. Why virtualism paves the way to political impotence: a reply to Daniel Miller's critique of 'The Laws of the Markets'. *Economic Sociology European Electronic Newsletter*, 6(2): 3–20.

Callon, M. 2006. What does it mean to say that economics is performative? *Working Paper* 5. Paris: Centre de Sociologie de l'Innovation, Ecole des Mines.

Callon, M., Lascoumes, P. and Barthe, Y. 2001. *Agir dans un Monde Incertain: Essai sur la Démocratie Technique*. Paris: Seuil.

Callon, M., Méadel, C. and Rabeharisoa, V. 2002. The economy of qualities. *Economy and Society*, 31: 194–217.

Callon, M. and Rip, A. 1992. Humains, non-humains: morale d'une coexistence, in *La Terre Outragée: les Experts sont Formels!* Edited by J. Theys and B. Kalaora. Paris: Editions Autrement, 140–56.

Cameron Brown, C.A. 1936. *Refrigeration for the Farm and Dairy*. Oxford: Oxford University, Institute for Research in Agricultural Economics.

Campkin, B. 2007. Degradation and regeneration: theories of dirt and the contemporary city, *Dirt: New Geographies of Cleanliness and Contamination*, edited by B. Campkin and R. Cox. London: I.B. Tauris, 68–79.

Carpenter, K.J. 1998. Early ideas on the nutritional significance of lipids. *Journal of Nutrition*, 128, 423S–6S.

Castree, N. 1995. The nature of produced nature: materiality and knowledge construction in Marxism. *Antipode*, 27: 12–48.

Castree, N. 2001. Socializing nature: theory, practice, and politics, in *Social Nature: Theory, Practice, and Politics*, edited by N. Castree and B. Braun. Malden, MA: Blackwell, 1–19.

Castree, N. 2006. A congress of the world. *Science as Culture*, 15: 159–70.

Chabot, P. 2003. *La Philosophie de Simondon*. Paris: Vrin.

Chabot, P. 2005. The philosophical August 4th: Simondon as a reader of Bergson. *Angelaki*, 10: 103–8.

Chalmers, C.H. 1945. *Bacteria in Relation to the Milk Supply*. London: Arnold.

Chatriot, A., Chessel, M.-E. and Hilton, M. (eds) 2004. *Au Nom du Consommateur: Consommation et Politique en Europe et aux Etats-Unis au XXe Siècle*. Paris: La Découverte.

Chemist 1831. *The Domestic Chemist: Comprising Instructions for the Detection of Adulteration in Numerous Articles*. London: Bumpus and Griffin.

Chessel, M.-E. 2006. Consumers' Leagues in France: a transatlantic perspective, in *The Expert Consumer: Associations and Professionals in Consumer Society*, edited by A. Chatriot, M.-E. Chessel and M. Hilton. Aldershot: Ashgate, 53–69.

Chevallier, A. 1844. Observations sur la vente du lait. *Annales d'Hygiène Publique et de Médecine Légale*, 1(31): 453–8.

Chevallier, A. 1850. Nouvelle falsification du lait par un liquide contenant de la dextrine. *Journal de Chimie Médicale*, 3(6): 29–30.

Chevallier, A. 1854. Falsification du lait – loi du 27 Mars 1851 – délit. Cour impériale de Metz. *Journal de Chimie Médicale*, 3(10): 765–8.

Chevallier, A. 1856. Sur le commerce du lait pour l'alimentation de la population Parisienne, *Journal de Chimie Médicale, de Pharmacie et de Toxicologie*, 4(2): 633–46.

Chevallier, A. and Henry, O. 1839. Mémoire sur le lait. *Journal de Pharmacie et de Sciences Accessoires*, 25, 6: 333–55, 401–21.

Chevallier, A. and Reveil, O. 1856. Notice sur le lait: falsifications qu'on lui fait subir. Instructions sur les moyens à employer pour les reconnaître. *Journal de Chimie Médicale, de Pharmacie et de Toxicologie*, 4(2): 342–64, 402–16.

Chevallier, J.G.A. 1812. *Le Conservateur de la Vue*. Paris: Chevallier and Le Normant.

Chevallier, J.G.A. 1819. *Essai sur l'Art de l'Ingénieur en Instrumens de Physique Expérimentale en Verre*. Paris: Chevallier, Huzard, Delaunay and Pillet.

Chick, H., Hume, M. and Macfarlane, M. 1971. *War on Disease: A History of the Lister Institute*. London: Andre Deutsch.

Chirnside, R.C. and Hamence, J.H. 1974. *The 'Practising Chemists': A History of the Society for Analytical Chemistry 1874–1974*. London: Society for Analytical Chemistry.

Clausnitzer, F. and Mayer, A. 1879. Bestimmung des fetts in der milch. *Forschungen auf dem Gebiete der Viehhaltung*, 2: 265.

Clayton, E.G. 1908. *Arthur Hill Hassall: Physician and Sanitary Reformer*. London: Baillière.

Cleaveland S., Laurenson M.K. and Taylor L.H. 2001. Diseases of humans and their domestic mammals: pathogen characteristics, host range and the risk of emergency. *Philosophical Transactions of the Royal Society B: Biological Sciences*, 356: 991–9.

Cleland, J. 1816. *Annals of Glasgow*. Glasgow: J. Hedderwick.

Clifton, G.C. 1995. *Directory of British Scientific Instrument Makers, 1550–1851*. London: Zwemmer.

Coe, S.D. and Coe, M.D. 2003. *The True History of Chocolate*. London: Thames and Hudson.

Cohen, L. 2003. *A Consumers' Republic: The Politics of Mass Consumption in Postwar America*. New York: Knopf.

Coit, H.L. 1909. The Medical Milk Commission on the American continent: its origin, its purpose and its growth. *Public Health*, 23: 93–7.

Colebrook, C. 2002. *Gilles Deleuze*. London: Routledge.

Collins, E.J.T. 1993. Food adulteration and food safety in Britain in the 19th and early 20th centuries. *Food Policy*, 18: 95–109.

Collins, E.J.T. 2000. The Great Depression, 1875–1896, in *The Agrarian History of England and Wales, VII: 1850–1914*, edited by E.J.T. Collins, 138–207.

Collins, H.M. and Evans, R.J. 2002. The third wave of science studies: studies of expertise and experience. *Social Studies of Science*, 32: 235–96.

Collins, H.M. and Evans, R.J. 2003. King Canute meets the beach boys: responses to the third wave. *Social Studies of Science*, 33: 435–52.

Collins, H.M. and Evans, R.J. 2007. *Rethinking Expertise*. Chicago: Chicago University Press.

Collins, H.M. and Yearley, S. 1992. Epistemological chicken, in *Science as Practice and Culture*, edited by A. Pickering. Chicago: University of Chicago Press, 301–26.

Colquhoun, P. 1829. *A Treatise on the Police and Crimes of the Metropolis*. London: Longman, Rees, Orme, Brown, and Green.

Combes, M. 1999. *Simondon Individu et Collectivité: pour une Philosophie du Transindividual*. Paris: Presses Universitaires de France.

Conford, P. 2001. *The Origins of the Organic Movement*. Edinburgh: Floris Books.

Cook, I. 2004. Follow the thing: papaya. *Antipode*, 36: 642–64.

Cook, I., Crang, P. and Thorpe, M. 1998. Biographies and geographies: consumer understandings of the origins of foods. *British Food Journal*, 100: 162–7.

Cook, I., Crang, P. and Thorpe, M. 2004. Tropics of consumption: 'getting with the fetish' of 'exotic fruit', *Geographies of Commodities*, edited by A. Hughes and S. Reimer. London: Routledge, 173–92.

Corbin, A. 1986. *The Foul and the Fragrant: Odour and the French Social Imagination*. Leamington Spa: Berg.

Cornish, W.R. and Clark G. de N. 1989. *Law and Society in England 1750–1950*. London: Sweet and Maxwell.

Court of General Sessions in and for the City and County of New York 1881. *The People vs Daniel Schrumpf; Misdemeanor, Adulteration of Milk; Record, Testimony and Proceedings*. New York: M.B. Brown.

Cowgill, G.R. 1964. Jean Baptiste Boussingault – a biographical sketch. *Journal of Nutrition*, 84: 1–9.

Crease, R. and Selinger, E. (eds) 2006. *The Philosophy of Expertise*. New York: Columbia University Press.

Crewe, L. 2004. Unravelling fashion's commodity chains, in *Geographies of Commodity Chains*, edited by A. Hughes and S. Reimer. London: Routledge, 195–214.

Cribb, C.H. and Moor, C.G. 1899. The statistics of adulteration. *British Food Journal*, 1: 224–6.

Crighton-Browne, Sir J. 1918. Milk and health. *Journal of State Medicine*, 26: 140–50, 178–86.

Cronon, W. 1991. *Nature's Metropolis: Chicago and the Great West*. New York: Norton.

Cronon, W. 1995. The trouble with wilderness, *Uncommon Ground: Toward Reinventing Nature*, edited by W. Cronon. New York: Norton, 69–90.

Crossley, E.L. 1958. The British Dairy Institute and the University of Reading. *Journal of the British Dairy Farmers' Association*, 62: 5–8.

Crowther, C. 1939. Some problems of animal nutrition, in *Agriculture in the Twentieth Century*, edited by H.E. Dale. Oxford: Clarendon Press, 361–95.

Cumming, R.H. and Mattick, A.T.R. 1920. An enquiry concerning the state of cleanliness of empty milk churns. *Journal of Hygiene*, 19: 84–6.

Curwen, J.C. 1809. *Hints on Agricultural Subjects*. London: J. Johnson and B. Crosby.

Dale, H.E. 1956. *Daniel Hall: Pioneer in Scientific Agriculture*. London: Murray.

Dant, T. 1999. *Material Culture and the Social World: Values, Activities and Lifestyles*. Buckingham: Open University Press.

Dant, T. 2005. *Materiality and Society*. Maidenhead: Open University Press.

Dant, T. 2008. The 'pragmatics' of material interaction. *Journal of Consumer Culture*, 8: 11–33.

Dasgupta, P. 1988. Trust as a commodity, in *Trust: Making and Breaking Cooperative Relations*, edited by D. Gambetta. Oxford: Blackwell, 49–72.

Daston, L. 1994. Historical epistemology, in *Questions of Evidence: Proof, Practice, and Persuasion across the Disciplines*, edited by J. Chandler, A.I. Davidson and H. Harootunian. Chicago: University of Chicago Press, 282–9.

Daston, L. 2000. Preternatural philosophy, in *Biographies of Scientific Objects*, edited by L. Daston. Chicago: University of Chicago Press, 15–41.

Daston, L. 2007. The history of emergencies. *Isis*, 98: 801–8.

Daston, L. and Galison, P. 2007. *Objectivity*. New York: Zone Books.

Daston, L. and Vidal, F. (eds) 2004. *The Moral Authority of Nature*. Chicago: University of Chicago Press.

Daunton, M. and Hilton, M. (eds) 2001. *The Politics of Consumption: Material Culture and Citizenship in Europe and America*. Oxford: Berg.

David, P. 2007. Path dependence: a foundational concept for historical social science. *Cliometrica*, 1: 91–114.

Davies, B. 1926. Milk: its hygiene and technics. *The Milk Industry*, 6(9): 69–80.

Davies, B. 1930. *The Food Council: The Consumers' Council Bill and Milk Distribution*. London: np.

Davies, B. 1933. *The Nation's Milk Supply: its Hygienic Production and Control*. London: United Dairies Laboratory Department.

Davis, J. 2004. Baking for the common good: a reassessment of the assize of bread in medieval England. *Economic History Review*, 57: 465–502.

Davis, J.G. 1947. Chemical standards and analytical methods for milk. *British Medical Bulletin*, 5: 206–7.

Davis, J.G. 1950. *A Dictionary of Dairying*. London: Leonard Hill.

Davis, J.G. 1951. *Milk Testing: The Laboratory Control of Milk*. London: Dairy Industries Ltd.

Davis, J.G. 1952. The chemical composition of milk between 1900 and 1950. *Analyst*, 77: 499–524.

Davis, J.G. 1983. Personal recollections of developments in dairy bacteriology over the last fifty years. *Journal of Applied Bacteriology*, 55: 1–12.

Davis, P. 2002. *The Victorians*. Oxford: Oxford University Press.

Day, F.E. and Grimes, M. 1918. The graduation and calibration of Gerber new milk butyrometers. *Analyst*, 43: 123–33.

Deakin, S. and Wilkinson, F. 2005. *The Law of the Labour Market: Industrialization, Employment, and Legal Evolution*. Oxford: Oxford University Press.

Deakin, S. 2006a. Les conventions du marché du travail et l'évolution du droit, in *L'Economie des Conventions, Méthodes et Résultats. Tome I: Débats*, edited by F. Eymard-Duvernay. Paris: La Découverte, 231–47.

Deakin, S. 2006b. 'Capacitas': contract law and the institutional preconditions of a market economy. *European Review of Contract Law*, 2: 317–41.

Dean, M. 1999. *Governmentality: Power and Rule in Modern Society*. London: Sage.

De Beistegui, M. 2005. Science and ontology: from Merleau-Ponty's 'reduction' to Simondon's 'transduction'. *Angelaki*, 10: 109–22.

De Certeau, M. 1984. *The Practice of Everyday Life*. Berkeley, CA: University of California Press.

De Certeau, M., Giard, L. and Mayol, P. 1998. *The Practice of Everyday Life, Volume 2: Living and Cooking*. Minneapolis: University of Minnesota Press.

De Landa, M. 2000. *A Thousand Years of Nonlinear History*. New York: Swerve.

De Landa, M. 2006. *A New Philosophy of Society: Assemblage Theory and Social Complexity*. London: Continuum.

Delépine, S. 1909. Report to the Local Government Board on investigations in the Public Health Laboratory of the University of Manchester upon the prevalence and sources of tubercle bacilli in cow's milk, Appendix B, no. 5, Medical Officer of Health to the Local Government Board *Annual Report for 1908–9*, PP 1909 (Cd.4935) xxviii.777–880.

Delépine, S. 1921. The milk question and 'The Milk Industry'. *The Milk Industry*, 1(7): 7–10, 1(8): 7–10, 1(9), 1(10): 5–7.

Deleuze, G. 2006. *The Fold: Leibniz and the Baroque*. London: Continuum.

Deleuze, G. and Guattari, F. 1987. *A Thousand Plateaus: Capitalism and Schizophrenia*. Minneapolis: University of Minnesota Press.

Demeritt, D. 1998. Science, social constructivism and nature, in *Remaking Reality: Nature at the Millennium*, edited by B. Braun and N. Castree. London: Routledge, 173–93.

Demeritt, D. 2001. Being constructive about nature, in *Social Nature: Theory, Practice, and Politics*, edited by N. Castree and B. Braun. Malden, MA: Blackwell, 22–40.

Demeritt, D. 2002. What is the 'social construction of nature'? A typology and sympathetic critique. *Progress in Human Geography*, 26: 767–90.

Department of Health for Scotland 1945. *The Freezing Point-Hortvet-Test of Milk: Report of a Sub-Committee of the Science Advisory Committee*. Edinburgh: Department of Health for Scotland.

De Planhol, X. and Claval, P. 1994. *An Historical Geography of France*. Cambridge: Cambridge University Press.

Derrida, J. 1976. *Of Grammatology*. Baltimore, MD: Johns Hopkins University Press.

Desrosières, A. 1998. *The Politics of Large Numbers: A History of Statistical Reasoning*. Cambridge, MA: Harvard University Press.

DeSilvey, C. 2006. Observed decay: telling stories with mutable things, *Journal of Material Culture*, 11: 318–38.

De Sousa, I.S.F and Busch, L. 2006. Standards and state building: the construction of soybean standards in Brazil, in *Agricultural Standards: The Shape of the Global Food and Fibre System*, edited by J. Bingen and L. Busch. Dordrecht: Springer, 125–35.

Dewey, J. 1910. *The Influence of Darwin on Philosophy*. New York: Holt.

Deyeux, C. 1793. Examen comparatif du lait de deux vaches nourris successivement avec le fourrage ordinaire et celui de maïs, ou bled de Turkie, *Annales de Chimie*, 17: 320–32.

Dicey, A.V. 1905. *Lectures on the Relation Between Law and Public Opinion in England during the Nineteenth Century*, London: Macmillan.

Dinocourt, H. 1846. *Instruction pour l'Usage du Galactomètre Centésimal et du Lactomètre, Instruments Propres à Faire Reconnaître la Pureté du Lait de Vaches*. Paris: A. Bailly.

Direction Générale de la Concurrence, de la Consommation et de la Répression des Fraudes (ed.) 2007. *La Loi du 1er Août 1905*. Paris: La Documentation Française.

Divers Dr 1868. On London milk. *British Medical Journal*, i: 356–7.

Dodd, F.L. 1905. Municipal milk supply and public health. *Fabian Tract* No. 122. London: Fabian Society.

Dodd, G. 1856. *The Food of London*. London: Longmans, Brown, Green and Longmans.

Dodge, M. and Kitchin, R. 2005. Code and the transduction of space. *Annals of the Association of American Geographers*, 95: 162–80.

Donné, A. 1839. Mémoire sur le lait. *Journal des Débats Politiques et Littéraires* 27 September.

Donné, A. 1841. Rapport sur un Mémoire de M. Donné. *Comptes Rendus Hebdomadaires des Séances de l'Académie des Sciences*, 17: 585–98.

Donné, A. and Foucault, J.-B.L. 1844. *Cours de Microscopie Complémentaire des Etudes Médicales*. Paris: Baillière.

Donnelly, J. 1994. Consultants, managers, testing slaves: changing roles for chemists in the British alkali industry, 1850–1920. *Technology and Culture*, 35: 100–28.

Dornbusch, D. 1998. An analysis of media coverage of the BSE crisis, in *The Mad Cow Crisis: Health and the Public Good*, edited by S.C. Ratzan. London: UCL Press, 138–52.

Douglas, L.M. 1904. *Refrigeration in the Dairy*. London: Wm Douglas and Sons.

Douglas, L.M. 1908. Geneva International Food Congress. *Journal of the Royal Society of Arts*, 56: 1059–61.

Douglas, M. 1966. *Purity and Danger*. London: Routledge and Kegan Paul.

Douny, L. 2007. The materiality of domestic waste: the recycled cosmology of the Dogon of Mali, *Journal of Material Culture*, 12: 309–31.

Liquid Materialities

Dumas, J.-B., Boussingault, J.-B. and Payen, A. 1843. Recherches sur l'engraissement des bestiaux et la formation du lait. *Annales de Chimie*, series 3(8): 63–114.

Dumouchel, P. 1992. Gilbert Simondon's plea for a philosophy of technology. *Inquiry*, 35: 407–21.

DuPuis, M. 2002. *Nature's Perfect Food: How Milk Became America's Drink.* New York: New York University Press.

Dwork, D. 1987. *War is Good for Babies and Other Young Children.* London: Tavistock.

Dyer, B. and Mitchell, C.A. 1932. *The Society of Public Analysts and Other Analytical Chemists: some Reminiscences of its First Fifty Years and a Review of its Activities.* Cambridge: Heffer.

Eastwood, A. 1909. Report on American methods for the control and improvement of the milk supply. *Reports to the Local Government Board on Public Health and Medical Matters*, new series: 1. London: HMSO.

Eckles, C.H. and Shaw, R.H. 1913. The influence of breed and individuality on the composition and properties of milk. *Bulletin, Bureau of Animal Industry*: 155. Washington, DC: United States Department of Agriculture.

Eden, S., Bear, C. and Walker, G. 2008a. Understanding and (dis)trusting food assurance schemes: Consumer confidence and the 'knowledge fix'. *Journal of Rural Studies*, 24: 1–14.

Eden, S., Bear, C. and Walker, G. 2008b. Mucky carrots and other proxies: problematising the knowledge-fix for sustainable and ethical consumption, *Geoforum*, 39: 1044–57.

Edensor, T. 2005. *Industrial Ruins: Space, Aesthetics and Materiality.* Oxford: Berg.

Edmond, G. and Mercer, D. 2004. Daubert and the exclusionary ethos: the convergence of corporate and judicial attitudes towards the admissibility of expert evidence in tort litigation. *Law and Policy*, 26: 231–57.

Egan, H. 1976. A century of food analysis, in *Food Quality and Safety, a Century of Progress: Proceedings of the Symposium Celebrating the Centenary of the Sale of Food and Drugs Act 1875, London, October 1975*, chaired by Lord Zuckerman. London: HMSO, 105–22.

Elden, S. 2001. *Mapping the Present: Heidegger, Foucault and the Project of a Spatial History.* London: Continuum.

Elden, S. 2005. Genealogy, ontology and the political: three conceptual questions to Engin Isin. *Political Geography*, 24: 355–9.

Ellis, K.A., Innocent, G., Grove-White, D., Cripps, P., McLean, W.G., Howard, C.V. and Mihm, M. 2006. Comparing the fatty acid composition of organic and conventional milk. *Journal of Dairy Science*, 89: 1938–50.

Ellis, K.A., Monteiro, A., Innocent, G., Grove-White, D., Cripps, P., McLean, W.G., Howard, V. and Mihm, M. 2007. Investigation of the vitamins A and E and β-carotene content in milk from UK organic and conventional dairy farms. *Journal of Dairy Research*, 74: 484–91.

Elias, N. 1939. *Über den Prozess der Zivilisation*. Basel: Haus zum Falken.

Elsdon, G.D. and Stubbs, J.R. 1927. The immersion refractometer and its value in milk analysis. *Analyst*, 52: 193–214.

Elsdon, G.D. and Stubbs, J.R. 1930. The freezing point of milk as a means of detecting added water. *Analyst*, 55: 423–32.

Elsdon, G.D. and Stubbs, J.R. 1933. The freezing-point of pasteurised and sterilised milks. *Analyst*, 58: 7–10.

Elsdon, G.D. and Stubbs, J.R. 1934. The technique of the freezing-point test for milk. *Analyst*, 59: 585–93.

Embrey, G. 1893. The Lister-Babcock milk tester: with some suggestions for extending its use. *Analyst*, 18: 118–25.

Enock, A.G. 1943. *This Milk Business: From 1895 to 1943*. London: H.K. Lewis.

Erdman, H.E. 1921. *Marketing of Whole Milk*. New York: Macmillan.

Ericsson, K.A., Charness, N., Feltovich, P.J. and Hoffman, R.R. (eds) 2006. *The Cambridge Handbook of Expertise and Expert Performance*. Cambridge: Cambridge University Press.

Estcourt, C. 1887. Analysis of decomposed milk. *Analyst*. 12: 224–5.

European Commission 2006. Risk issues, *Special Eurobarometer* 238/Wave 64.1. Brussels: European Commission.

Evans, F.J. 1976. Enforcing the law, in *Food Quality and Safety, a Century of Progress: Proceedings of the Symposium Celebrating the Centenary of the Sale of Food and Drugs Act 1875, London, October 1975*, chaired by Lord Zuckerman. London: HMSO, 127–37.

Evans, R. 2008. The sociology of expertise: the distribution of social fluency. *Social Compass*, 2: 281–98.

Eymard-Duvernay, F. 1989. Conventions et qualité et formes de coordination. *Revue Économique*, 40: 329–59.

Eymard-Duvernay, F. (ed.) 2006. *L'Economie des Conventions, Méthodes et Résultats*. Paris: La Découverte.

Faber, H. 1887. A new method of ascertaining the amount of fat in milk. *Analyst*, 12: 6–11.

Faber, H. 1889. On condensed milk and the estimation of casein and lactalbumen. *Analyst*, 14: 141–7.

Fagan, B. 2005. Globalization, the WTO and the Australia-Philippines 'banana war', in *Cross-Continental Food Chains*, edited by N. Fold and B. Pritchard. London: Routledge, 207–22.

Fairer, J.A. 1930. Cleaner milk production in Leicestershire. *Medical Officer*, 44: 125–6.

Farrington, E.H. and Woll, F.W. 1901. *Testing Milk and its Products: A Manual for Dairy Students, Creamery and Cheese-Factory Operators and Dairy Farmers*. Madison, WI: Mendota Book Co.

Favereau, O. and Lazega, E. (eds) 2002. *Conventions and Structures in Economic Organization: Markets, Networks and Hierarchies*. Cheltenham: Elgar.

Favereau, O., Biencourt, O. and Eymard-Duvernay, F. 2002. Where do markets come from? From (quality) conventions! In *Conventions and Structures in Economic Organization: Markets, Networks and Hierarchies*, edited by O. Favereau and E. Lazega. Cheltenham: Elgar, 213–52.

Faÿs-Sallois, F. 1980. *Les Nourrices à Paris au XIXe Siècle*. Paris: Histoire Payot.

Federation of United Kingdom Milk Marketing Boards 1972. *United Kingdom Dairy Facts and Figures*. Thames Ditton: Federation.

Ferguson, P. 2005. Eating orders: markets, menus, and meals. *Journal of Modern History*, 77: 679–700.

Ferrières, M. 2006. *Sacred Cow, Mad Cow: A History of Food Fears*. New York: Columbia University Press.

Filby, F.A. 1934. *A History of Food Adulteration and Analysis*. London: Allen and Unwin.

Fildes, V. 1988. *Wet Nursing: A History from Antiquity to the Present*. Oxford: Blackwell.

Fine, B. 2003. Callonistics: a disentanglement. *Economy and Society*, 32: 478–84.

Fine, B. and Leopold, E. 1993. *The World of Consumption*. London: Routledge.

Flandrin, J.-L. and Montanari, M. (eds) 1999. *Food: A Culinary History*. New York: Columbia University Press.

Fleck, L. 1935. *Genesis and Development of a Scientific Fact*. Chicago: University of Chicago Press.

Fleischmann, W. 1885. Beiträge zur kenntnis des wesens der milch. *Journal für Landwirtschaft*, 33: 251–69.

Fleischmann, W. [1896]. *The Book of the Dairy: A Manual of the Science and Practice of Dairy Work*. London: Blackie.

Fleischmann, W. and Morgen, A. 1882. Ueber die beziehungen welche zwischen dem specifischen gewicht der mich einerseits und dem procentischen gehalt derselben an fett und trokensubstanz andererseits bestehen. *Journal für Landwirtschaft*, 30: 293–309.

Foot, P. 1794. *General View of the Agriculture of Middlesex*. London: J. Nichols.

Forrester, R.B. 1927. The fluid milk market in England and Wales. *Ministry of Agriculture and Fisheries, Economic Series* No. 16. London: HMSO.

Foucaud, E. 1841. *Les Artisans Illustres*. Paris: Béthune et Plon.

Foulerton, A.G.R. 1899. The influence on health of chemical preservatives in food. *Lancet*, ii: 1427–32, 1577–83.

Fox, P.F. (ed.) 1992. *Advanced Dairy Chemistry. Volume 1: Proteins*. London: Chapman and Hall.

Fox, P.F. (ed.) 1995. *Advanced Dairy Chemistry. Volume 2: Lipids*. London: Chapman and Hall.

Fox, P.F. and McSweeney, P.L.H. 1998. *Dairy Chemistry and Biochemistry*. London: Blackie.

Frank, L.C. and Moss, F.J. 1932. *The Extent of Pasteurization and Tuberculin Testing in American Cities of 10,000 Population and Over in 1927 and 1931*. Washington, DC: United States Public Health Service.

Franklin, Sir R. 1953. *Ministry of Agriculture: Report of the Working Party on Quality Milk Production*. London: HMSO.

Freear, K., Buckley, W. and Williams, R.S. 1919. *A Study of Two Types of Commercial Milk*. Cambridge: Cambridge University Press.

Freidberg, S. 2003. Not all sweetness and light: new cultural geographies of food. *Social and Cultural Geography*, 4: 3–6.

Freidberg, S. 2004a. *French Beans and Food Scares: Culture and Commerce in an Anxious Age*. New York: Oxford University Press.

Freidberg, S. 2004b. The ethical complex of corporate food power. *Environment and Planning D: Society and Space*, 22: 513–31.

French, M. and Phillips, J. 2000. *Cheated not Poisoned? Food Regulation in the United Kingdom, 1875–1938*. Manchester: Manchester University Press.

French, M. and Phillips, J. 2003. Sophisticates or dupes? Attitudes toward food consumers in Edwardian Britain. *Enterprise and Society*, 4: 442–70.

Fressoz, J.B. 2007. Beck back in the 19th century: towards a genealogy of risk society. *History and Technology*, 23: 333–50.

Fruton, J.S. 1988. The Liebig research group – a reappraisal. *Proceedings of the American Philosophical Society*, 132(1): 1–66.

Fukuyama, F. 1995. *Trust: The Social Virtues and the Creation of Prosperity*. New York: Free Press.

Fullwiley, D. 2007. The molecularization of race: institutionalizing human difference in pharmacogenetics practice. *Science as Culture*, 16: 1–30.

Furlough, E. 1991. *Consumer Cooperation in France: The Politics of Consumption, 1834–1930*. Ithaca, NY: Cornell University Press.

Furlough, E. and Strikwerda, C. (eds) 1999. *Consumers against Capitalism? Consumer Cooperation in Europe, North America, and Japan, 1840–1990*. Lanham, MD: Rowman and Littlefield.

Fussell, G.E. 1954. Goat's milk for Londoners: a nineteenth century enterprise. *Agriculture*, 61: 450–51.

Fussell, G.E. 1966. *The English Dairy Farmer, 1500–1900*. London: Routledge.

Garnier, J. and Harel, C. 1844. *Des Falsifications des Substances Alimentaires et des Moyens Chimique de les Reconnaître*. Paris: Baillière.

Gaultier de Claubry, H.-F. 1842. Sur la sophistication du lait au moyen de la matière cérébrale. *Annales d'Hygiène Publique et de Médecine Légale*, 1(27): 287–95.

Gerber, N. 1892. Die Acid-Butyrometrie als Universal-Fettbestimmungs-Methode für Milch und alle flüssigen und festen Molkerei-Produkte, sowie Oleomargarine, etc. *Schweizerische Landwirtschaftlichen Zeitschrift*, 11: 50–52.

Gibbon, P. and Ponte, S. 2005. *Trading Down: Africa, Value Chains, and the Global Economy*. Philadelphia: Temple University.

Gibbs-Smith, E.G. and Hobday, F.T.G. 1928. Milk bottles as a possible cause of epidemics. *Journal of State Medicine*, 36: 660–64.

Giddens, A. 1999. *Runaway World: How Globalization is Reshaping Our Lives.* London: Profile.

Giles, R.F. 1976. The development of food legislation in the United Kingdom, in *Food Quality and Safety, a Century of Progress: Proceedings of the Symposium Celebrating the Centenary of the Sale of Food and Drugs Act 1875, London, October 1975*, chaired by Lord Zuckerman. London: HMSO, 4–13.

Giovannucci, D. and Ponte, S. 2005. Standards as a new form of social contract? Sustainability initiatives in the coffee industry. *Food Policy*, 30: 284–301.

Girard, C. 1904. *Analyse des Matières Alimentaires et Recherché de Leurs Falsifications.* Paris: Veuve Charles Dunod.

Golan, T. 2004. *Laws of Men and Laws of Nature.* Cambridge, MA: Harvard University Press.

Gooday, G.J.N. 2004. *The Morals of Measurement: Accuracy, Irony, and Trust in Late Victorian Electrical Practice.* Cambridge: Cambridge University Press.

Goodman, D. 1999. Agro-food studies in the 'age of ecology': nature, corporeality, bio-politics. *Sociologia Ruralis*, 39: 17–38.

Goodman, D. 2001. Ontology matters: the relational materiality of nature and agro-food studies. *Sociologia Ruralis*, 41: 182–200.

Goodman, D. 2003. The quality 'turn' and alternative food practices: reflections and agenda. *Journal of Rural Studies*, 19: 1–7.

Goodman, D. and Redclift, M. 1991. *Refashioning Nature: Food, Ecology, and Culture* London: Routledge.

Goodman, D.E, Sorj, B. and Wilkinson, J. 1987. *From Farming to Biotechnology.* Oxford: Blackwell.

Goodman, M. 2004. Reading fair trade: political ecological imaginary and the moral economy of fair trade foods. *Political Geography*, 23: 891–915.

Gorman, M. 2002. Levels of expertise and trading zones. *Social Studies of Science*, 32: 933–8.

Gottlieb, E. 1892. Eine bequeme Methode zur Bestimmung von Fett in Milch. *Landwirtschaftlichen Versuchsstationen*, 40: 1–27.

Goubert J.-P. 1986. *La Conquête de l'Eau: l'Avènement de la Santé à l'Age Industriel.* Paris: Robert Laffont.

Goubert, J.-P. 1988. The development of water and sewerage systems in France, 1850–1950, in *Technology and the Rise of the Networked City in Europe and America*, edited by J.A. Tarr and G. Dupuy. Philadelphia: Temple University Press, 116–36.

Grabher, G. 2009. Yet Another Turn? The Evolutionary Project in Economic Geography. *Economic Geography*, 85: 119–27.

Graham, J.N. 1935. Ten years' work on improving a county's milk supply. *Home Farmer*, 2(2): 30.

Graham, S. and Marvin, S. 2001. *Splintering Urbanism: Networked Infrastructures, Technological Mobilities and the Urban Condition.* London: Routledge.

Granovetter, M. 1973. The strength of weak ties. *American Journal of Sociology*, 78: 360–80.

Granovetter, M. 1985. Economic action and social structure: the problem of embeddedness. *American Journal of Sociology*, 91: 481–510.

Gronow, J. 1997. *The Sociology of Taste*. London: Routledge.

Gronow, J. 2004. Standards of taste and varieties of goodness: the (un)predictability of modern consumption, in *Qualities of Food*, edited by M. Harvey, A. McMeekin and A. Warde. Manchester: Manchester University Press, 38–60.

Gronow, J. and Warde, A. (eds) 2001. *Ordinary Consumption*. London: Routledge.

Grotenfelt, G. 1894. *The Principles of Modern Dairy Practice from a Bacteriological Point of View*. New York: Wiley.

Grummer, R.R. 1991. Effect of feed on the composition of milk fat. *Journal of Dairy Science*, 74: 3244–57.

Guersent, M. 1818, Maladies laiteuses, *Dictionnaire des Sciences Medicales*, 30: 270–90.

Guillaume, P. 2006. Combattre la fraude sur le lait: entre économie, hygiène et politique, in *Fraude, Contrefaçon et Contrebande de l'Antiquité à Nos Jours*, edited by G. Béaur, H. Bonin and C. Lemercier. Geneva: Droz, 579–92.

Gurney, P. 1996. *Cooperative Culture and the Politics of Consumption in England, 1870–1930*. Manchester: Manchester University Press.

Guthman, J. 2002. Commodified meanings and meaningful commodities: re-thinking production-consumption links through the organic system of provision. *Sociologia Ruralis*, 42: 295–311.

Guthman, J. 2007. The Polanyian way? Voluntary food labels as neoliberal governance. *Antipode*, 39: 456–78.

Hacking, I. 1990. *The Taming of Chance*. Cambridge: Cambridge University Press.

Hacking, I. 1999. *The Social Construction of What?* Cambridge, MA: Harvard University Press.

Hacking, I. 2002. *Historical Ontology*. Cambridge, MA: Harvard University Press.

Haggard, H.R. 1911. *Rural Denmark and its Lessons*. London: Longmans, Green and Co.

Halewood, M. 2005. On Whitehead and Deleuze: the process of materiality. *Configurations*, 13: 57–76.

Hall, H.S. [1951]. The relative importance of utensil sterilization and milk cooling. *National Institute for Research in Dairying*, Paper No. 1063.

Hall, J.S. and Buckett, M. 1969. Milk quality, *Department of Agriculture and Fisheries for Scotland, Advisory Bulletin*, No. 7. Edinburgh: HMSO.

Haller, A. von 1779. *First Lines of Physiology*. Edinburgh: Macfarquhar and Elliot.

Halsey, M. 2006. *Deleuze and Environmental Damage: Violence of the Text*. Aldershot: Ashgate.

Hamburger, P.A. 1989. The development of the nineteenth-century consensus theory of contract. *Law and History Review*, 7: 241–329.

Hamlin, C. 1986. Scientific method and expert witnessing: Victorian perspectives on a modern problem. *Social Studies of Science*, 16: 485–513.

Hamlin, C. 1990. *A Science of Impurity: Water Analysis in Nineteenth Century Britain*. Berkeley: University of California Press.

Hamlin, C. 2004. Wanklyn, James Alfred (1834–1906), in *Dictionary of National Biography*, edited by C. Matthew. Oxford: Oxford University Press. Available at: http://www.oxforddnb.com/view/article/36723 [accessed: 7 May 2009].

Hammond, P.W. 1992. 150 years of the Laboratory of the Government Chemist. *Analytical Proceedings*, 29: 311–14.

Hammond, P.W. and Egan, H. 1992. *Weighed in the Balance: A History of the Laboratory of the Government Chemist*. London: HMSO.

Hand, M. and Shove, E. 2007. Condensing practices: ways of living with a freezer. *Journal of Consumer Culture*, 7: 79–104.

Hanley, J.A. 1939. Agricultural education in college and county, in *Agriculture in the Twentieth Century*, edited by H.E. Dale. Oxford: Clarendon Press, 87–122.

Harbers, H., Mol, A. and Stollmeyer, A. 2002. Food matters: arguments for an ethnography of daily care. *Theory, Culture and Society*, 19: 207–26.

Hardin, R. 2002. *Trust and Trustworthiness*. New York: Russell Sage Foundation.

Harding, F. 1995. Adulteration of milk, in *Milk Quality*, edited by F. Harding. London: Blackie, 60–74.

Hardy, A. 1999. Food, hygiene, and the laboratory: a short history of food poisoning in Britain, *circa* 1850–1950. *Social History of Medicine*, 12: 293–311.

Harley, W. 1828–9. On feeding milk cows, as practised at Willow Bank Dairy, Glasgow. *Quarterly Journal of Agriculture*, 1: 170–74.

Harley, W. 1829. *The Harleian Dairy System*. London: Ridgway.

Harman, G. 2007. *Heidegger Explained: from Phenomenon to Thing* Peru, IL: Open Court.

Harman, G. 2009. *Prince of Networks: Bruno Latour and Metaphysics*. Prahran: re.press.

Harris, J.A. 1915. Standard dairy score cards. *Science*, 8 October: 503–5.

Hartley, R.M. 1842. *Essay on Milk as Article of Sustenance*. New York: Jonathan Leavitt.

Harvey, D. 1996. *Justice, Nature and the Geography of Difference*. Oxford: Blackwell.

Harvey, M., McMeekin, A. and Warde, A. (eds) 2004. *Qualities of Food*. Manchester: Manchester University Press.

Harvey, W.C. and Hill, H. 1936. *Milk Production and Control*. London: H.K. Lewis.

Hassall, A.H. 1855. *Food and its Adulterations: Comprising Reports of the Analytical Sanitary Commission of 'The Lancet' for the Years 1851–1854*. London: Longman, Brown, Green, and Longmans.

Hassall, A.H. 1857. *Adulterations Detected*. London: Longman, Brown, Green, Longmans, and Roberts.

Hassall, A.H. (ed.) 1871. *Food, Water and Air in Relation to the Public Health*. London: Wyman and Sons.

Hassall, A.H. 1876. *Food: its Adulterations, and the Methods for their Detection*. London: Longmans, Green and Co.

Hassall, A.H. 1893. *The Narrative of a Busy Life: An Autobiography*. London: Longmans, Green and Co.

Hassard, C. 1905. The milk trade from within. *Economic Review*, 15: 74–84, 203–11.

Hatanaka, M., Bain, C. and Busch, L. 2005. Third-party certification in the global agrifood system. *Food Policy*, 30: 354–69.

Hatanaka, M., Bain, C. and Busch, L. 2006. Differentiated standardization, standardized differentiation: the complexity of the global agrifood system, in *Between the Local and the Global: Confronting Complexity in the Contemporary Agri-Food Sector*, edited by T. Marsden and J. Murdoch. Amsterdam: Elsevier, 39–68.

Hayes, K. 2008. *Milk and Melancholy*. Cambridge, MA: MIT Press.

Hehner, O. 1882a. On some points in milk analysis. *Analyst*, 7: 60–64.

Hehner, O. 1882b. On the relation between the specific gravity, the fat, and the solids not fat in milk. *Analyst*, 7: 129–35.

Hehner, O. 1891. Remarks in discussion of T. Eustace Hill, The Werner-Schmid method of milk analysis. *Analyst*, 16: 73.

Hehner, O. 1892. Note on the Leffmann-Beam method of determining fat in milk. *Analyst*, 17: 102–4.

Hehner, O. and Richmond, H.D. 1888. On the relation of specific gravity, fat and solids-not-fat in milk, upon the basis of the Society of Public Analysts' method. *Analyst*, 13: 26–36.

Heinemann, P.G. 1921. *Milk*. Philadelphia: W.B. Saunders.

Heisch, C. 1882. Presidential address. *Analyst*, 7: 13–15.

Hennion, A. 2005. Pragmatics of taste, in *The Blackwell Companion to the Sociology of Culture*, edited by M.D. Jacobs and N.W. Hanrahan. Oxford: Blackwell, 131–44.

Herreid, E.O. 1942. The Babcock Test: a review of the literature. *Journal of Dairy Science* 25: 335–70.

Hess, M. 2004. 'Spatial' relationships? Towards a reconceptualization of embeddedness. *Progress in Human Geography*, 18: 165–86.

Hetherington, K. 2003. Spatial textures: place, touch, and praesentia. *Environment and Planning A*, 35: 1933–44.

Hetherington, K. 2007. *Capitalism's Eye: Cultural Spaces of the Commodity*. London: Routledge.

Hierholzer, V. 2007. The 'war against food adulteration': municipal food monitoring and citizen self-help associations in Germany, 1870s–1880s, in *Food and the City in Europe since 1800*, edited by P.J. Atkins, P. Lummel and D.J. Oddy. Aldershot: Ashgate, 117–28.

Hildebrandt, M. 2007. The trial of the expert: épreuve and prevue. *New Criminal Law Review*, 10: 78–101.

Hill, A.H. 1876. Milk standards. *Analyst*, 1: 40–46.

Hill, A.H. 1899. Antiseptics in food. *Public Health*, 11: 527–38.

Hill, T.E. 1891. The Werner-Schmid method of milk analysis. *Analyst*, 16: 67–73.

Hilton, M. 2003. *Consumerism in Twentieth-Century Britain: The Search for a Historical Movement*. Cambridge: Cambridge University Press.

Hilton, M. 2006. The organized consumer movement since 1945, in *The Expert Consumer: Associations and Professionals in Consumer Society*, edited by A. Chatriot, M.-E. Chessel and M. Hilton. Aldershot: Ashgate, 187–203.

Hilton, M. 2008. The banality of consumption, in *Citizenship and Consumption*, edited by K. Soper and F. Trentmann. Basingstoke: Palgrave Macmillan, 87–103.

Hinchliffe, S. 2001. Indeterminacy in-decisions – science, policy and politics in the BSE (Bovine Spongiform Encephalopathy) crisis. *Transactions of the Institute of British Geographers*, 26: 182–204.

Hinchliffe, S. 2007. *Geographies of Nature: Societies, Environments, Ecologies*. London: Sage.

Hinchliffe, S., Kearnes, M., Degen, M. and Whatmore, S. 2005. Urban wild things: a cosmopolitical experiment. *Environment and Planning D: Society and Space*, 23: 643–58.

Hinchliffe, S. and Whatmore, S. 2006. Living cities: towards a politics of conviviality. *Science as Culture*, 15: 123–38.

Hinrichs, C.C. 2000. Embeddedness and local food systems: notes on two types of direct agricultural market. *Journal of Rural Studies*, 16: 295–303.

Hirschman, A.O. 1970. *Exit, Voice and Loyalty: Responses to Decline in Firms, Organizations and States*. Cambridge, MA: Harvard University Press.

Hiscox, E.R. 1923. A study of a cause of 'blowing' in tins of sweetened condensed milk. *Annals of Applied Biology*, 10, 3–4: 370–77.

Hobday, F.T.G. 1927. Clean milk. *Lancet*, ii: 738–40.

Hobson, G. 1930. *History of British Friesian Cattle*. Lewes: Baxter.

Hodson, H.V. 1765. *The Annual Register* 5. London: Longmans.

Holderness, B.A. 2000. Dairying, in *The Agrarian History of England and Wales, VII: 1850–1914*, edited by E.J.T. Collins. Cambridge: Cambridge University Press, 472–8.

Hole, T. 1924. Growth of certified and Grade A (Tuberculin Tested) Milk. *The Dairyman, The Cowkeeper and Dairyman's Journal*, June: 533–5.

Hole, T.A. 1926. Growth of milk grading in England and Wales. *The Dairyman, The Cowkeeper and Dairyman's Journal*, May: 533–4.

Holmes, O.W. 1881. *The Common Law*. Boston, MD: Little, Brown.

Holt, J. 1795. *General View of the Agriculture of the County of Lancaster*. London: Nicol.

Holt, J.C. 1977. *The University of Reading: The First Fifty Years*. Reading: Reading University Press.

Hopkins, F.G. 1912. Feeding experiments illustrating the importance of accessory factors in normal dietaries. *Journal of Physiology*, 44: 425–60.

Hopkins, F.G. 1920. Note on the vitamine content of milk. *Biochemical Journal*, 14: 721–4.

Hortvet, J. 1921. The cryoscopy of milk. *Journal of Industrial and Engineering Chemistry*, 13(3): 198–208.

Horwitz, M.J. 1977. *The Transformation of American Law*. Cambridge, MA: Harvard University Press.

Hoskins, J. 2006. Agency, biography and objects, in *Handbook of Material Culture*, edited by C. Tilley, W. Keane, S. Küchler, M. Rowlands and P. Spyer. London: Sage, 74–84.

Houston, A.C. 1905. *Report on the Bacteriological Examination of Milk*. London: London County Council.

Howell, J.B. 1934. Bacteriological testing of milk. *Lancet*, ii: 1073–4.

Hoy, W.A. 1923. How to steam dairy utensils with an ordinary farm copper. *The Dairyman, The Cowkeeper and Dairyman's Journal*, April: 378–80.

Hoy, W.A. 1924. Can the ordinary farmer produce clean milk? *Journal of the British Dairy Farmers' Association*, 36: 103–10.

Hoy, W.A. and Hickson, P. 1924. The sterilization of dairy utensils. *The Dairyman, The Cowkeeper and Dairyman's Journal*, June: 526–30.

Hoy, W.A. and Williams, R.S. 1921. *Report on a Simple Steam Sterilizer*. London: Dairy Supply Co.

Hughes, A. 2005. Responsible retailers: ethical trade and the strategic re-regulation of cross-continental food supply chains, in *Cross-Continental Food Chains*, edited by N. Fold and B. Pritchard. London: Routledge, 141–54.

Hughes, A. 2006. Geographies of exchange and circulation: transnational trade and governance. *Progress in Human Geography*, 30: 635–43.

Hughes, E.B. 1960. Pure food for the people: the manufacturers' contribution, *Pure Food and Pure Food Legislation*, edited by A.J. Amos. London: Butterworths, 21–39.

Hughes, R.E. 2000. Vitamin C, in *The Cambridge World History of Food*, edited by K.F. Kiple and K.C. Ornelas. New York: Cambridge University Press, 754–63.

Hughes, T.P. 1983. *Networks of Power: Electrification in Western Society, 1880–1930*. Baltimore, MD: Johns Hopkins University Press.

Husson, A. 1856. *Les Consommations de Paris*. Paris: Guillaumin.

Ilbery, B. and Kneafsey, M. 1998. Product and place: promoting quality products and services in the lagging regions of the European Union. *European Urban and Regional Studies*, 5: 329–41.

Ilbery, B. and Kneafsey, M. 2000. Producer constructions of quality in regional speciality food production: a case study from south west England. *Journal of Rural Studies*, 16: 217–30.

Ingold, T. 2007. Materials against materiality. *Archaeological Dialogues*, 14: 1–16.

Institute of Chemistry of Great Britain and Ireland 1879. Report of a conference on the adulteration of food. *Proceedings of the Institute of Chemistry*, 3: B025–53

Isin, E.F. 2004. The neurotic citizen. *Citizenship Studies*, 8: 217–35.

Jackson, P. 2000. Rematerializing social and cultural geography. *Social and Cultural Geography*, 1: 9–14.

Jackson, P. 2005. *Reflections: A History of De Laval*. Stockholm: Alfa Laval.

Jackson, P., Ward, N. and Russell, P. 2006. Mobilising the commodity chain concept in the politics of food and farming. *Journal of Rural Studies*, 22: 129–41.

Jago, W. 1909. *A Manual of Forensic Chemistry*. London: Stevens and Haynes.

Jasanoff, S. 2002. Breaking the waves in science studies. *Social Studies of Science*, 33: 389–400.

Jeffcock, W.P. 1937 *Agricultural Politics 1915–1935*. Ipswich: Harrison.

Jenkins, A. 1970. *Drinka Pinta: The Story of Milk and the Industry that Serves It*. London: Heinemann.

Jenness, R. 1956. Progress in the basic chemistry of milk. *Journal of Dairy Science*, 39: 651–6.

Jenness, R. 1974. The composition of milk, in *Lactation: A Comprehensive Treatise. Volume III: Nutrition and Biochemistry of Milk/Maintenance*, edited by B.L. Larson and V.R. Smith. New York: Academic Press, 3–106.

Jenness, R. 1999. Composition of milk, in *Fundamentals of Dairy Chemistry*, edited by N.P. Wong, R. Jenness, M. Keeney, and E.H. Marth. Gaithersburg, MD: Aspen, 1–38.

Jensen, C.O. 1909. *Essentials of Milk Hygiene*. Philadelphia: Lippincott.

Johnson, C. 1817. On a lactometer. *Annals of Philosophy*, 10, 58: 304–5.

Johnson, J. [pseud. of Bruno Latour] 1988. Mixing humans and nonhumans together: the sociology of a door-closer. *Social Problems*, 35: 298–310.

Johnston, V.J. 1985. *Diet in Workhouses and Prisons*. London: Garland.

Jones, P., Comfort, D. and Hillier, D. 2003. Retailing fair trade food products in the UK. *British Food Journal*, 105: 800–10.

Jones, W.E.D. [1924]. *The Importance of Clean Milk Production*. [no place: no publisher].

Joyce, P. 2003. *The Rule of Freedom: Liberalism and the Modern City*. London: Verso.

Kaika, M. 2005. *City of Flows: Modernity, Nature, and the City*. London: Routledge.

Kaika, M. and Swyngedouw, E. 2000. Fetishizing the modern city: the phantasmagoria of urban technological networks. *International Journal of Urban and Regional Research*, 24: 120–38.

Karpik, L. 1989. L'économie de la qualité, *Revue Française de Sociologie*, 30: 187–210.

Kassim, L. 2001. The co-operative movement and food adulteration in the nineteenth century. *Manchester Region History Review* 15: 9–18.

Kay, H.D. 1944. Milk. *Proceedings of the Nutrition Society*, 2: 121–3.

Kay, H.D. 1947. Presidential Address on the compositional quality of milk. *Journal of the Royal Sanitary Institute*, 67: 515–8.

Kay, H.D. 1950. Safe milk in Great Britain. *Nature*, 165: 144.

Kay, H.D. 1968. Rupert Edward Cecil Lee Guinness, second Earl of Iveagh, 1874–1967: elected FRS 1964. *Biographical Memoirs of Fellows of the Royal Society*, 14: 287–307.

Kearnes, M.B. 2003. Geographies that matter – the rhetorical deployment of physicality? *Social and Cultural Geography*, 4: 139–52.

Kearnes, M. 2006. Chaos and control: nanotechnology and the politics of emergence. *Paragraph*, 29(2): 57–80.

Kearnes, M. and Macnaghten, P. 2006. Introduction: (re)imagining nanotechnology, *Science as Culture*, 15: 279–90.

Kelly, E., Cook, L.B. and Gamble, J.A. 1916. Milk and cream contents. *United States Department of Agriculture, Bulletin*, No. 356. Washington, DC: Government Printing Office.

Kelly, E. and Posson, R.J. 1926. How to conduct milk and cream contests. *United States Department of Agriculture, Departmental Circular*, No. 384. Washington, DC: Government Printing Office.

Kelly, E. and Taylor, G.B. 1919. Milk and cream contests. *United States Department of Agriculture, Departmental Circular*, No. 53. Washington, DC: Government Printing Office.

Kete, K. 1988. La rage and the bourgeoisie: the cultural context of rabies in the French nineteenth century. *Representations*, 22: 89–107.

Kiple, K. and Ornelas, O. (eds) 2000. *The Cambridge World History of Food*. Cambridge: Cambridge University Press .

Kitzinger, J. and Reilly, J. 1997. The rise and fall of risk reporting: media coverage of human genetics research, 'false memory syndrome' and 'mad cow disease'. *European Journal of Communication*, 12: 319–50.

Kjærnes, U. 2006. Trust and distrust: cognitive decisions or social relations? *Journal of Risk Research*, 9: 911–32.

Kjærnes, U., Dulsrud, A. and Poppe, C. 2006. Contestation over food safety: the significance of consumer trust, in *What's the Beef?: The Contested Governance of European Food Safety*, edited by C.K. Ansell and D. Vogel. Cambridge, MA: MIT Press, 61–79.

Kjærnes, U., Harvey, M. and Warde, A. 2007. *Trust in Food: A Comparative and Institutional Analysis*. Basingstoke: Palgrave Macmillan.

Klein, U. 2003. *Experiments, Models, Paper Tools: Cultures of Organic Chemistry in the Nineteenth Century*. Stanford, CA: Stanford University Press.

Klein, U. 2005. Technoscience avant la lettre. *Perspectives on Science*, 13: 226–66.

Klein, U. and Lefèvre, W. 2007. *Materials in Eighteenth-Century Science: a Historical Ontology*. Cambridge, MA: MIT Press.

Kline, J.N. 2007. Eat dirt: CpG DNA and immunomodulation of asthma. *Proceedings of the American Thoracic Society*, 4: 283–8.

Knight, E.G., Freear, K. and Williams, R.S. 1920. *A Study of Factors Concerned in the Production of Clean Milk. Part I*. London: P.S. King and Son.

Knight, E.H. 1884. *Knight's New Mechanical Dictionary*. Boston: Houghton, Mifflin.

Knorr-Cetina, K. 1981. *The Manufacture of Knowledge: An Essay on the Constructivist and Contextual Nature of Science*. Oxford: Pergamon.

Knorr-Cetina, K. 1983. The ethnographic study of scientific work: towards a constructivist interpretation of science, in *Science Observed: Perspectives on the Social Study of Science*, edited by K. Knorr-Cetina and M. Mulkay. London: Sage, 115–40.

Kopytoff, I. 1986. The cultural biography of things: commoditization as process, in *The Social Life of Things: Commodities in Cultural Perspective*, edited by A. Appadurai. Cambridge: Cambridge University Press, 64–91.

Kroen, S. 2004. A political history of the consumer. *Historical Journal*, 47: 709–36.

Kubler, G. 1962. *The Shape of Time: Remarks on the History of Things*. New Haven: Yale University Press.

Kurlansky, M. 1997. *Cod: A Biography of the Fish that Changed the World*. New York: Walker.

Kurlanksy, M. 2002. *Salt: A World History*. New York: Walker

La Berge, A.F. 1991. Mothers and infants, nurses and nursing: Alfred Donné and the medicalization of child care in nineteenth-century France, *Journal of the History of Medicine and Allied Sciences*, 46: 20–43.

Laclau, E. 1990. *New Reflections on the Revolution of Our Time*. London: Verso.

Laird, J. 1905. A pure milk supply. *Public Health*, 17: 437–44.

Lane, C.B. and Weld, I.C. 1907. A city milk and cream contest as a practical method of improving the milk supply. *Bureau of Animal Industry, Circular*, No. 117. Washington, DC: United States Department of Agriculture.

Laporte, D.-G. 2000. *The History of Shit*. Cambridge, MA: MIT Press.

Larsen, C. and White, W. 1913. *Dairy Technology*. New York: Wiley.

Lash, S. 2006. Life (vitalism). *Theory, Culture and Society*, 23(2–3): 323–49.

Laszlo, P. 2007. *Citrus: A History*. Chicago: University of Chicago Press.

Lásztity, R. and Abonyi, T. 2009. Prediction of wheat quality – past, present, future: a review. *Food Reviews International*, 25: 126–41.

Latham, A. and McCormack, D.P. 2004. Moving cities: rethinking the materialities of urban geographies, *Progress in Human Geography*, 28: 701–24.

Latour, B. 1987. *Science in Action: How to Follow Scientists and Engineers through Society*. Cambridge, MA: Harvard University Press.

Latour, B. 1988. *The Pasteurization of France*. Cambridge MA: Harvard University Press.

Latour, B. 1990. Drawing things together, *Representation in Scientific Practice*, edited by M. Lynch and S. Woolgar. Cambridge, MA: MIT Press, 19–68.

Latour, B. 1993. *We Have Never Been Modern*. New York: Harvester Wheatsheaf.

Latour, B. 2003. The promises of constructivism, in *Chasing Technoscience: Matrix for Materiality*, edited by D. Ihde and E. Selinger. Bloomington: University of Indiana Press, 27–46.

Latour, B. 2004. *La Fabrique du Droit: une Ethnographie du Conseil d'Etat*. Paris: La Découverte.

Latour, B. 2007. A plea for earthly sciences. Available at: http://www.bruno-latour.fr/articles/article/102-BSA-GB.pdf [accessed 10 May 2009].

Law, J. and Mol, A. 1995. Notes on materiality and sociality. *Sociological Review*, 43: 274–94.

Law, M. 2003. The origin of state pure food regulation. *Journal of Economic History*, 63: 1103–30.

Leach, A.E. 1904. *Food Inspection and Analysis* New York: Wiley.

Le Canu, L.-R. 1839. Concernant l'analyse du lait. *Journal de Pharmacie et des Sciences Accessoires*, 25: 201–5.

Lefebvre, H. 1991. *The Production of Space*. Oxford: Blackwell.

Leffmann, H. and Beam, W. 1892. A rapid and accurate method of determining fat in milk. *Analyst*, 17: 83–4.

Leffmann, H. and Beam, W. 1893. *Analysis of Milk and Milk Products*. London: Kegan Paul Trench Trubner.

Lehmann, C.G. 1854. *Physiological Chemistry*. London: Cavendish Society.

Leighton, G. 1929. The supervision of the food supply. *Journal of State Medicine*, 37: 92–9.

Lévy, T. 2002. The theory of conventions and a new theory of the firm, in *Intersubjectivity in Economics*, edited by E. Fullbrook. London: Routledge, 254–72.

L'Héritier, S.D. 1842. *Traité de Chimie Pathologique*. Paris: Baillière.

Liebowitz, S.J. and Margolis, S.E. 1990. The fable of the keys. *Journal of Law and Economics*, 33: 1–25.

Ling, E.R. 1944. *A Text Book of Dairy Chemistry*. London: Chapman and Hall.

Liversage, V. 1926. Cost of production of tuberculin tested milk. *The Dairyman, The Cowkeeper and Dairyman's Journal*, May: 528–31.

Liverseege, J.F. 1899. The comparative amount of adulteration in large towns. *British Food Journal*, 1: 101–2.

Liverseege, J.F. 1932. *Adulteration and Analysis of Food and Drugs*. London: J. and A. Churchill.

Llewellyn, K.N. 1936. On warranty of quality, and society. *Columbia Law Review*, 36: 699–744.

Llewellyn, K.N. 1939. Across sales on horseback. *Harvard Law Review*, 52: 725–46.

Lockie, S. and Goodman, M. 2006. Neoliberalism and the problem of space: competing rationalities of governance in fair trade and mainstream agri-environmental networks, in *Between the Local and the Global: Confronting Complexity in the Contemporary Agri-Food Sector*, edited by T. Marsden and J. Murdoch. Amsterdam: Elsevier, 95–117.

Lorimer, H. 2005. Cultural geography: the busyness of being 'more-than-representational', *Progress in Human Geography*, 29: 83–94.

Loudon, J.C. 1825. *An Encyclopaedia of Agriculture*. London: Longman.

Luhmann, N. 1979. *Trust and Power*. Chichester: Wiley.

Luhmann, N. 1988. Familiarity, confidence, trust: problems and alternatives, in *Trust: Making and Breaking Cooperative Relations*, edited by D. Gambetta. Oxford: Blackwell, 94–107.

Lupton, D. 1999. *Risk*. London: Routledge.

Lynch, M. and McNally, R. 2003. 'Science', 'common sense', and DNA evidence: a legal controversy about the public understanding of science. *Public Understanding of Science*, 12: 83–103.

Lyon, S. 2006. Evaluating fair trade consumption: politics, defetishization and producer participation. *International Journal of Consumer Studies*, 30: 452–64.

McCormack, D.P. 2007. Molecular affects in human geographies. *Environment and Planning A*, 39: 359–77.

MacDonagh, O.O.G.M. 1958. The nineteenth-century revolution in government: a reappraisal. *Historical Journal*, 1: 56–67.

MacFarlane, A. 2004. *Empire of Tea: The Remarkable History of the Plant That Took over the World*. New York: Penguin.

Macgregor, A.S.M. 1890. *The Milk Supply of Copenhagen*. Edinburgh: Scott Ferguson Burness.

Macgregor, A.S.M. 1930. The place of pasteurization in a scheme of milk distribution. *British Medical Journal*, ii: 463–65.

McHugh, F.G. 1943. The milk supply of the future. *Journal of the Royal Sanitary Institute*, 63: 111–16

Mackenzie, A. 2002. *Transductions: Bodies and Machines at Speed*. London: Continuum.

Mackenzie, A. 2005. Problematising the technological: the object as event? *Social Epistemology*, 19: 381–99.

Mackenzie, A. 2006. The strange meshing of impersonal and personal forces in technological action. *Culture, Theory and Critique*, 47: 197–212.

MacKenzie, D. 1990. *Inventing Accuracy: A Historical Sociology of Nuclear Missile Guidance Systems*. Cambridge, MA: MIT Press.

Mackenzie, L. 1899. The hygienics of milk. *Edinburgh Medical Journal*, 5: 372–8, 563–76.

Mackintosh, J. 1922. How to produce clean milk. *Journal of the Ministry of Agriculture*, 29: 17–29.

Mackintosh, J. 1939. The evolution of milk production, in *Agriculture in the Twentieth Century*, edited by H.E. Dale. Oxford: Clarendon Press, 397–422.

Mackintosh, J. and Whitby, L.E.H. 1931. The bacteriological control of milk supplies. *Lancet*, ii: 147–50.

McKone, H.T. 1991. The history of food colorants before aniline dyes. *Bulletin of Food Chemistry*, 10: 25–31.

MacLeod, R.M. 1967. The frustration of state medicine, 1880–1899. *Medical History*, 11: 15–40.

MacLeod, R.M. 1968. Treasury control and social administration. *Occasional Papers on Social Administration*: 23. London: Bell.

MacLeod, R.M. 1976. Science and the Treasury: principles, personalities, and policies, 1870–1885, in *The Patronage of Science in the Nineteenth Century*, edited by G.L'E. Turner. Leyden: Noordhoff International, 115–72.

MacLeod, R. (ed.) 2003. *Government and Expertise: Specialists, Administrators, and Professionals*. Cambridge: Cambridge University Press.

MacNutt, J.S. 1917. *The Modern Milk Problem in Sanitation, Economics and Agriculture* New York: Macmillan.

Maddock, C. 1926. Production and distribution of Grade A (Tuberculin Tested) milk. *Journal of State Medicine*, 34: 537–44.

Maggs, J.H. 1924. The organization of United Dairies (Ltd), in *Proceedings of the World's Dairy Congress, Washington, DC, October 2, 3, Philadelphia PA, October 4, Syracuse NY, October 5, 6, 8, 9, 10*, 1923, edited by L.A. Rogers and K.D. Lenoir. Washington, DC: Government Printing Office, volume I, 235–41.

Magnello, B. 2000. *A Century of Measurement: An Illustrated History of the National Physical Laboratory*. Bath: Canopus.

Maine, H.S. 1861. *Ancient Law: its Connection with the Early History of Society, and its Relation to Modern Ideas*. London: Murray.

Mansfield, B. 2003a. Fish, factory trawlers, and imitation crab: the nature of quality in the seafood industry. *Journal of Rural Studies*, 19: 9–21.

Mansfield, B. 2003b. Spatializing globalization: a 'geography of quality' in the seafood industry. *Economic Geography*, 79: 1–16.

Mansfield, B. 2003c. From catfish to organic fish: making distinctions about nature as cultural economic practice. *Geoforum*, 34: 329–42.

Mansfield, B. 2003d. 'Imitation crab' and the material culture of commodity production. *Cultural Geographies*, 10: 176–95.

Marchand, E. 1854. Note sur une nouvelle méthode de dosage du beurre dans le lait, *Journal de Chimie Médicale, de Pharmacie et de Toxicologie*, 3rd series, 11: 641–42.

Marsili, R. (ed.) 2002. *Flavour, Fragrance and Odour Analysis*. Boca Raton, FL: CRC Press.

Marsili, R. (ed.) 2006. *Sensory-Directed Flavour Analysis* Boca Raton, FL: CRC Press.

Martin, R. and Sunley, P. 2006. Path dependence and regional economic evolution. *Journal of Economic Geography*, 6: 395–437.

Marx, K. 2007 (original 1867). *Capital: A Critique of Political Economy. Volume 1 – Part 1*. New York: Cosimo.

Massumi, B. 2002. *Parables for the Virtual: Movement, Affect, Sensation*. Durham, NC: Duke University Press.

Matthews, S. 2000. The administration of the livestock census of 1866. *Agricultural History Review*, 48: 223–8.

Mattick, A.T.R. 1921. The sterilization of empty milk churns by steam under pressure. *Journal of Hygiene*, 20: 165–72.

Mattick, A.T.R. 1944. Bacteriological aspects of milk processing and distribution. *Proceedings of the Nutrition Society*, 2: 141–9.

Mattick, E.C.V. 1927. *The Production and Distribution of Clean Milk*. London: The Dairyman Ltd.

Max Planck Institute for the History of Science 2006. *The Shape of Experiment.* Berlin: Max Planck Institute for the History of Science. Available at: http://www.mpiwg-berlin.mpg.de/Preprints/P318.PDF [accessed: 4 May 2009].

Mayhew, H. 1861. *London Labour and the London Poor. Volume I: The Street Folk.* London: Griffin, Bohn.

Mayhew, M. 1988. The 1930s nutrition controversy. *Journal of Contemporary History*, 23: 445–64.

Mennell, S. 1996. *All Manners of Food: Eating and Taste in England and France from the Middle Ages to the Present.* Urbana: University of Illinois Press.

Mercier, L.-S. 1783. *Le Tableau de Paris.* Amsterdam: n.p.

Merriman, P. 2005. Materiality, subjectification, and government: the geographies of Britain's Motorway Code, *Environment and Planning, D: Society and Space*, 23: 235–50.

Meskell, L. 2004. *Object Worlds in Ancient Egypt: Material Biographies Past and Present.* Oxford: Berg.

Middleton, J. 1e 1798, 2e 1807. *View of the Agriculture of Middlesex.* London: Nicol.

Milk Control Board 1919. *Memorandum on the Milk Supply.* London: Ministry of Food.

Milk Marketing Board 1934. *Scheme to Establish a Roll of Accredited Producers.* London: Milk Marketing Board.

Millán-Verdú, C., Garrigós-Oltra, L., Blanes-Nadal, G. and Domingo-Beltrán, M. 2003. The history of optical analysis of milk: the development and use of lactoscopes. *Journal of Chemical Education*, 80: 762–7.

Miller, D. 1987. *Material Culture and Mass Consumption.* Oxford: Blackwell.

Miller, D. 1998. Why some things matter, in *Material Culture: Why Some Things Matter*, edited by D. Miller. London: University of Chicago Press, 3–21.

Miller, D. 2002. Turning Callon the right way up. *Economy and Society*, 31: 218–33.

Miller, D. 2005. Materiality: an introduction, in *Materiality*, edited by D. Miller. Durham, NC: Duke University Press, 1–50.

Miller, D. 2007. Stone age or plastic age, *Archaeological Dialogues*, 14: 23–7.

Miller, P. 2001. Governing by numbers: why calculative practices matter. *Social Research*, 68, 379–96.

Miller, W.I. 1997. *The Anatomy of Disgust.* Cambridge, MA: Harvard University Press.

Millington, G. 2005. Meaning, materials and melancholia: understanding the Palace Hotel, *Social and Cultural Geography*, 6: 531–49.

Minchin, H. 1860. Means of determining the quality of milk. *Chemical News*, 2: 135–7.

Mingay, G.E. 1982. *British Friesians: An Epic of Progress.* Rickmansworth: British Friesian Cattle Society.

Ministry of Agriculture and Fisheries 1924. Guide to the conduct of clean milk competitions. *Miscellaneous Publication*: 43. London: HMSO.

Mintz, S. 1985. *Sweetness and Power: The Place of Sugar in Modern History.* New York: Viking.

Mitchell, J. 1848. *Treatise on the Falsifications of Food and the Chemical Means Employed to Detect Them.* London: Baillière.

Mitchell, P. 2001. The development of quality obligations in sale of goods. *Law Quarterly Review,* 117: 645–63.

Mitchell, T. 2002. *Rule of Experts: Egypt, Techno-politics, Modernity.* Berkeley: University of California Press

Mol, A. 1999. Ontological politics: a word and some questions, in *Actor Network Theory and After,* edited by J. Law and J. Hassard. Oxford: Blackwell, 74–89.

Mol, A. 2002. *The Body Multiple: Ontology in Medical Practice.* Durham, NC: Duke University Press.

Mol, A. and Mesman, J. 1996. Neonatal food and the politics of theory: some questions of method. *Social Studies of Science,* 26: 419–44.

Molotch, H. 2003. *Where Stuff Comes From: How Toasters, Toilets, Cars, Computers and Many Other Things Come to be as They Are* London: Routledge.

Molotiu, A. 2000. Focillon's Bergsonian rhetoric and the possibility of deconstruction, *Invisible Culture* 3. Available at: http://www.rochester.edu/in_visible_culture/issue3/molotiu.htm [accessed: 4 May 2009]

Monfalcon, J.-B. and de Polinière, A.P.I. 1846. *Traité de la Salubrité dans les Grandes Villes.* Paris: Baillière.

Monier, F., Chesney, F. and Roux, E. 1909. *Traité Théorique et Pratique sur les Fraudes et les Falsifications.* Paris: Sirey.

Monier-Williams, G.W. 1914. Report to the Local Government Board on the freezing point of milk considered in its relation to the detection of added water, *Food Report to the Local Government Board,* 103.

Moore, S.G. 1921. *This Concerns You.* London: St Catherine's Press.

Moore, T.S. and Philip, J.C. 1947. *The Chemical Society, 1841–1941.* London: Chemical Society.

Moran, W. 1993. The wine appellation as territory in France and California. *Annals of the Association of American Geographers,* 83: 694–717.

Morgan, B. 1964. *Express Journey, 1864–1964: A Centenary History of the Express Dairy Co. Ltd.* London: Newman Neame.

Morgan, K., Marsden, T. and Murdoch, J. 2006. *Worlds of Food: Place, Power, and Provenance in the Food Chain.* Oxford: Oxford University Press.

Morris, C. and Young, C. 2000. 'Seed to shelf', 'teat to table', 'barley to beer' and 'womb to tomb': discourses of food quality and quality assurance schemes in the UK. *Journal of Rural Studies,* 16: 103–115.

Morris, C. and Young, C. 2004. New geographies of agro-food chains: an analysis of UK quality assurance schemes, in *Geographies of Commodity Chains,* edited by A. Hughes and S. Reimer. London: Routledge, 83–101.

Morrison-Low, A.D. 1998. Hydrometer, in *Instruments of Science: An Historical Encyclopedia,* edited by R. Bud and D.J. Warner. London: Garland, 311–13.

Morton, J.C. 1865. On London milk. *Journal of the Society of Arts,* 14: 65–78.

Müldner, G. and Richards, M.P. 2006. Diet in medieval England: the evidence from stable isotopes, in *Food in Medieval England: Diet and Nutrition*, edited by C.M. Woolgar, D. Serjeantson and T. Waldron. Oxford: Oxford University Press, 228–38.

Mullaly, J. 1853. *The Milk Trade in New York and Vicinity*. New York: Fowlers and Wells

Muradian, R. and Pelupessy, W. 2005. Governing the coffee chain: the role of voluntary regulatory systems. *World Development*, 33: 2029–44.

Murdoch, J. 1998. The spaces of actor-network theory. *Geoforum*, 29: 357–74.

Murdoch, J. and Miele, M. 1999. 'Back to nature': changing 'worlds of production' in the food sector. *Sociologia Ruralis*, 39: 465–83.

Murdoch, J. and Miele, M. 2004. A new aesthetic of food? Relational reflexivity in the 'alternative' food movement, in *Qualities of Food*, edited by M. Harvey, A. McMeekin and A. Warde. Manchester: Manchester University Press, 156–75.

Murray, A.H. 1922. Copenhagen's milk supply. *The Milk Industry*, 3(6): 87–90.

Musselin, C. and Paradeise, C. 2005. The concept of quality: a brief historical review in the French social sciences and three questions, in *Quality: A Debate*, edited by C. Musselin and C. Paradeise. *Sociologie du Travail*, 47: S90–94.

Mutersbaugh T 2005a. Just-in-space: certified rural products, labour of quality, and regulatory spaces. *Journal of Rural Studies*, 21: 389–402.

Mutersbaugh, T. 2005b. Fighting standards with standards: harmonization, rents, and social accountability in certified agro-food networks. *Environment and Planning A*, 37: 2033–51.

Mutersbaugh, T., Klooster, D., Renard, M.-C. and Taylor, P. 2005. Certifying rural spaces: quality-certified products and rural governance. *Journal of Rural Studies*, 21: 381–8.

Mythen, G. 2004. *Ulrich Beck: A Critical Introduction to the Risk Society*. London: Pluto Press.

Nabhan, G.P. 2004. *Why Some Like it Hot: Food, Genes and Cultural Diversity*. Washington, DC: Island Press.

National Institute for Research in Dairying 1924. Studies concerning the handling of milk, *Ministry of Agriculture, Miscellaneous Publications*, No. 41. London: HMSO.

National Institute for Research in Dairying 3e 1929, 4e 1931. Studies concerning the handling of milk. *Ministry of Agriculture, Research Monograph*: No. 1. London: HMSO.

Nayak, A. 2008. On the way to theory: a processual approach. *Organization Studies*, 29, 173–90.

Newman, G. 1903. *Report on the Milk Supply of Finsbury*. London: Finsbury Borough Council.

Newman, G. 1926. Report on public health. *British Medical Journal*, ii: 566.

Nicholson, W. 1813. *A Journal of Natural Philosophy, Chemistry and the Arts*: 35. London: Sherwood, Neely and Jones.

Nimmo, R. 2008a. Auditing nature, enacting culture: rationalization as disciplinary purification in early twentieth-century British dairy farming. *Journal of Historical Sociology*, 21: 272–302.

Nimmo, R. 2008b. Governing nonhumans: knowledge, sanitation and discipline in the late 19th and early 20th-century British milk trade. *Distinktion*, 16: 77–97.

Nimmo, R. 2010. *Milk, Modernity and the Making of the Human: Purifying the Social*. London: Routledge.

Niven, J. 1896. *On the Improvement of the Milk Supply of Manchester*. Manchester: Heywood.

Niven, J. 1923. *Observations on the History of Public Health Effort in Manchester*. Manchester: Heywood.

Normandy, A. 1850. *The Commercial Hand-book of Chemical Analysis*. London: George Knight and Sons.

North, C.E. 1918. *Farmers' Clean Milk Book*. New York: Wiley.

North, C.E. 1922. Safeguarding milk. *The Dairyman, the Cowkeeper and Dairyman's Journal*, November: 132–8.

Novo, M., Reija, B. and Al-Soufi, W. 2007. Freezing point of milk: a natural way to understand colligative properties. *Journal of Chemical Education*, 84: 1673–5.

Nowotny, H., Scott, P. and Gibbons M. 2001. *Re-thinking Science: Knowledge and the Public in an Age of Uncertainty*. Cambridge: Polity Press.

O'Connell, J. 1993. The creation of universality by the circulation of particulars. *Social Studies of Science*, 23: 129–73.

Oddy, D.J. 2007. Food quality in London and the rise of the public analyst, in *Food and the City in Europe since 1800*, edited by P.J. Atkins, P. Lummel and D.J. Oddy. Aldershot: Ashgate, 91–103.

Oddy, D.J., Atkins, P.J. and Amilien, V. (eds) 2009. *The Rise of Obesity in Europe: A Twentieth Century Food History*. Aldershot: Ashgate.

Oestigaard, T. 2004. Approaching material culture: a history of changing epistemologies. *Journal of Nordic Archaeological Science*, 14: 79–87.

Olsen, B. 2003. Material culture after text: re-membering things. *Norwegian Archaeological Review*, 36(2): 87–104.

Olsen, B. 2007. Keeping things at arm's length: a genealogy of asymmetry. *World Archaeology*, 39: 79–88.

O'Malley, P. 2000. Uncertain subjects: risks, liberalism and contract. *Economy and Society*, 29: 460–84.

Oosterveer, P. 2002. Reinventing risk politics: reflexive modernity and the European BSE crisis. *Journal of Environmental Policy and Planning*, 4: 215–29.

Orland, B. 2003. Turbo-cows: producing a competitive animal in the nineteenth and early twentieth century, in *Industrializing Organisms: Introducing Evolutionary History*, edited by S.R. Schrepfer and P. Scranton. London: Routledge, 167–89.

Orr, T. 1908. *Report on an Investigation as to the Contamination of Milk*. Beverley: Joint Committee of the East and West Ridings and of the County Boroughs of Leeds, Bradford, Hull, Rotherham and Sheffield.

Orr, T. 1925. Raw versus pasteurised milk. *The Milk Industry*, 5(8): 33–5.

O'Sullivan, A., O'Connor, B., Kelly, A. and McGrath, M.J. 1999. The use of chemical and infrared methods for analysis of milk and dairy products. *International Journal of Dairy Technology*, 52: 139–48.

Otter, C. 2006. The vital city: public analysis, dairies and slaughterhouses in nineteenth-century Britain. *Cultural Geographies*, 13: 517–37.

Pacey, A. and Payne, P. (eds) 1985. *Agricultural Development and Nutrition*. London: Hutchinson.

Parker, H.N. 1917. *City Milk Supply*. New York: McGraw Hill.

Parmentier, A.-A. 1817. Lait, in *Nouveau Dictionnaire d'Histoire Naturelle Appliquée aux Arts, à l'Agriculture, à l'Economie Rurale et Domestique, à la Médecine, etc.*, edited by J.E. Sève. Paris: Déterville, volume 17: 218–36.

Parmentier, A.-A. and Deyeux, N. 1799. Mémoire … sur … déterminer, par l'examen comparé des propriétés physiques et chimiques, la nature des laits de femme, de vache, de chèvre, d'ânesse, de brebis et de jument, *Recueils de la Société Royal de Médecine*, 9.

Parmentier, A.-A. and Deyeux, N. 1790. *Précis d'Expériences et Observations sur les Différentes Espèces de Lait*. Strasbourg: F.G. Levrault.

Parrott, N., Wilson, N. and Murdoch, J. 2002. Spatializing quality: regional protection and the alternative geography of food. *European Urban and Regional Studies*, 9: 241–61.

Paulus, I. 1974. *The Search for Pure Food*. London: M. Robertson.

Payen, A. 1828. Du lait de plusieurs femmes et du lait de chèvre. *Journal de Chimie Médicale*, series 1(4): 118–26.

Payen, A. 1865. *Des Substances Alimentaires et des Moyens de les Améliorer, de les Conserver et d'en Reconnaître les Altérations*. Paris: Hachette.

Pearmain, T.H. and Moor, C.G. 1897. *The Analysis of Food and Drugs. Part I: Milk and Milk Products*. London: Baillière, Tindall and Cox, London.

Péligot, E.-M. 1836. Mémoire sur la composition chimique du lait d'anesse. *Annales de Physiques et de Chimiques*, 62: 432.

Pelouze, T-J. and Frémy, E. 1865. *Traité de Chimie Générale, Analytique, Industrielle et Agricole*. Paris: Victor Masson et Fils.

Pels, D. 1996. The politics of symmetry. *Social Studies of Science*, 26: 277–304.

Pennington, T.H. 2003. *When Food Kills: BSE, E. Coli, and Disaster Science*. Oxford: Oxford University Press.

People's League of Health 1932. *Report of a Special Committee appointed by the People's League of Health (Inc.) to Make a Survey of Tuberculosis of Bovine Origin in Great Britain*. London: People's League of Health.

People's League of Health 1934. *A Safe Milk Supply and the Report of the Committee on Cattle Diseases (Economic Advisory Council)*. London: People's League of Health.

People's League of Health 1936. *What is the Quality of the Milk Supplied to School Children?* London: People's League of Health.

Pereira, J. 1843. *A Treatise on Food and Diet* London: Longman, Brown, Green and Longmans.

Perkin, H. 1989. *The Rise of Professional Society: England since 1880*. London: Routledge.

Perren, R. 1978. *The Meat Trade in Britain 1840–1914*. London: Routledge and Kegan Paul.

Pertzoff, V.A. and Bell, B. 1932. Un historique des premières recherches sur la composition de la caséine. *Le Lait*, 12: 161–72.

Phillips, A.D.M. 2004. Rebuilding rural England: farm building provision, 1850–1900, in *Home and Colonial: Essays on Landscape, Ireland, Environment and Empire*, edited by A.R.H. Baker. London: Historical Geography Research Group, 39–51.

Philo, C. 2000. More words, more worlds: reflections on the 'cultural turn' and human geography, in *Cultural Turns/Geographical Turns: Perspectives on Cultural Geography*, edited by I. Cook, D. Crouch, S. Naylor and J. Ryan. Harlow: Prentice Hall, 26–53.

Pickering, A. 1995. *The Mangle of Practice: Time, Agency and Science*. Chicago: University of Chicago Press.

Pickering, A. 2005. Decentring sociology: synthetic dyes and social theory. *Perspectives on Science*, 13: 352–405.

Pilcher, R.B. 1914. *The Institute of Chemistry of Great Britain and Ireland ... History of the Institute: 1877–1914*. London: [Bradbury, Agnew, & Co. Ltd., printers].

Pilcher, R.B. 1919. *The Profession of Chemistry*. London: Constable.

Pitt, J.C. 2003. Against the perennial: small steps toward a Heraclitian philosophy of science. *Techné*, 7: 1–13.

Plassmann, H., O'Doherty, J., Shiv, B. and Rangel, A. 2008. Marketing actions can modulate neural representations of experienced pleasantness. *Proceedings of the National Academy of Sciences*, 105: 1050–54.

Pocock, W.D. 1933. United Dairies Ltd, in *The Milk Trade: A Comprehensive Guide to the Development of the Dairy Industry*, edited by C. Raison. London: Virtue and Co., 11–20.

Poggiale, A.B. 1849. Dosage du sucre de lait par la méthode des volumes, et détermination de la richesse du lait. *Comptes Rendus Hebdomadaires des Séances de l'Académie des Sciences*, 28(16): 505–7.

Polanyi, K. 1944. *The Great Transformation*. Boston: Beacon.

Ponte, S. and Gibbon, P. 2005. Quality standards, conventions and the governance of global value chains. *Economy and Society*, 34: 1–31.

Poovey, M. 1998. *A History of the Modern Fact: Problems of Knowledge in the Sciences of Wealth and Society*. Chicago: University of Chicago Press.

Popp, M. 1904. The Gottlieb-Röse Method for determining Fat in Milk, *Zeitschrift für Untersuchung der Nahrungs-und Genussmittel*, 7: 6–12.

Poppe, C. and Kjærnes, U. 2003. *Trust in Food in Europe: A Comparative Analysis*. Oslo: National Institute for Consumer Research.

Porter, T.M. 1986. *The Rise of Statistical Thinking, 1820–1900*. Princeton, NJ: Princeton University Press.

Porter, T.M. 1995. *Trust in Numbers: The Pursuit of Objectivity in Science and Public Life*. Princeton, NJ: Princeton University Press.

Posner, R.A. (ed.) 1992. *The Essential Holmes*. Chicago: Chicago University Press.

Powell, D. 2001. Mad cow disease and the stigmatization of British beef, in *Risk, Media and Stigma: Understanding Public Challenges to Modern Science and Technology*, edited by J. Flynn, P. Slovic, and H. Kunreuther. London: Earthscan, 219–28.

Powell, D. and Leiss, W. 1997. *Mad Cows and Mother's Milk: The Perils of Poor Risk Communication*. Montreal: McGill-Queen's University Press.

Power, M. 1997. *The Audit Society: Rituals of Verification*. Oxford: Oxford University Press.

Probyn, E. 2000. *Carnal Appetites: Foodsexidentities*. London: Routledge.

Proctor, F. and Hoy, W.A. 1925. The influence of different methods of cleaning dairy utensils on the bacterial content of churns and on the keeping properties of milk. *Journal of Hygiene*, 24: 419–26.

Prout, W. 1834. *Chemistry, Meteorology and the Function of Digestion*. London: Pickering.

Provan, A.L. and Jenkins, D.I. 1949. Trends in milk quality. *Journal of the Society of Dairy Technology*, 2: 88–94.

Pure Food Society of Great Britain [1915]. *Milk under Seal: Special Rules made by the Milk Certification Committee of the Pure Food Society of Great Britain (Incorporated)*. [London: n.p.].

Quevenne, T.-A. 1841. Mémoire sur le lait. *Annales d'Hygiène Publique et de Médecine Légale*, series 1(26): 5–125, 257–380.

Quevenne, T-A. 1842. Falsifications du lait. *Annales d'Hygiène Publique et de Médecine Légale*, 1(27): 241–86.

Raalte, A. van 1929. The freezing point of milk. *Analyst*, 54: 266–8.

Rabier, C. 2007. *Fields of Expertise: A Comparative History of Expert Procedures in Paris and London, 1600 to Present*. Newcastle: Cambridge Scholars Press.

Rabinow, P. and Rose, N. 2006. Biopower today. *Biosocieties*, 1: 195–217.

Race, J. 1918. *The Examination of Milk for Public Health Purposes*. New York: Wiley.

Raison, C. (ed.) 1933. *The Milk Trade*. London: Virtue and Co.

Raw, N. 1937. *The Control of Bovine Tuberculosis in Man*. London: Baillière, Tindall and Cox.

Recknagel, G. 1883. Über eine physikalische Eigenschaft der Milch. *Milch-Zeitung*, 12: 419–22, 437–8.

Renard, M.-C. 1999. The interstices of globalization: the example of Fair Coffee. *Sociologia Ruralis*, 39: 494–500.

Renard, M.-C. 2003. Fair trade: quality, market and conventions. *Journal of Rural Studies*, 19: 87–96.

Renard, M.-C. 2005. Quality certification, regulation and power in fair trade. *Journal of Rural Studies*, 21: 419–31.

Rennes, J. 1923. *Le Lait qui Tue. Le Lait qui Sauve*. Lyon: Collection le Lait.

Renney, H. 1906. Desirability of licensing dairies, cowsheds and milkshops in lieu of registration. *Journal of the Royal Sanitary Institute*, 27: 627–32.

Renting, H., Marsden, T.K. and Banks, J. 2003. Understanding alternative food networks: exploring the role of short food supply chains in rural development? *Environment and Planning* A, 35: 393–412.

Rew, R.H. 1892. An inquiry into the statistics of the production and consumption of milk and milk products in Great Britain. *Journal of the Royal Statistical Society*, 55: 244–86.

Rheinberger, H.-J. 1997. *Toward a History of Epistemic Things: Synthesizing Proteins in the Test Tube*. Stanford, CA: Stanford University Press.

Richmond, H.D. 1888. An instrument for calculating milk results. *Analyst*, 13: 65.

Richmond, H.D. 1889. Fat extraction from milk solids. *Analyst*, 14: 121–31.

Richmond, H.D. 1892. Leffmann and Beam's method of fat estimation in milk, part I. *Analyst*, 17: 144–52.

Richmond, H.D. 1893a. Leffmann and Beam's method of fat estimation in milk, part II. *Analyst*, 18: 130–34.

Richmond, H.D. 1893b. The discrimination between abnormal and adulterated milks. *Analyst*, 18: 270–9.

Richmond, H.D. 1898. The calculation of 'added water' in adulterated milks. *Analyst*, 23: 169–74.

Richmond, H.D. 1e 1899, 2e 1914, 3e 1920, 4e 1942, 5e 1953. *Dairy Chemistry*. London: Griffin.

Richmond, H.D. 1904. The composition of milk. *Analyst*, 29: 180–90.

Richmond, H.D. 1920. An improved slide rule for dairy calculations. *Analyst*, 45: 218–20.

Richmond, H.D. and Boseley, L.K. 1893a. The Leffmann-Beam method for fat-estimation in milk, part III. *Analyst*, 18: 62–9.

Richmond, H.D. and Boseley, L.K. 1893b. Note on the detection of adulteration of fresh milk by diluted condensed milk. *Analyst*, 18: 174–80.

Richmond, H.D. and Boseley, L.K. 1897. The detection of mixtures of diluted condensed or sterilized milk with fresh milk. *Analyst*, 22: 95–7.

Richmond, H.D. and Miller, E.H. 1906. The method of analysis of milk used in the Government Laboratory for samples referred under the Sale of Food and Drugs Acts. *Analyst*, 31: 317–35.

Rideal, S. and Foulerton, A.G.R. 1899. On the use of boric acid and formic aldehyde as milk preservatives. *Public Health*, 11: 554–68.

Rip, A. 2003. Constructing expertise in a third wave of science studies? *Social Studies of Science*, 33: 419–34.

Roberts, C.G. 1872. Report on the trials of implements at Cardiff. *Journal of the Royal Agricultural Society of England*, 2(8): 404–79.

Roberts, G.K. 1979. The Royal College of Chemistry (1845–53): a social history of chemistry in early-Victorian England, unpublished PhD thesis, Johns Hopkins University.

Rocke, A.J. 1992. Berzelius's animal chemistry: from physiology to organic chemistry (1805–1814), in *Enlightenment Science in the Romantic Era: The Chemistry of Berzelius and its Cultural Setting*, in E.M. Melhade and T. Frängsmyr. Cambridge: Cambridge University Press, 107–31.

Rocke, A.J. 2000. Organic analysis in comparative perspective: Liebig, Dumas, and Berzelius, 1811–1837, *Instruments and Experimentation in the History of Chemistry*, edited by F.L. Holmes and T.H. Levere. Cambridge, MA: MIT Press, 273–310.

Roe, E. 2006a. Material connectivity, the immaterial and the aesthetic of eating practices: an argument for how genetically modified foodstuff becomes inedible, *Environment and Planning A*, 38: 465–81.

Roe, E. 2006b. Things becoming food and the embodied, material practices of an organic food consumer. *Sociologia Ruralis*, 46: 104–21.

Rolet, A. 1908. *Le Lait Hygiénique: Production et Vente*. Paris: Lucien Laveur.

Romagnoli, D. 1999. 'Mind your manners': etiquette at the table, in *Food: A Culinary History from Antiquity to the Present*, edited by J.-L. Flandrin and M. Montanari. New York: Columbia University Press, 328–38.

Röse, B. 1888. Studien aus dem Gebiete der Chemie der Lebensmittel und der Verbrauchsgegenstände. Zur Analyse der Milch. Fettbestimmung. *Zeitschrift fur Angewandte Chemie*, 1: 100.

Rose, N. 1999. *Governing the Soul: The Shaping of the Private Self*. London: Free Association Books.

Rose, N. 2006. *The Politics of Life Itself: Biomedicine, Power, and Subjectivity in the Twenty-First Century*. Princeton, NJ: Princeton University Press.

Rose, N. and Miller, P. 1992. Political power beyond the state: problematics of government. *British Journal of Sociology*, 43: 173–205.

Rosenau, M.J. 1912. *The Milk Question*. Boston: Houghton Mifflin.

Ross, H.E. 1910. *A Dairy Laboratory Guide*. New York: Orange Judd Co.

Rothschild, H. de 1901. *Bibliographia Lactaria*. Paris: Octave Doin, with supplement 1902.

Rowlands, A. and Hoy, W.A. 1949. Approved hypochlorites in milk production. *Agriculture*, 56(7): 304–9.

Rozier, F. (ed.) 1806. *Cours Complet d'Agriculture, Théorique, Pratique, Economique et de Médecine* Paris: Chez Delalain, volume 12.

Rugg, H.H. [1850]. *Observations on London milk*. London: Bailey and Moon.

Russell, C.A. 2003. *Edward Frankland: Chemistry, Controversy and Conspiracy in Victorian England*. Cambridge: Cambridge University Press.

Russell, C.A., Coley, N.G., Roberts, G.K. 1977. *Chemists by Profession: The Origins and Rise of the Royal Institute of Chemistry*. Milton Keynes: Open University Press.

Russell, Sir E.J. 1966. *A History of Agricultural Science in Great Britain, 1620–1954*. London: Allen and Unwin.

Ryan, M. 1831. *A Manual of Medical Jurisprudence and State Medicine*. London: Sherwood, Gilbert and Piper.

Salaman, R.N. 1949. *The History and Social Influence of the Potato*. Cambridge: Cambridge University Press.

Sassatelli, R. 2004. The political morality of food: discourses, contestation and alternative consumption, in *Qualities of Food*, edited by M. Harvey, A. McMeekin and A. Warde. Manchester: Manchester University Press, 176–91.

Sassatelli, R. 2006. Virtue, responsibility and consumer choice: framing critical consumerism, in *Consuming Cultures: Global Perspectives, Historical Trajectories, Transnational Exchanges*, edited by J. Brewer and F. Trentmann. Oxford: Berg, 219–50.

Sassatelli, R. 2007. *Consumer Culture: History, Theory and Politics*. Los Angeles: Sage.

Savage, W.G. 1912. *Milk and the Public Health*. London: Macmillan.

Savage, W.G. 1929. *The Prevention of Human Tuberculosis of Bovine Origin*. London: Macmillan.

Sayer, A. 2001. For a critical cultural political economy. *Antipode*, 33: 687–708.

Sayer, A. 2006. Approaching moral economy, in *The Moralization of the Markets*, edited by N. Stehr, C. Henning and B. Weiler. New Brunswick, NJ: Transaction Publishers, 77–97.

Schaffer, S. 1995. Accurate measurement is an English science, in *The Values of Precision*, edited by M.N. Wise. Princeton, NJ: Princeton University Press, 135–72.

Schickore, J. 2002. (Ab)using the past for present purposes: exposing contextual and trans-contextual features of error. *Perspectives on Science*, 10: 433–56.

Schickore, J. 2003. Cheese mites and other delicacies: the introduction of test objects into microscopy. *Endeavour*, 27: 134–8.

Schickore, J. 2006. Misperception, illusion and epistemological optimism: vision studies in early nineteenth-century Britain and Germany. *British Journal for the History of Science*, 39: 383–405.

Schmid, W. 1888. Bestimmung des Fettgehaltes in Milch, Rahm und dergleichen, *Zeitschrift fur Analytsche Chemie*, 27: 464.

Schmidgen, H. 2005. Thinking technological and biological beings: Gilbert Simondon's philosophy of machines, *Revista do Departamento de Psicologia – UFF*, 17(2): 11–18.

Schmidt, F. 1878. Über Fettbestimmung in der Milch mittelst des Lactobutyrometers. *Journal für Landwirthschaft*, 26: 361–400.

Scholliers, P. 2007. Twenty-five years of studying *un phénomène social total*: food history writing on Europe in the nineteenth and twentieth centuries, *Food, Culture and Society*, 10: 449–71.

Scott, J.C. 1985. *Weapons of the Weak: Everyday Forms of Peasant Resistance*. New Haven: Yale University Press.

Scott, W.L. 1861. On food: its adulterations, and the methods of detecting them. *Journal of the Society of Arts*, 9: 153–66.

Scurfield, H. 1923. The milk problem. *Journal of State Medicine*, 31: 28–40.

Sedgwick, W.T. and Batchelder, J.L. 1892. A bacteriological examination of the Boston milk supply. *Boston Medical and Surgical Journal*, 126: 25–8.

Serventi, S. and Sabban, F. 2002. *Pasta: The Story of a Universal Food*. New York: Columbia University Press.

Shanks, M. 2007. Symmetrical archaeology. *World Archaeology*, 39: 89–96.

Shapin S. 1991. 'A scholar and a gentleman': the problematic identity of the scientific practitioner in early modern England. *History of Science*, 29: 279–327.

Shapin, S. 2003. Trusting George Cheyne: scientific expertise, common sense, and moral authority in early eighteenth-century dietetic medicine. *Bulletin of the History of Medicine*, 77: 263–97.

Shaviro, S. 2007. Deleuze's encounter with Whitehead. Available at: http://www.dhalgren.com/Othertexts/DeleuzeWhitehead.pdf [accessed 4 May 2009].

Shaw, R.H. 1917. *Chemical Testing of Milk and Cream*. Washington, DC: Government Printing Office.

Sheail, P. 2003. *Wilfred Buckley of Moundsmere and the Clean Milk Campaign*. Winchester: Hampshire County Council.

Shelmerdine, A. 1921. Municipal milk supply. *Journal of State Medicine*, 29: 297–304.

Sherbon, J.W. 1999. Physical properties of milk, in *Fundamentals of Dairy Chemistry*, edited by N.P. Wong, R. Jenness, M. Keeney and E.H. Marth. Gaithersburg, MD: Aspen, 409–60.

Shiga, J. 2006. Translations: artifacts from an actor-network perspective. *Artifact*, 1(1): 40–55.

Shove, E., Watson, M., Hand, M. and Ingram, J. 2007. *The Design of Everyday Life*. Oxford: Berg.

Shutt, F.T. 1892. On the Babcock method of milk analysis. *Analyst*, 17: 227–9.

Simon, J.F. 1845. *Animal Chemistry* London: Sydenham Society.

Simoons, F.J. 1978. The geographic hypothesis and lactose malabsorption: a weighing of the evidence. *American Journal of Digestive Diseases*, 23: 963–80.

Simoons, F.J. 1994. *Eat Not This Flesh: Food Avoidances from Prehistory to the Present*. Madison: University of Wisconsin Press.

Simpson, A.W.B. 1975. Innovation in nineteenth century contract law. *Law Quarterly Review*, 91: 247–78.

Simpson, E.S. 1957. The Cheshire grass-dairying region. *Transactions of the Institute of British Geographers*, 23: 141–62.

Singh, H. and Bennett, R.J. 2002. Milk and milk processing, in *Dairy Microbiology Handbook*, edited by R.K. Robinson. New York: Wiley-Interscience, 1–38

Sismondo, S. 2004. *An Introduction to Science and Technology Studies*. Malden, MA: Blackwell.

Sklar, K.K. 1999. The Consumers' white label campaign of the National Consumers' League, 1898–1918, in *Getting and Spending: European and American Consumer Societies in the Twentieth Century*, edited by S. Strasser, C. McGovern and M. Judt. Cambridge: Cambridge University Press, 17–35.

Smith, A. 1776. *An Inquiry into the Nature and Causes of the Wealth of Nations*. London: Methuen.

Smith, A.F. 2002. *Peanuts: The Illustrious History of the Goober Pea*. Urbana, IL: University of Illinois Press.

Smith, B. and Brogaard, B. 2003. Sixteen days. *Journal of Medicine and Philosophy*, 28: 45–78.

Smith, B. and Smith, D.W. (eds) 1995. *The Cambridge Companion to Husserl*. Cambridge: Cambridge University Press.

Smith, D.F. 1997. Nutrition science and the two World Wars, in *Nutrition in Britain: Science, Scientists and Politics in the Twentieth Century*, edited by D.F. Smith. London: Routledge, 142–65.

Smith, D.F. 2000. The Carnegie Survey: background and intended impact, in *Order and Disorder: The Health Implications of Eating and Drinking in the Nineteenth and Twentieth Centuries*, edited by A. Fenton. East Linton: Tuckwell Press, 64–80.

Smith, E. 1864. *Report on the Food of the Poorer Labouring Classes in England*, Appendix no. 5 to the 6th Report of the Medical Officer to the Privy Council, P.P. 1864 (3416), xxviii.

Smith, N. 1984. *Uneven Development: Nature, Capital and the Production of Space*. Oxford: Blackwell.

Smith, S.D. 2001. Coffee, microscopy, and the Lancet's Analytical Sanitary Commission. *Social History of Medicine*, 14: 171–97.

Smith-Gordon, L. 1919. *The Irish Milk Supply*. Dublin: Cooperative Reference Library.

Smithers, H. 1825. *Liverpool, its Commerce, Statistics and Institutions*. Liverpool: Thomas Kaye.

Smollett, T. 1771. *The Expedition of Humphry Clinker*. London: Bumbus.

Snyder, H. 1906. *Dairy Chemistry*. New York: Macmillan.

Sommerfeld P. von 1909. *Handbuch der Milchkunde*. Wiesbaden: Bergmann.

Sorensen, B.H. 1982. Consumer protection and the development of public law in the nineteenth century: the regulation of food quality, unpublished MLitt. thesis, University of Oxford.

Spary, E.C. 2005. Ways with food, *Journal of Contemporary History*, 40: 763–71.

Spencer, A.J. 1904. *Dixon's Law of the Farm*. London: Stevens and Sons.

Spencer, H. 1872. *Social Statics*. New York: D. Appleton.

Spottiswoode Cameron, J. 1904. The sophistication of foods. *British Food Journal*, 6: 2–4.

Stanford, J.K. 1956. *British Friesians: A History of the Herd*. London: Max Parrish.

Stanziani, A. 2003a. *La Qualité des Produits en France, XVIIIe–XXe Siècles*. Paris: Belin.

Stanziani A. 2003b. La fraude dans l'agroalimentaire, genèse historique: la falsification du vin en France, 1880–1905. *Revue d'Histoire Moderne et Contemporaine*, 2: 154–186.

Stanziani, A. 2003c. Action économique et contentieux judiciaires le cas du plâtrage du vin en France, 1851–1905. *Genèses*, 50: 71–90.

Stanziani, A. 2005. Products, norms, and historical dynamics, in *Quality: A Debate*, edited by C. Musselin and C. Paradeise. *Sociologie du Travail*, 47: S114–23.

Stanziani, A. 2006. Qualité des produits et règles de droit dans une perspective historique, in *L'Economie des Conventions, Méthodes et Résultats*, edited by F. Eymard-Duvernay. Paris: La Découverte, volume 2: 61–74.

Stanziani, A. 2007a. Negotiating innovation in a market economy: foodstuffs and beverages adulteration in nineteenth-century France. *Enterprise and Society*, 8: 375–412.

Stanziani A. 2007b. A l'origine du service de la répression des fraudes: concurrence, expertise et qualité des produits en France, 1789–1914, in *La Loi du 1er Août 1905: Cent Ans de Protection des Consommateurs*, edited by Direction Générale de la Concurrence, de la Consommation et de la Répression des Fraudes. Paris: *La Documentation Française*, 209–27.

Stanziani, A. 2009. Information, quality and legal rules: wine adulteration in nineteenth century France. *Business History*, 51: 268–91.

Starr, P. 1987. The sociology of official statistics, in *The Politics of Numbers*, edited by W. Alonso and P. Starr. New York: Russell Sage Foundation, 7–57.

Stassart, P. and Whatmore, S. 2003. Metabolising risk: food scares and the un/re-making of Belgian beef. *Environment and Planning A*, 35: 449–62.

Stengers, I. 2002. *Penser avec Whitehead: une Libre et Sauvage Création de Concepts*. Paris: Seuil.

Stengers, I. 2004. Résister à Simondon. *Multitudes*, 18.

Stiegler, B. 1998. *Technics and Time, 1: The Fault of Epimetheus*. Stanford, CA: Stanford University Press.

Stokes, A.W. 1887. Allowances for decomposed milk. *Analyst*, 12: 226–34.

Stokes, A.W. 1889. The Werner-Schmid method of determining fat in milk and cream. *Analyst*, 14: 29–32.

Stokes, A.W. 1892. Estimation of fat in milk by the Babcock method. *Analyst*, 17: 127–30.

Stoljar, S.J. 1952. Conditions, warranties and descriptions of quality in sale of goods. *Modern Law Review*, 15: 425–45, 16 (1953): 174–197.

Storper, M. and Salais, R. 1997. *Worlds of Production: The Action Frameworks of the Economy*. Cambridge, MA: Harvard University Press.

Strachan, D.P. 1989. Hay fever, hygiene and household size. *British Medical Journal*, 299: 1259–60.

Straete, E.P. 2008. Modes of qualities in development of speciality food, *British Food Journal*, 110: 62–75.

References 325

Sumner, J. 2001. John Richardson, saccharometry and the pounds-per-barrel extract: the construction of a quantity. *British Journal for the History of Science*, 34: 255–73.

Super, J.C. 2002. Food and history. *Journal of Social History*, 36: 165–78.

Sussman, C. 2000. *Consuming Anxieties: Consumer Protest, Gender, and British Slavery*. Stanford, CA: Stanford University Press.

Sussman, G.D. 1982. *Selling Mothers' Milk: The Wet-Nursing Business in France, 1715–1914*. Urbana: University of Illinois Press.

Sweeney, J. 1904. *At Scotland Yard*. London: Grant Richards.

Swithinbank, H. and Newman, G. 1903. *Bacteriology of Milk*. London: Murray.

Swyngedouw, E. 2004. *Social Power and the Urbanization of Water: Flows of Power*. Oxford: Oxford University Press.

Swyngedouw, E. 2006. Circulations and metabolisms: (hybrid) natures and (cyborg) cities. *Science as Culture*, 15: 105–21.

Sykes, J.F.J. 1887–8. On the supervision of dairies, cowsheds and milkshops. *Transactions of the Sanitary Institute of Great Britain*, 9: 180–92.

Sylvander, B. and Biencourt, O. 2006. Negotiating standards for animal products: a procedural approach applied to raw milk, in *Agricultural Standards: The Shape of the Global Food and Fibre System*, edited by J. Bingen and L. Busch. Dordrecht: Springer, 95–109.

Szabadváry, F. 1966. *History of Analytical Chemistry*. London: Pergamon Press.

Tallontire, A. 2007. CSR and regulation: towards a framework for understanding private standards initiatives in the agri-food chain. *Third World Quarterly*, 28: 775–91.

Taylor, D. 1976. The English dairy industry, 1860–1930. *Economic History Review*, 2nd series 29: 585–601.

Taylor, D. 1987. Growth and structural change in the English dairy industry, c.1860–1930, *Agricultural History Review*, 35: 47–64.

Téchoueyres, I. 2007. Food markets in the City of Bordeaux – from the 1960s until today: historical evolution and anthropological aspects, in *Food and the City in Europe since 1800*, edited by P.J. Atkins, P. Lummel and D.J. Oddy. Aldershot: Ashgate, 239–49.

Teeven, K.M. 1990. *A History of the Anglo-American Common Law of Contract*. New York: Greenwood Press.

Teil, G. and Hennion, A. 2004. Discovering quality or performing taste? A sociology of the amateur, in *Qualities of Food*, edited by M. Harvey, A. McMeekin and A. Warde. Manchester: Manchester University Press, 19–37.

Thévenot, L. 2002. Which road to follow? The moral complexity of an 'equipped' humanity, in *Complexities: Social Studies of Knowledge Practices*, edited by J. Law and A. Mol. Durham, NC: Duke University Press, 53–87.

Thévenot, L. 2006. Convention school, in *International Encyclopedia of Economic Sociology*, edited by J. Beckert and M. Zafirovski. London: Routledge, 111–5.

Thomas, S.B. 1963. The history of the Society for Applied Bacteriology from 1931 to 1945. *Journal of Applied Bacteriology*, 26: 66–8.

Thorburn Burns, D. 2007. Alexander Wynter Blyth (1844–1921): a pioneering and innovative public analyst. *Journal of the Association of Public Analysts*, 35: 17–29.

Thorpe, T.E. 1905. The analysis of milk referred to the Government Laboratory in connexion with the Sale of Food and Drugs Act, *Journal of the Chemical Society, Transactions*, 87: 206–25.

Tishkoff, S.A. et al. 2006. Convergent adaptation of human lactase persistence in Africa and Europe. *Nature Genetics*, 39: 31–40.

Tobin, W. 2006. Alfred Donné and Léon Foucault: the first applications of electricity and photography to medical illustration. *Journal of Visual Communication in Medicine*, 29(1): 6–13.

Tocher, J.F. 1925. *Variations in the Composition of Milk*. Edinburgh: HMSO.

Tollens, B. and Grote, F. 1879. Zur Fettbestimmung in der Milch mittelst des Marchand'schen Lactobutyrometers. *Journal für Landwirthschaft*, 27, 145–52.

Toscano, A. 2006. *The Theatre of Production: Philosophy and Individuation between Kant and Deleuze*. Basingstoke: Palgrave Macmillan.

Trentmann, F. 2001. Bread, milk and democracy: the reconfiguration of consumption and citizenship in twentieth-century British popular politics, in *The Politics of Consumption*, edited by M. Daunton and M. Hilton. Oxford: Berg, 129–63.

Trentmann, F. 2006a. Knowing consumers – histories, identities, practices, in *The Making of the Consumer: Knowledge, Power and Identity in the Modern World*, edited by F. Trentmann. Oxford: Berg, 1–27.

Trentmann, F. 2006b. The modern genealogy of the consumer: meanings, identities and political synapses before affluence, in *Consuming Cultures: Global Perspectives, Historical Trajectories, Transnational Exchanges*, edited by J. Brewer and F. Trentmann. Oxford: Berg, 19–69.

Trentmann, F. 2007. Before 'fair trade': empire, free trade, and the moral economies of food in the modern world. *Environment and Planning D: Society and Space*, 25: 1079–102.

Tuckett, J.D. 1816. *A History of the Past and Present State of the Labouring Population*. London: Longman.

Tully, J. 1995. *Strange Multiplicity*. Cambridge: Cambridge University Press.

Turner, F.M. 1980. Public science in Britain, 1880–1919. *Isis*, 71: 589–608.

Turner, J. 2004. *Spice: The History of a Temptation*. London: HarperCollins.

Turner, M.E. 1981. Arable in England and Wales: estimates from the 1801 crop returns. *Journal of Historical Geography*, 7: 291–302.

Turner, M. 1998. Counting sheep: waking up to new estimates of livestock numbers in England, c. 1800. *Agricultural History Review*, 46: 142–61.

Turner, M.E., Beckett, J.V. and Afton, B. 2001. *Farm Production in England 1700–1914*. Oxford: Oxford University Press.

Tustin, P.B. 1929. Retail aspects of milk hygiene. *Journal of the Royal Sanitary Institute*, 50: 312–5, 322–3.

Tyler, C. 1956. The development of feeding standards for livestock. *Agricultural History Review*, 4: 97–107.

Valverde, M. 2003. *Law's Dream of a Common Knowledge*. Princeton, NJ: Princeton University Press.

Van Vliet, B., Chappells, H. and Shove, E. 2005. *Infrastructures of Consumption: Environmental Innovation in the Utility Industries*. London: Earthscan.

Vaughan, P., Cook, M. and Trawick, P. 2007. A sociology of reuse: deconstructing the milk bottle, *Sociologia Ruralis*, 47: 120–34.

Ventura, F. and Milone, P. 2000. Theory and practice of multi-product farms: farm butcheries in Umbria. *Sociologia Ruralis*, 40: 452–65.

Vernois, M. and Becquerel, A. 1853. *Du Lait Chez la Femme dans l'Etat de Santé et dans l'Etat de Maladie*. Paris: Baillière.

Vernon, K. 1997. Science for the farmer? Agricultural research in England, 1909–1936. *Twentieth Century British History*, 8: 310–33.

Vernon, K. 2000. Milk and dairy products, in *The Cambridge World History of Food*, edited by K. Kiple and K. Ornelas. Cambridge: Cambridge University Press, 692–702.

Vieth, P. 1889. The methods for determining fat in milk. *Analyst*, 14: 86–89.

Vieth, P. 1892. The average composition of milk. *Analyst*, 17: 84–9.

Vigarello, G. 1988. *Concepts of Cleanliness: Changing Attitudes in France Since the Middle Ages*. Cambridge: Cambridge University Press.

Vincent, J. 2006. The moral expertise of the British consumer, c. 1900: a debate between the Christian Social Union and the Webbs, in *The Expert Consumer: Associations and Professionals in Consumer Society*, edited by A. Chatriot, M.-E. Chessel and M. Hilton. Aldershot: Ashgate, 37–51.

Voelcker, J.C.A. 1867. London milk: a report, with analysis of the milk sold in the various districts of London. *British Medical Journal*, ii: 479–80.

Wagner, P. 1994. Action, coordination and institution in recent French debates. *Journal of Political Philosophy*, 3: 270–89.

Wagner, P. 2001. *A History and Theory of the Social Sciences: Not all That is Solid Melts into Air*. London: Sage.

Waley-Cohen, C. 1933. Standards in the milk industry: joint responsibility of Medical Officers, producers, distributors, and the veterinary sciences. *Journal of the Royal Sanitary Institution*, June: 670–83.

Wallace, R.H. 1898. *Adulteration of Dairy Produce*. Edinburgh: Anderson.

Walstra, P., Geurts, T.J., Noomen, A, Jellema, A. and van Boekel, M.A.J.S. 1999. *Dairy Technology: Principles of Milk Properties and Processes*. New York: Marcel Dekker.

Walton, J.K. 1979. Mad dogs and Englishmen: the conflict over rabies in late Victorian England. *Journal of Social History*, 13: 219–39.

Walton, J.R. 1984. The diffusion of the improved shorthorn breed of cattle in Britain during the eighteenth and nineteenth centuries. *Transactions of the Institute of British Geographers*, 9: 22–36.

Walton, J.R. 1986. Pedigree and the national cattle herd, circa 1750–1950. *Agricultural History Review*, 34: 149–70.

Walton, J.R. 1999. Pedigree and productivity in the British and North American cattle kingdoms before 1930. *Journal of Historical Geography*, 25: 441–62.

Wanklyn, J.A. 1874. *Milk Analysis: A Practical Treatise on the Examination of Milk and its Derivatives, Cream, Butter, and Cheese*. London: Trübner.

Ward, A.G. 1976. Advising on food standards in the United Kingdom. 1: the changing role of the Food Standards Committee, in *Food Quality and Safety, a Century of Progress: Proceedings of the Symposium Celebrating the Centenary of the Sale of Food and Drugs Act 1875, London, October 1975*, chaired by Lord Zuckerman. London: HMSO, 22–40.

Ward, S. 1996. *Reconfiguring Truth: Postmodernism, Science Studies and the Search for a New Model of Knowledge*. Lanham, MD: Rowman and Littlefield.

Washer, P. 2006. Representations of mad cow disease. *Social Science and Medicine*, 62: 457–66.

Weatherell, C., Tregear, A. and Allinson, J. 2003. In search of the concerned consumer: UK public perceptions of food, farming and buying local. *Journal of Rural Studies*, 19: 233–44.

Webmore, T. 2007. What about 'one more turn after the social' in archaeological reasoning? Taking things seriously. *World Archaeology*, 39: 563–78.

Weibull, M. 1898. Gottlieb's method for estimating fat in milk. *Chemiker-Zeitung*, 22: 632.

Whatmore, S. 2002. *Hybrid Geographies: Natures, Cultures, Spaces*. London: Sage.

Whetham, E.H. 1978. *The Agrarian History of England and Wales. Volume VIII: 1914–39*. Cambridge: Cambridge University Press.

White, E. 1928. Purchasing milk on a quality basis, in *World's Dairy Congress – 1928: Report of Proceedings, Great Britain, June 26th–July 12th*. London: World's Dairy Congress Committee, 323–8.

Whitehead, A.N. 1920. *The Concept of Nature*. Cambridge: Cambridge University Press.

Whittaker, S. 2005. *Liability for Products: English Law, French Law, and European Harmonization*. Oxford: Oxford University Press.

Wigner, G. 1883. The milk supply of London. *Analyst*, 8: 243–5.

Wigner, G.W. 1879. Milk preservatives. *Analyst*, 4: 88–90.

Wilkinson, J. 1997. A new paradigm for economic analysis? *Economy and Society*, 26: 305–39.

Wilkinson, J. 2006. Network theories and political economy: from attrition to convergence? In *Between the Local and the Global: Confronting Complexity in the Contemporary Agri-Food Sector*, edited by T. Marsden and J. Murdoch. Amsterdam: Elsevier, 11–38.

[Williams, J.D.] 1830. *Deadly Adulteration or Slow Poisoning: or, Disease and Death in the Pot and Bottle*. London: Sherwood, Gilbert and Piper.

Williams, R. 1980. Ideas of nature, in *Problems in Materialism and Culture: Selected Essays*, by R. Williams. London: Verso, 67–85.

Williams, R.S. 1917. The wastage of milk. *Journal of the Royal Agricultural Society of England*, 78: 24–34.

Williams, R.S. 1925. Bacteriological standards and a pure milk supply. *British Medical Journal*, ii: 241–4.

Williams, R.S. 1928. Education and advisory work amongst milk producers and the handling of milk at the farm, in *World's Dairy Congress – 1928: Report of Proceedings, Great Britain, June 26th–July 12th*. London: World's Dairy Congress Committee, 302–9.

Williams, R.S. and Cornish, E.C.V. 1917. *The Milk Supply: A Suggestion*. Cambridge: Cambridge University Press.

Williams, R.S. and Hoy, W.A. 1928. The milk supply: what shall our policy be? *Journal of State Medicine*, 36: 63–78.

Williams, R.S. and Mattick, A.T.R. 1922. Report concerning the present position and possible developments of graded milk. *Medical Officer*, 27: 223–5.

Williams, R.S. and Mattick, A.T.R. 1923. Pasteurization of milk – No. II. *Modern Farming*, 6(10): 11.

Williams, R.S. and Mattick, E.C.V. 1931. The importance of a complete study of the nutritional value of milk. *Journal of State Medicine*, 39: 141–56.

Wilson, G.S. 1935. The bacteriological grading of milk. *Medical Research Council, Special Report Series*, 206.

Wilson, G.S. 1936. The modified, methylene blue reduction test for the grading of raw milk on the basis of bacterial cleanliness. *Veterinary Record*, 16: 494–7.

Wilson, G.S. 1942. *The Pasteurization of Milk*. London: Arnold.

Wilson, G.S. and Miles, A. 1975 *Topley and Wilson's Principles of Bacteriology, Virology and Immunity*. London: Arnold.

Wing, H.H. 1913. *Milk and its Products, a Treatise Upon the Nature and Qualities of Dairy Milk and the Manufacture of Butter and Cheese*. London: Macmillan.

Witmore, C.L. 2007. Symmetrical archaeology: excerpts of a manifesto. *World Archaeology*, 39: 546–62.

Wohl, A.S. 1983. *Endangered Lives: Public Health in Victorian Britain*. London: Dent.

Wohlert, K. 1983. Voraussetzungen und Strategien der schwedischen Separatorgesellschaft Alfa-Laval bis zum Ersten Weltkrieg. *Zeitschrift für Unternehmensgeschichte*, 28: 188–213.

Wolff, H.W. 1912. *Cooperation in Agriculture*. London: P.S. King and Son.

Wood, J. 1928. Our milk supply: the essential conditions for securing a pure and good milk for co-operators. *The Producer*, 12: 315–16, 342.

Woolsey Biggart, N. and Beamish, T.D. 2003. The economic sociology of conventions: habit, custom, practice, and routine in market order. *Annual Review of Sociology*, 29: 443–64.

Worboys, M. 2004. Delépine, Auguste Sheridan (1855–1921), in *Oxford Dictionary of National Biography*. Oxford: Oxford University Press. Available at: http://www.oxforddnb.com/view/article/57113 [accessed: 11 May 2009].

Wright, R.F. and Huck, P. 2002. Counting cases about milk: our 'most nearly perfect' food, 1860–1940. *Law and Society Review*, 36: 51–112.

Wynne, B. 2003. Seasick on the third wave? Subverting the hegemony of propositionalism. *Social Studies of Science*, 33: 401–17.

Wynter, A. 1854. The London commissariat. *Quarterly Review*, 190: 271–308.

Parliamentary Papers

Astor Committee. *Departmental Committee on Production and Distribution of Milk* [Chairman: Waldorf Astor], First Interim Report, PP 1917–18 (Cd.8608) xvi.1003; Second Interim Report, PP 1917–18 (Cd.8886) xvi.1011; Report to the Food Controller PP 1918 (Cd.9095) xii.125; Third Interim Report PP 1919 (Cmd.315) xxv.615; Final Report PP 1919 (Cmd.483) vvx.645.

Cook Committee. *Interdepartmental Committee on Milk Composition in the United Kingdom* [Chairman: T.W. Cook], PP 1960 (Cmnd.147) xix.119.

Foster Committee. *Select Committee on Food Products Adulteration* [Chairman: Sir Walter Foster], PP 1894 (253) xii.1, 1895 (363) x.73.

Read Committee. *Select Committee into the Adulteration of Food Act (1872)* [Chairman: Clare Sewell Read], PP 1874 (262) vi.243.

Russell Committee. *Select Committee on Food Products Adulteration* [Chairman T.W. Russell], PP 1896 (288) ix.483.

Scholefield Committee. *Select Committee into the Adulteration of Food, Drinks and Drugs* [Chairman: William Scholefield], PP 1854–5 (432, 480, 480–I) viii.221, 373, PP 1856 (379) viii.1.

Sclater Booth Committee. *Select Committee on Sale of Food and Drugs Act (1875) Amendment Bill* [Chairman: George Sclater-Booth], PP 1878–9 (155) x.1.

Mackenzie Committee. *Inter-Departmental Committee on the Laws, Regulations and Procedure Governing the Sale of Milk in Scotland* [Chairman: Sir Leslie Mackenzie], PP 1922–II (Cmd.1749) ii.835.

Maxwell Committee. *Departmental Committee into the Use of Preservatives and Colouring Matters in the Preservation and Colouring of Food* [Chairman: Herbert Maxwell], PP 1902 (Cd.833) xxxiv.579.

Wenlock Committee. *Departmental Committee to Inquire into the Desirability of Regulations, under Section 4 of the Sale of Food and Drugs Act, 1899, for Milk and Cream* [Chairman: Lord Wenlock], PP 1901 (Cd.491, Cd.484) xxx.371.

Willis Committee. *Departmental Committee on the Composition and Description of Food (Composition and Description)* [Chairman: Frederick James Willis], PP 1933–4 (Cmd.4564) xii.159.

Index

adulteration xv, xviii, 11, 27, 39–40, 44–5,
  49, 59, 60, 62, 64, 67–70, 73, 79, 82,
  85, 89, 93, 96, 99, 101–2, 104–6,
  110, 119–22, 125, 127, 135–6, 138,
  141–4, 146, 150–51, 157–8, 160–71,
  174, 178–82, 187, 188–93, 195–202,
  204–15, 219, 223–4, 231, 237
Aylesbury Dairy Company 84, 120, 122,
  151, 174–5, 181, 186, 199, 238

Babcock, Stephen Moulton 72, 77, 78, 79,
  81, 82, 83, 86
bacteria 228, 230, 232–3, 240–41, 247–9,
  251–4, 256, 258, 262–5, 271–2, 274–5
bacteriology 139, 223, 243, 247, 248, 275
bad practice 102, 151, 233
Barham, George xvii, xviii, 123, 145, 146,
  147, 192, 207
Barlow Committee 241, 254, 255
biopolitics 14, 160, 161, 176, 226
bovine tuberculosis xiv, 18, 127, 162, 223,
  254, 273
Britain *see* United Kingdom
British Dairy Farmers' Association 102,
  122, 238
British Standards Institution 65, 85, 159
BSE xiv, 3, 7–8, 49, 91, 144, 156
butter xvi, xvii, 13, 26, 59, 65–6, 76–7, 79,
  89, 110, 120, 122, 128, 137, 166, 171,
  181, 183, 188, 199, 224, 248

certification xx, 16, 18, 115, 117–18, 151,
  233, 237–8, 254–5, 261, 270
cheese xvi, xvii, 26, 59, 65, 66, 110, 122,
  166–7, 171, 181, 199, 247, 265
chemical analysis 41, 60, 73, 106, 139,
  148, 249
chemistry 18, 35, 40, 45, 49, 56–60, 68,
  74–5, 79, 81, 83, 87, 89–90, 93–4, 99,
  101, 104, 120, 144, 156, 178, 196,
  201, 253, 278
  animal 67
  dairy 18, 45, 49, 56, 59, 81, 87, 89, 120

food 74, 99, 201
  gravimetric 75
  industrial 57, 90
  organic 18, 40, 49, 59, 60, 74, 81, 83, 94,
    156, 278
constructionism 15, 22, 29, 31, 36, 50,
  52–3, 56, 57, 90
  social 15, 31, 52–3, 90
contamination 19, 53, 92, 139, 145, 174,
  178, 225, 227–8, 230–31, 233, 237,
  240–41, 249–52, 258
Cook Committee 85, 126, 128, 129, 176,
  193, 270
Co-operative Society 149–51
cream xvii, xviii, xx, 39, 42, 58, 61–4,
  76–7, 89, 93, 100, 102, 122, 137, 143,
  169, 171, 180–81, 187, 199, 214, 271

dairy farmers/farming xvi, 26, 110, 116,
  123, 131, 183, 224, 266
  see also dairymen, milk producers
dairy industry 26, 52, 61, 64, 79, 145, 257,
  262
dairymen xviii, 94, 122, 146, 181, 184–6,
  189, 192, 205–6, 208, 233, 237, 242,
  255, 259
  see also dairy farmers/farming, milk
    producers
dirt xvi, xix–xx, 9, 20, 27, 73, 86, 137,
  139–40, 162, 182, 223–28, 230, 232,
  240, 241, 242, 244, 245, 247, 250,
  265, 270, 272, 278
  see also milk, dirty
disease xvi, xviii, xix–xx, 3, 4, 9, 19, 22,
  26–7, 52, 60, 92, 110, 112, 117, 126,
  131, 137, 162, 164, 174, 186, 223,
  224–7, 236, 248–9, 254–5, 257–8,
  271, 273
  infectious 92, 223, 236

ethics 9, 74, 149, 152
European Union (EU) 24, 131, 155
Excise laboratory *see* Somerset House

expertise 13, 36, 40, 42, 49, 91–113, 120,
    136, 138, 139, 144–6, 148, 149, 154,
    157, 175, 188, 195, 212, 219, 269
Express Dairies xvii, xviii, 146, 147, 175,
    192, 207, 238

fair trade 118, 138, 149, 150, 152
First World War xvii, 60, 111–12, 141,
    150–51, 164, 189, 195, 223, 227, 230,
    233, 237, 239, 247–50, 253, 274–5
food
    safety 4, 7–8, 16, 119, 142, 152, 154
    scares 3, 4, 91
Food and Drugs Acts (various years) 27,
    95, 100, 163, 180, 192, 196, 202, 203,
    209, 218
Foster Committee xviii, 84, 96, 99, 101–2,
    105–9, 121–2, 180–81, 186, 188–9,
    198, 205, 208, 210
France xvii, xviii, 3, 15, 62–3, 92, 94, 116,
    155–6, 159, 167, 176, 179, 185, 200,
    226
fraud xviii, 37, 50, 52, 59, 64, 93, 96, 99,
    102, 104, 107, 136, 157–8, 167, 171,
    180, 182, 189, 192, 197, 201, 205–8,
    210, 224

geography xiii, xiv, 7, 15, 25, 28, 32, 33,
    113, 138, 153, 226
    economic 25, 33
    historical 7, 113, 138
    human 28
Germany 3, 40, 77, 84, 112, 150, 154, 168
grades/grading 115, 116, 126, 234, 237,
    255, 257, 259, 260, 261, 262, 264,
    270, 273

history xiii, xiv, xx, 3–5, 7, 8, 11, 12, 14,
    23, 28–30, 32, 33, 35–103, 118, 135,
    152–3, 160, 169, 198, 200, 202, 208,
    217, 219, 223, 225–6, 279
    agricultural xiii
    diet-disease 3
    economic xiii, 5, 32, 153
    epistemological 160, 226
    food xiii, xiv, xx, 1, 3–5, 8, 23, 28, 30,
        35, 49, 225
    natural 12

political 5
social 5
'thing' 9, 14
hygiene xix, 138, 152, 162, 223, 225–6,
    230–33, 235, 245, 247, 268, 274
    food 223, 225, 233
hygienism 163, 226, 278

Inland Revenue Laboratory *see* Somerset
    House

knowability 40, 53, 55–90
knowledges 22, 24, 211

labelling xx, 36, 115–17, 127, 130, 151
lactometers 39–45, 55, 56, 60–66, 82,
    88–90, 130, 167, 168
Latour xix, 10, 28, 144, 203
Latour, Bruno xix, 10, 28–30, 36, 52, 79,
    88, 144, 165, 196, 203
    actor network theory 8, 10, 26, 28–30,
        36, 90
legislation 4, 68, 93, 97–8, 100, 102–3,
    124, 129, 144, 159, 162–3, 169, 180,
    196, 198–9, 202, 208, 218–19, 224,
    256, 258
    anti-adulteration 144, 169, 180
Local Government Board 98, 103–5, 107,
    121–2, 135, 143, 163, 171, 173, 178,
    181, 186–92, 198, 205, 207, 209, 237,
    249, 254–6, 258–9

maceration 70, 73, 79, 99
Mackenzie Committee 124–6, 206, 210
Mackintosh, James 112, 238–44, 256, 265
materialism 11, 15, 28, 36, 52
material quality xx, 11, 16–17, 23, 49, 137,
    219, 220, 277
Medical Officers of Health 64, 103–4, 109,
    125, 240, 244, 250, 267, 279
medicine xx, 12, 31, 32, 163
Metropolitan Dairymen's Society 102, 122
Midland Counties Dairy Ltd 126, 146, 269,
    270, 271
milk
    butter- xvii, 188
    certified 27, 200, 233, 235, 238, 254,
        257–8, 261–2, 264, 270

clean milk 126, 137–8, 166, 182, 196,
220, 223–4, 231, 233–45, 247,
253–8, 261, 264–7, 270, 272, 275
condensed 27, 89, 171, 180, 240
dirty xiv, 110, 113, 225–45, 247, 261,
264, 272
drinking-xvi, xviii, xx, 27, 240
'invalid' and 'infant' xviii, 27
'knowability' of 55–90
liquid xvi, xvii, 65, 116, 120, 131, 148,
166, 240, 277
natural 18, 39,–41, 59, 130, 196, 211
quality 42, 60, 99, 127–8, 166, 200,
223–4, 236, 267, 270, 272–4
raw 61, 87, 235, 265, 274
skimmed xvii
skim xvii, xviii, 27, 58, 70, 143, 180,
188, 201, 211
skimming of xvii, 41, 64, 89, 137, 166,
180
unpasteurized 50, 164
watering of xv, 41–5, 63, 64, 70, 81, 82,
100, 102, 119, 122, 127, 129, 137,
143, 144, 157, 166, 178, 181, 187,
192, 196, 208
whole xvii, 27, 41, 59, 76, 100, 130,
143–44, 180, 187, 192, 201, 207
milk composition 19, 39, 57, 60, 66, 83,
97–8, 112–13, 119, 124–5, 128–9,
131, 175, 180, 188, 193, 199, 201, 216
milk constituents
butterfat xviii, 18, 58, 67, 69, 76, 77,
90, 110, 120, 122–3, 126, 128, 175,
193, 271, 278
fat xvi, xx, 18, 37, 57–8, 62–3, 65–73,
75–9, 81–7, 89, 90, 100, 101, 112,
120–31, 170, 176–7, 179–81, 188,
199, 212–15, 232,–3, 237, 252–3,
270
solids non-fat 69, 71, 120–21
solids-not-fat 18, 67, 69–70, 73, 81–3,
85, 120, 122–3, 126–9, 131, 176,
179, 193
Milk Marketing Board 116, 126, 128, 131,
176, 183, 240, 243, 267, 272–3
milk producers 26, 123, 125, 207, 240,
256, 266, 268, 272
see also dairy farmers/farming, dairymen

milk trade xvii, xix, 26, 39, 44–5, 59, 64,
123, 137, 147, 151, 167–8, 175, 185,
196, 207, 209, 214, 237, 240, 272
Ministry of Agriculture 116, 127, 183–4,
230, 242–3, 259, 264, 267, 273–4
modernity xiii, 53, 91, 140, 146, 159–60,
163, 179, 226, 279

National Clean Milk Society 224, 254,
257–8, 265
National Institute for Research in Dairying
(NIRD) 27, 126, 224, 239–40, 242–4,
260, 268, 270
networks xx, 23–4, 26–7, 29, 51, 118, 149,
154, 217
nutrition 4, 35, 87, 112, 128, 148, 161

ontogenesis 10, 14, 31, 32, 36, 89, 195, 277
ontologies/ontology 7, 8, 9, 12, 14, 22,
31–2, 36, 53, 75, 90, 130, 196, 202
legal 195–220

Pasteur, Louis 29, 223, 226, 247
pasteurization xix, 27, 87–8, 127, 223,
235–6, 240, 245, 262, 278

physics 45, 49, 57, 60, 65, 67–8, 81, 120,
144, 178
dairy 45, 49
Pickering, Andrew 36, 49, 53, 55–7, 62,
90, 156, 158
pollution 139, 162, 166, 224, 226
preservatives 102, 137, 141, 151, 180,
184–7, 223, 228
public health xviii, xix, 8–9, 50, 138, 142,
163, 170, 175–6, 223, 226, 230, 232,
236, 249, 252, 254, 273, 275

Read Committee xvi, xviii, 59, 64, 65, 69,
95–6, 99, 104–5, 121, 147, 189, 192,
198, 201, 207, 210, 214–15
Royal Agricultural Society xvi, xviii, 121,
265
rules xvi, 19, 25–6, 82, 94, 96, 116,
118–19, 130, 149, 202–3, 213, 217,
243, 265
Russell Committee 71, 101–2, 107, 122,
179, 181, 187

Sale of Food and Drugs Acts (various
    years) 27, 95–6, 100–102, 104–5, 122,
    163, 171, 174, 180, 192, 196–8, 202,
    208, 210, 213, 218–19
Sale of Milk Regulations 50, 124, 126,
    130, 199, 211, 215
salmonella xiv, 3, 7, 156
sampling 18, 98, 104, 107, 125, 157, 161,
    165, 174, 178, 187, 189, 198, 205,
    216, 233, 256, 272
sanitation xix, 109, 139, 151, 163, 233
Scholefield, William 93, 95, 104, 168, 181,
    187, 197, 200
Schrumpf, Daniel 39–45, 49, 56, 88, 94,
    144, 195
Second World War 127, 129, 135, 150,
    182, 245, 252, 270, 274
Simondon, Gilbert 10, 33, 34, 36, 37, 89,
    179, 196, 216, 278
Society of Public Analysts (SPA) 68–71,
    86, 95, 97, 98, 101, 105, 120, 163, 232
sociology 11, 25, 28, 33, 49, 58, 75, 147,
    155, 211
Somerset House 72, 95–102, 121, 186,
    198, 214
specific gravity 39, 41, 42, 43, 55, 56, 61,
    62, 63, 64, 65, 76, 81, 82, 83, 84, 93,
    167, 168
standards xvii, 4–5, 12, 19, 24, 40, 43, 58,
    65, 68, 81, 85, 92, 94, 101–3, 106, 113,
    115–31, 138, 146, 149, 151–3, 155,
    159, 161, 174, 179–80, 193, 199–200,
    204, 211, 213, 220–31, 233, 235–6,
    240, 259, 261–3, 270, 271, 274, 278

sterilization 241, 242, 261, 266

technology xvi, 12, 29, 35–6, 39, 41, 57–8,
    61–2, 65, 74, 80, 85, 88, 90, 94, 131,
    139, 192, 217, 245, 275, 278
traceability 3, 16, 24
trust xvii, 18, 26, 51, 57, 80, 91, 115,
    117, 119, 136, 138–9, 143, 148, 149,
    152–6, 164–5, 178, 209, 211, 219,
    279

United Dairies 141, 146, 240, 245, 256,
    266, 270
United Kingdom xiii, xiv, xv, xvi, xviii,
    xx, 24, 26, 52, 65, 68, 72, 85, 92, 94,
    103–5, 111, 116, 118, 124, 126–8,
    130–31, 135–6, 138, 148, 151, 157,
    159–62, 168, 170, 176, 180–81, 184,
    196, 198, 200, 202, 216, 219, 234–8,
    243–4, 249, 254–5, 273–4, 277
United States of America xvii, xviii, 23, 50,
    62, 65, 72, 77–8, 81, 112, 140, 161,
    179, 181, 202, 216, 219, 232–3, 236,
    240, 244, 254, 256
urbanization xvi, 51, 119, 166, 201, 226

variability 44, 60, 85, 97, 107
vitamins xx, 19, 87, 164, 279

Wanklyn, Alfred 64–5, 68–70, 72, 79, 82,
    84, 95, 120, 184, 188
Willis Committee 127

yoghurt xx, 131

For Product Safety Concerns and Information please contact our EU
representative GPSR@taylorandfrancis.com
Taylor & Francis Verlag GmbH, Kaufingerstraße 24, 80331 München, Germany